1
Plant Cell Monographs

Series Editor: David G. Robinson

Plant Endocytosis

Volume Editors:
Jozef Šamaj · František Baluška · Diedrik Menzel

With 39 Figures and 3 Tables

Springer

Volume Editors:

Dr. Jozef Šamaj
Dr. František Baluška
Professor Dr. Diedrik Menzel

Institute of Cellular and Molecular Botany
Rheinische Friedrich-Wilhelms-University Bonn
Department of Plant Cell Biology
Kirschallee 1
D-53115 Bonn
Germany

Series Editor:

Professor Dr. David G. Robinson
Ruprecht-Karls-University of Heidelberg
Heidelberger Institute for Plant Sciences (HIP)
Department Cell Biology
Im Neuenheimer Feld 230
D-69120 Heidelberg
Germany

ISSN 1861-1370
ISBN-10 3-642-42535-6 Springer Berlin Heidelberg New York
ISBN-13 978-3-642-42535-6 Springer Berlin Heidelberg New York
DOI 10.1007/b103851

Springer is a part of Springer Science+Business Media

springeronline.com

Editor: Dr. Christina Eckey, Heidelberg
Desk Editor: Anette Lindqvist, Heidelberg
Cover design: *Design & Production* GmbH, Heidelberg
Typesetting and Production: LE-TEX Jelonek, Schmidt & Vöckler GbR, Leipzig

Printed on acid-free paper 149/3152 YL – 5 4 3 2 1 0

Editors

Jozef Šamaj received his Ph.D. degree from the Comenius University in Bratislava, Slovakia. He completed three post-doctoral programmes supported by Eurosilva, the Alexander von Humboldt Foundation, and the EU Marie Curie Programme in well-recognized laboratories in Toulouse, Bonn, and Vienna. His scientific work has focused on the cell biology of somatic embryogenesis, lignification in tree species, arabinogalactan proteins, the cytoskeleton, and signalling proteins. Jozef Šamaj has co-authored more than 60 research papers, reviews, and book chapters. Currently he is a senior lecturer and associate researcher at the Institute of Cellular and Molecular Botany in Bonn, Germany. His main research interest centers on the role of stress-induced MAP-kinase cascades in relation to vesicular trafficking and the cytoskeleton during stress responses in *Arabidopsis thaliana*.

František Baluška studied at the Comenius University in Bratislava, Slovakia. After his doctorate he worked at the Institute of Botany, Slovak Academy of Sciences (Bratislava, Slovakia), and completed two post-doctoral programmes supported by the Royal Society (London) and the Alexander von Humboldt Foundation (Bonn) in Bristol and Bonn. During that time, he focused on the cell biology of the plant cytoskeleton, especially as related to growth and polarity of cells in the root apex. He now is lecturer and associate researcher at the Institute of Cellular and Molecular Botany in Bonn, Germany. His current main research interest is the signal-mediated interactions between membrane trafficking and the cytoskeleton in the two plant model species *Zea mays* and *Arabidopsis thaliana*.

Diedrik Menzel graduated at the Free University of Berlin (Germany) in 1982. He was a predoctoral research fellow at the Department of Biochemistry, University of Melbourne, and a postdoctoral research associate at the Departments of Botany and Zoology of the University of California at Berkeley. In 1988 he habilitated at the Faculty of Biology, University of Heidelberg, and became a leader of the Plant Cytoskeleton Group at the Max-Planck-Institute for Cell Biology in Ladenburg. In 1996 he was appointed full professor in Plant Cell Biology at the University of Bonn. His work is related to cell architecture and morphogenesis in higher plants and algae. The emphasis of his current work is on actin cytoskeleton and endomembrane dynamics, as well as the molecular architecture of plant myosins. His other interests include gene expression and intracellular transport of mRNA in the unicellular green alga *Acetabularia*.

Preface

In all eukaryotic cells, endocytosis is an ancient and essential process necessary for the internalization of molecules from the plasma membrane and the extracellular environment. Endocytosis is also needed for plasma membrane recycling, uptake and degradation of signal molecules, as well as for selective sorting and delivery of these components to various intracellular destinations. In some special cases, endocytosis is involved in the internalization of pathogenic or symbiotic microorganisms. Significantly, recent data in various systems (yeast, plants and animals) revealed that endocytic and secretory pathways are very closely related, being dependent on each other, together constituting a dynamic and signal-responsive system of endomembrane flow in eukaryotic cells.

Skeptics had argued that rigid cell walls and high turgor pressure might preclude endocytosis in plant cells, and consequently the study of endocytosis in higher plants has remained dormant up to the recent time. Despite these negative predictions, plants are emerging as very useful model system for studying endocytic vesicle recycling, cell wall turnover and cell-to-cell communication based on signalling endosomes. Availability of *Arabidopsis thaliana*, with its sequenced genome, mutants, and strong genetics, will foster plants to conquer the forefront of this fascinating field of cell biology. Recent convincing evidence reveal that turgid plant cells not only accomplish endocytosis but also use it in order to regulate signalling pathways, abundance of important cell wall molecules, plasma membrane lipids, transport proteins and receptors. Surprisingly, plant endocytosis even satisfies the nutritional demands of plant cells. Last but not least, it is also emerging that endocytosis is involved in other physiological and developmental processes such as gravitropism, gas exchange and stomatal movements, cell growth, cell wall morphogenesis and cytokinesis. Despite the growing body of evidence, there is still no definitive proof of receptor-mediated endocytosis in plants, even though it was shown that some receptors indeed cycle via endocytosis in plants.

Other key questions have been addressed with growing confidence and amongst the answers that are beginning to emerge are the following:

1. cell wall composition is regulated through endocytosis;
2. recycling of plasma membrane transporters involves BFA-sensitive regulators of small GTPases;

3. the trans Golgi network (TGN) should be considered as a part of the endocytic network.

Intriguing similarities between the roles of endocytosis in plants and animals are emerging, e.g. during cytokinesis. Additionally, some results suggest that plants, like animals, are capable of synaptic cell-to-cell communication involving rapid recycling of endocytic vesicles.

These recent data, obtained from plant experimental model systems, support the view that endocytosis has an immense and ancient importance in all eukaryotic cells tracing back to their evolutionary origin. Several of the crucial molecular players involved in endocytosis-related processes were characterized and thus help to draw a picture as detailed as never before. Many of these will be described in this book.

The Editors June 2005

Contents

Plant Cell Monogr (1)
J. Šamaj · F. Baluška · D. Menzel: Plant Endocytosis
DOI 10.1007/7089_002/Published online: 22 September 2005

Methods and Molecular Tools for Studying Endocytosis in Plants—an Overview

Jozef Šamaj[1,2]

[1]Institute of Cellular and Molecular Botany,
University of Bonn, Kirschallee 1, 53115 Bonn, Germany
jozef.samaj@uni-bonn.de

[2]Institute of Plant Genetics and Biotechnology,
Slovak Academy of Sciences, Akademicka 2, 949 01 Nitra, Slovakia
jozef.samaj@uni-bonn.de

Abstract Proteins of the endocytosis machinery in plants, such as clathrin and adaptor proteins, were isolated and characterized using combinations of molecular biological (cloning and tagging) and biochemical methods (gel filtration, pull-down assays, surface plasmon resonance and immunoblotting). Other biochemical methods, such as cell fractionation and sucrose density gradients, were applied in order to isolate and further characterize clathrin-coated vesicles and endosomes in plants. Endocytosis was visualized in plant cells by using both non-fluorescent and fluorescent markers, and by employing antibodies raised against endosomal proteins or green fluorescent protein-tagged endocytic proteins in combination with diverse microscopic techniques, including confocal laser scanning microscopy and electron microscopy. Genetic and cell biological approaches were used together to address the role of a few proteins potentially involved in endocytosis. Additionally, biochemical and/or biophysical/electrophysiological methods were occasionally combined with microscopic methods (including both in situ and in vivo visualization) in plant endocytosis research.

1 Introduction

A variety of methods have been used to study endocytosis in isolated protoplasts, suspension cells and intact cells organized within tissues and organs. Among them, microscopic, biophysical/electrophysiological, biochemical, molecular and genetic methods and their combinations have been very helpful in revealing the diversity of the endocytic pathways and molecules involved (reviewed by Holstein, 2002; Geldner, 2004; Šamaj et al., 2004, 2005; Murphy et al., 2005).

2 Biochemical and Molecular Biological Methods

2.1 Isolation of Clathrin-Coated Vesicles

Plant clathrin-coated vesicles (CCVs) were isolated from cucumber and zucchini hypocotyls (Depta et al., 1991; Holstein et al., 1994). CCV components were protected against proteolysis using homogenization media composed of 0.1 M MES (pH 6.4), 1 mM EGTA, 3 mM EDTA, 0.5 mM $MgCl_2$, a mixture of proteinase inhibitors and 2% (w/v) fatty-acid-free BSA (Holstein et al., 1994). The crude CCV fraction (40 000–120 000 g pellet) was further purified by centrifugation in Ficoll/sucrose according to Campbell et al., (1983) and then by isopycnic centrifugation in a sucrose density gradient using a vertical rotor (160 000 g, 2.5 h, Depta et al., 1991). CCV-enriched fractions (collected at 40–45% sucrose) were removed, pooled and pelleted. CCV fractions were stored at – 80 °C for further use. Immunoblotting was performed using monoclonal antibodies against mammalian adaptins and clathrin. Confirmation of the presence of a β-type adaptin in plants was provided by dot and Southern blotting experiments using genomic DNA from zucchini hypocotyls and a β-adaptin cDNA clone from human fibroblasts (Holstein et al., 1994).

2.2 Cloning, Tagging and Interactions between Plant Clathrin and Adaptor Proteins

A full-length cDNA clone for *Arabidopsis* clathrin light chain was isolated and tagged with GST-myc epitopes. It was shown that this construct specifically interacts (binds) with the His-tagged hub region of mammalian clathrin heavy chain using Superose 12 gel filtration and immunoblotting (Scheele and Holstein, 2002). In a similar approach, *Arabidopsis* adaptor proteins AP180 and αC-adaptin were cloned and tagged with His or GST, respectively, and their binding requiring the plant-specific DPF motif was confirmed via pull-down assays and immunoblotting, or alternatively by surface plasmon resonance analysis (Barth and Holstein, 2004). It was also shown in this study using the same approach that AP180 binds to *Arabidopsis* clathrin heavy chain, and αC-adaptin binds several mammalian endocytic proteins such as amphiphysin, epsin and dynamin. AP180 promotes clathrin assembly into cages having almost uniform size and distribution. When the DLL domain was deleted from AP180, its clathrin assembly activity was abolished but its binding to triskelia was not affected, which suggests that this motif is not involved in clathrin binding (Barth and Holstein, 2004). These combined molecular biological and biochemical studies revealed that clathrin and adaptor proteins isolated from plants display the same structural and functional features as their mammalian counterparts.

2.3
Cell Fractionation and Isolation of Endosomes

Cell fractionation on sucrose gradients combined with immunoblotting with specific marker antibodies represents the most useful method for isolation of endomembranous compartments (e.g. Boonsirichai et al., 2003; Preuss et al., 2004; Fig. 1a). These methods have been applied to show that peripheral plasma membrane protein ARG1, which localizes to endocytic brefeldin A (BFA) compartments together with auxin efflux facilitator PIN2, cofractionates with the plasma membrane marker H-ATPase and with different en-

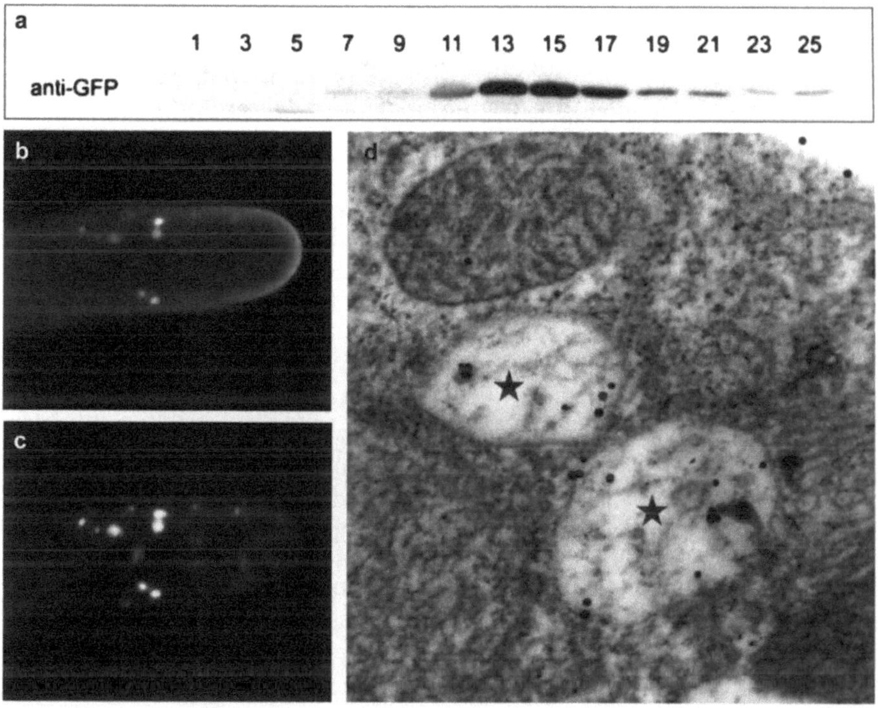

Fig. 1 Methods for studying endocytosis in plants. **a** Microsomal membranes were isolated from *Arabidopsis* plants stably transformed with endosomal marker RabF2a tagged to GFP and fractionated on sucrose gradients. Subsequently, they were subjected to immunoblotting with GFP antibody. Please note that RabF2a is enriched in endosomal fractions 13–17. **b, c** Confocal laser scanning microscopy imaging showing colocalization of endosomal tracer FM4-64 **b** with GFP-tagged endosomal molecular marker RabF2a **c** in actively growing root hairs of *Arabidopsis* roots. **d** Immunogold electron microscopy localization of arabinogalactan proteins (AGPs) within pre-vacuolar compartments (indicated by *stars*) of *Drosera* glandular cells using monoclonal antibody JIM13. AGPs represent cell wall cargo, which is internalized from plasma membrane and delivered via pre-vacuolar compartments to lytic vacuole for degradation and turnover

domembranes including endosomes labelled with PEP12 antibody (Boonsirichai et al., 2003). Membrane fractionation on sucrose density gradients also revealed that the small GTPase RabA4b localizes to a novel endomembrane compartment associated with tip growth of root hairs (Preuss et al., 2004), which turned out to be related to both secretion and endocytosis. Recently, this method was also used for the isolation and identification of prevacuolar compartments (PVCs) representing late endosomes in plants (Tse et al., 2004). Cell extracts from protoplasts of tobacco BY-2 cells were collected and loaded on discontinuous sucrose gradients (consisting of 5 mL each of 25, 40, 55 and 70% (w/v) sucrose solution in basic buffer). The gradient was centrifuged at 110 000 g for 2 h. Immunoblotting with antibodies against vacuolar sorting receptor (VSR) was used to identify PVC/endosomal fractions enriched with VSRs. Further, these fractions were pooled, diluted and separated on a second, continuous 25–50% (w/v) linear sucrose gradient. Each fraction (1 mL) of this gradient was subjected to immunoblotting with specific marker antibodies directed against VSR. In this way, PVCs/late endosomes were isolated and biochemically characterized. Moreover, immunogold electron microscopic (EM) labelling with VSR antibodies revealed that the PVC/late endosome-enriched fractions in fact possess multivesicular bodies (MVBs). Thus, MVBs were identified as PVCs/late endosomes in tobacco suspension BY-2 cells (Tse et al., 2004).

2.4 Isolation of Plasma Membrane Lipid Rafts

Recently, lipid raft plasma membrane domains were identified in plants based on their insolubility with the detergent Triton X-100 (Berczi and Horvath, 2003; Mongrand et al., 2004; Borner et al., 2005). First results obtained using thin-layer chromatography revealed that both quantitative and qualitative differences exist between the lipid composition of plant plasma membranes isolated from etiolated bean hypocotyls and green *Arabidopsis* leaves (Berci and Horvath, 2003). Later, protocols for the preparation of Triton X-100 detergent-resistant membranes (DRMs) from *Arabidopsis* callus were developed by Borner et al., (2005). Further, a proteomics approach using two-dimensional gel electrophoresis and liquid chromatography–tandem mass spectrometry revealed that the DRMs were highly enriched in specific proteins. Among them, eight glycosylphosphatidylinositol (GPI)-anchored proteins, several plasma membrane (PM) ATPases, multidrug resistance proteins and proteins of the stomatin/prohibitin/hypersensitive response family, were identified, suggesting that the DRMs originated from PM domains. Further analysis has shown that PM contains phytosterol and sphingolipid-rich lipid domains with a specialized protein composition. DRMs were prepared by low-temperature detergent extraction. According to Borner et al., (2005), membranes were resuspended in cold TNE (25 mM Tris-HCl, 150 mM NaCl, 5 mM EDTA, pH 7.5) containing 4–6 : 1 (detergent-to-protein, w/w) excess of Triton X-100 (no detergent was used in the control extractions). The final concentration of Triton X-100 was approximately 2%. Extractions were performed on ice with shaking at 100 rpm for 35 min. Extracts were adjusted to 1.8 M sucrose (Suc)/TNE by addition of 3 volumes of cold 2.4 M Suc/TNE. Extracts were overlaid with Suc step gradients 1.6–1.4–1.2–0.15 M and centrifuged at 240 000 g in a Beckman SW50.1 rotor for 18 h at 4 °C. DRMs were visible as off-white to white bands near the 1.2/1.4 and 1.4/1.6 M interfaces. Control fractions had a grey-green tinge. Fractions of 1 mL (0.5 mL above and 0.5 mL below the centre of the bands) were collected to harvest the DRM fractions and control fractions. Membranes were diluted with 4 volumes of cold TNE and pelleted at 100 000 g for 2 h in a Beckman 50Ti rotor.

3
Genetic Approaches

Site-directed mutagenesis of important amino acids and truncated versions resulted in mislocalization of the mutated Rab5 proteins, Ara6 and Ara7, which were preferentially localized either to the plasma membrane or to the tonoplast but not to endosomes (Ueda et al., 2001). This mutational analysis revealed that Ara6 requires N-terminal fatty acid acylation, nucleotide-binding and a C-terminal amino acid sequence for its correct targeting to endosomes (Ueda et al., 2001).

Stably transformed plants carrying constitutively active mitogen-activated protein kinase (MAPK) SIMK mutant (carrying a point mutation preventing dephosphorylation) are able to overcome root-hair growth inhibition caused by the MAPK inhibitor, UO126, which is linked to downregulated endo/exocytosis in inhibitor-treated control root hairs (Šamaj et al., 2002).

Point mutation within the catalytic Sec7 domain of the endosomal protein, GNOM, which is an explicitly BFA-sensitive guanosine exchange factor for ADP-ribosylation factor (ARF-GEF), renders this protein BFA-insensitive. Transgenic plants carrying such a mutated GNOM version show defects in endosomal recycling of the auxin efflux facilitator PIN1. Additionally, the inhibition of polar auxin transport upon BFA treatment is rescued, and endosomes show morphological changes in this mutant (Geldner et al., 2003).

ARF1, the reaction partner of ARF-GEF, is a small GTPase involved in vesicular trafficking, and constitutive active, GTP-locked (Q71L) mutant localized to both Golgi and endosomes (similarly to wild-type protein). The dominant negative, GDP locked form (T31N) was rather diffusely distributed throughout the cytoplasm and the nucleus (Xu and Scheres, 2005). Recently, it was reported that overexpression of constitutive active mutant of another small GTPase, RAC10, in *Arabidopsis* plants abolished normal endocytic uptake of FM dye into root hairs (Bloch et al., 2005).

4
Electrophysiological Methods

Electrophysiological experiments provide the means for directly measuring protoplast plasma membrane surface area (membrane capacitance), and thus helped to prove that exo/endocytic cycles are accompanied by membrane internalization via endocytosis (Carroll et al., 1998; see the chapter by Homann, this volume). This membrane recycling occurs rapidly within seconds to minutes in response to osmotically induced cell volume changes. Hypoosmotic treatment reversibly increases the volume of guard cell protoplasts and *Fucus* zygotes with subsequent internalization of membrane, which can be monitored by electrophysiological measurements and concomitant uptake of FM dyes (Homann, 1998). Similarly, endocytic uptake of plasma membrane during hyperosmotically induced shrinkage of protoplasts was measured using patch-clamp measurements (membrane capacitance) correlated with the internalization of FM1-43 (Kubitscheck et al., 2000). Further patch-clamp capacitance measurements revealed that osmotically induced surface area changes in guard cell protoplasts occur through exo- and endocytosis of 300-nm vesicles (Homan and Thiel, 1999), which contain active potassium inward rectifying channel KAT1 organized in clusters (Hurst et al., 2004). Endocytic uptake of KAT1 was confirmed using FM4-64 as a most reliable endocytic marker in intact guard cells (Meckel et al., 2004).

5
Inhibitors of Vesicular Trafficking

5.1
Brefeldin A

Brefeldin A (BFA) is a fungal metabolite used as an inhibitor of vesicular trafficking, which blocks exocytosis/secretion but allows endocytosis to continue (Šamaj et al., 2004) or even enhances the endocytic uptake (Emans et al., 2002). As a consequence, endocytic material accumulates in intracellular compartments (called BFA compartments in cells of intact roots), which are hybrid organelles composed of endosomes and trans-Golgi network (TGN) (Šamaj et al., 2004). Several plasma membrane (PM) molecules, such as auxin efflux facilitators PIN1 and PIN2, H-ATPase, the syntaxin KNOLLE and ARG1, cycle between the PM and endosomal compartment. In the presence of BFA, they accumulate in endosomal BFA-induced compartments (Geldner et al., 2001; Baluška et al., 2002; Boonsirichai et al., 2003; Grebe et al., 2003). In addition to PM proteins, cell wall pectins also undergo similar BFA-sensitive endosomal recycling and colocalize on BFA compartments with PIN proteins (Baluška et al., 2002, 2005; Šamaj et al., 2004). Molecular targets of BFA are ARF-GEF proteins, which regulate vesicle formation and secretory steps on Golgi, TGN and endosomes in eukaryotic cells. GNOM/EMB30 is a BFA-sensitive ARF-GEF protein located on endosomes, which regulates endosomal recycling of PIN1 but likely not that of H-ATPase and KNOLLE (Geldner et al., 2003). It was also shown that BFA stimulates endocytic uptake of FM1-43, but inhibits its delivery to the tonoplast (Emans et al., 2002).

5.2 Wortmannin

Wortmannin is a specific inhibitor of the phosphatidylinositol P(I) 3-kinase in mammalian cells resulting in the blockage of endocytosis. In plant cells, it also inhibits endocytic uptake of FM1-43 (Emans et al., 2002) and protein sorting to the vacuole through its action on both the P(I) 3- and P(I) 4-kinases. Wortmannin affects the morphology of endosomes labelled with a FYVE-domain green fluorescent protein (GFP)-fusion construct and induces vacuolization of late endosomes/PVCs/MVBs labelled with yellow fluorescent protein (YFP)-tagged BP80 (binding protein 80, a vacuolar sorting receptor) in a dose-dependent manner, but it has no effect on the Golgi labelled with a CONST1-YFP construct (Tse et al., 2004; Voigt et al., 2005a). Recently, it was also shown that wortmannin inhibits recycling of BP80 between PVC and TGN, which was accompanied by leakage of the corresponding ligand to the vacuole. This drug does not prevent receptor–ligand binding but rather limits levels of BP80 (daSilva et al., 2005).

5.3 Auxin Efflux Inhibitors: TIBA and NPA

Surprisingly, auxin efflux inhibitors *N*-1-naphthylphthalamic acid (NPA) and 2,3,5-triiodobenzoic acid (TIBA) were found to non-specifically inhibit vesicular recycling of plasma membrane-associated molecules in *Arabidopsis* roots (Geldner et al., 2001). In the presence of TIBA, BFA did not induce intercellular accumulation of several proteins, such as auxin efflux facilitator PIN1, PM H-ATPase and syntaxin KNOLLE, in the endosomal compartment. Additionally, recovery from the effect of BFA (normally this effect is fully reversible by washout of BFA) was blocked by washout in the presence of TIBA, resulting in persistent localization of PM molecules to endocytic BFA compartments upon BFA treatment.

6
Non-fluorescent Markers for Endocytosis

6.1
Metal Markers

Heavy metals such as lanthanum, cationized ferritin and gold-conjugated bovine serum albumin and lectins were reported to be internalized by plant cells via endocytosis (Hubner et al., 1985; Hillmer et al., 1986; Lazzaro and Thomson, 1992; Villanueva et al., 1993). Internalization of these endocytic tracer molecules was followed using electron microscopy.

6.2
Biotinylated Markers

Biotinylated proteins such as peroxidase and bovine serum albumin (bBSA) are internalized by plant cells and localize to endomembranes (Horn et al., 1990, 1992; Bahaji et al., 2001). Uptake of these markers is temperature-sensitive, saturable and competed by free biotin showing properties of receptor-mediated endocytosis (Bahaji et al., 2001). Dividing cells seem to have higher endocytic rates and, furthermore, bBSA uptake is inhibited by treatment with the microtubule-depolymerizing drug nocodazole (Bahaji et al., 2001). Additionally, salt and osmotic stress initially inhibit but later on activate the uptake of bBSA (Bahaji et al., 2003).

7
Fluorescent Markers for Endocytosis

7.1
Labelled Signalling Ligands

Fluorescently labelled lipochitooligosaccharides (LCOs) and lipopolysaccharides (LPSs) were used to monitor endocytic uptake of the signal molecules produced by symbiotic and pathogenic bacteria, such as *Rhizobia* and *Xanthomonas* (Timmers et al., 1998; Gross et al., 2004). For LPSs it was shown that these are internalized in an amantadine-sensitive and energy- and temperature-dependent manner suggesting that the uptake was receptor-mediated. Moreover, it was shown that they pass through the endosomal compartment because they colocalize with endosomal marker Ara6 (Gross et al., 2004).

7.2 Lucifer Yellow

Lucifer Yellow (LY) is a membrane-impermeable dye, which is useful for studies of fluid-phase endocytosis in plant cells. Baluška et al., (2004) reported on the endocytic internalization of LY from specific plasmodesmata-associated subcellular PM domains into small vacuoles within cortex cells of maize roots. This fluid-phase endocytosis was dependent on the intact actomyosin cytoskeleton and likely related to the nutritional demands of these cells. Recently it was shown that sucrose follows the same route of fluid-phase endocytic uptake as LY and other fluid-phase markers in suspension cell cultures and storage root cells (see the chapter by Baluška et al., this volume). Additionally, under the condition of sugar starvation, LY was used to identify autolysosomes which accumulate around the nuclei of cultured tobacco cells treated with cysteine protease inhibitors (Yano et al., 2004).

7.3 Styryl FM Dyes

Lipophilic styryl dyes are membrane-impermeable polar fluorochromes which are fluorescent only upon their intercalation to the outer leaflet of the plasma membrane. Subsequently, they are internalized from the PM via endocytosis, label the membranes of different endosomal populations, and finally end up in the tonoplast. FM1-43 labels endosomes and the vacuolar tonoplast in tobacco suspensions (Emans et al., 2002) as well as vesicles in the clear zone and vacuoles of pollen tubes (Camacho and Malho, 2003). It seems that in some cell types such as stomata, the most hydrophobic FM4-64 is the better endosomal marker while FM1-43 can occasionally label the mitochondria, although the reason for this effect is unknown (Meckel et al., 2004). FM4-64 does not label endoplasmatic reticulum (ER) and Golgi itself (Bolte et al., 2004; Tse et al., 2004), but it might label the TGN (Bolte et al., 2004), which is considered to be part of the endomembrane sorting system integrated with endosomes and vacuoles (Šamaj et al., 2004, 2005). FM1-43 and FM4-64 were successfully used as endocytic tracers in different plant cell types such as *Arabidopsis* protoplasts, tobacco BY-2 suspension cells or intact fungal and plant cells, e.g. fungal hyphae, pollen tubes, root epidermal cells and root hairs, stomata and leaf epidermal cells (Carroll et al., 1998; Parton et al., 2001; Ueda et al., 2001, 2004; Emans et al., 2002; Geldner et al., 2003; Shope et al., 2003; Meckel et al., 2004; Uemura et al., 2004; Walther and Wendland, 2004; Ovecka et al., 2005; Voigt et al., 2005; Xu and Scheres, 2005). In particular, FM4-64 was useful in identifying early and late endosomes/PVCs in plant cells (Geldner et al., 2003; Ueda et al., 2001, 2004; Tse et al., 2004; Uemura et al., 2004; Voigt et al., 2005). Additionally, FM dyes were used to study the morphology and dynamics of vacuoles (Emans et al., 2002; Ovecka et al., 2005) and for the identification of autolysosomes in plant cells (Yano et al., 2004). Importantly, FM dyes can be used for quantification of endocytosis using high-resolution imaging (Ryan et al., 1997; Emans et al., 2002).

7.4 Filipin

Antibiotic filipin binds to structural sterols and because of its fluorescent properties it can be used for the visualization of sterols. Additionally, filipin-complexed sterols can also be visualized on the ultrastructural level. Filipin was used recently to label structural sterols on the plasma membrane and to study their internalization and endosomal trafficking in epidermal cells of intact *Arabidopsis* roots (Grebe et al., 2003; see the chapter by Ovecka and Lichtscheidl, this volume). It was shown that early endosomal trafficking of structural sterols is actin-dependent, BFA-sensitive, involves endosomes enriched with Ara6, and is connected to polar sorting events, such as recycling of PIN2, an auxin efflux facilitator (Grebe et al., 2003).

8 GFP Technology for Tagging Endocytic Proteins

Tagging with fluorescent proteins such as GFP and/or its fluorescent analogues (YFP or cyan fluorescent protein, CFP) as well as with DsRed was widely used to study endosomal trafficking and recycling of various proteins associated with plasma membrane and/or endosomal compartments. Importantly, some of these tagged proteins were also mutated and used for functional studies. For example, three *Arabidopsis* Rab-GTPases, Ara6, Ara7 and Rha1, were identified as endosomal markers (Ueda et al., 2001, 2004; Sohn et al., 2003; Šamaj et al., 2004). Several SNAREs were localized to the PM and/or endosomes using YFP tagging and colocalization with FM4-64 (Uemura et al., 2004). GNOM, a BFA-sensitive ARF-GEF (see above), was localized on endosomes and endosomal BFA compartments together with PIN1 (Geldner et al., 2003). Plasma membrane protein LTI6a tagged with GFP was also found on endosomal BFA compartments (Grebe et al., 2003).

Plasma membrane receptors, such as receptor-like kinases (RLKs) including SERK1, brasinosteroid receptors composed of BRI1 and SERK3 as well as CRINKLY4, were tagged with CFP, YFP and GFP and localized to endosomes (Shah et al., 2002; Rusinova et al., 2004; Gifford et al., 2005). SERK1 localization to endosomes was dependent on associated protein phosphatase KAPP (Shah et al., 2002). Additionally, GFP-tagged vacuolar sorting receptor BP80 was localized on PVCs/late endosomes (as shown above, Tse et al., 2004).

Endocytic internalization and recycling of plasma membrane potassium channel KAT1 was also studied using GFP tagging (Meckel et al., 2004). Additionally, peptide constructs encompassing the PI(3)P binding motif, FYVE, were used to study both endosomal trafficking towards the vacuole and for endosomal movements in intact roots together with tagged Rha1 and Ara6 (Voigt et al., 2005).

9 Immunolocalization Methods

Immunolabelling protocols adapted to sectioned and whole-mount samples were used to study endosomal cycling of plasma membrane molecules, cell wall pectins, and for localization of endosomal proteins. These methods were also very useful in order to verify the identity of plant endosomal organelles on the submicroscopic level using both confocal laser scanning microscopy (CLSM) and electron microscopy (EM).

A pharmacological approach employing BFA combined with immunofluorescence labelling revealed endosomal recycling of plasma membrane proteins including auxin efflux facilitators PIN1 (Geldner et al., 2001), PIN2 (Boonsirichai et al., 2003), H^+-ATPase (Geldner et al., 2001; Baluška et al., 2002) and plasma membrane associated protein ARG1 (Boonsirichai et al., 2003). Moreover, regulatory proteins such as ARF1 (Baluška et al., 2002), the ARF-GEF protein, GNOM, which represents one of the BFA targets in plant cells (as shown above, Geldner et al., 2003) and cytokinesis-specific syntaxin KNOLLE (Geldner et al., 2001) were immunolocalized to endocytic BFA compartments. Surprisingly, immunolocalization experiments with antibodies specific against cell wall pectins crosslinked with boron and calcium have shown that these cell wall components also take a similar route of endocytic recycling and accumulation in BFA compartments (Baluška et al., 2002; Šamaj et al., 2004).

Additionally, immunogold EM labelling with an antibody specific against Lucifer Yellow confirmed the existence of fluid-phase endocytosis in plant cells, and revealed that this type of endocytic internalization occurs preferentially on specialized plasmodesmata domains of maize root cortex cells (Baluška et al., 2004).

Both immunofluorescence localization with vacuolar sorting receptor (VSR) and pre-vacuolar compartment (PVC) marker antibodies (AtSYP21 and 14G7), as well as immunogold positive and negative EM labelling with VSR antibodies, were used to study endomembranous compartments containing these molecules. This study identified multivesicular bodies (200–500 nm) as PVCs and these actually correspond to late endosomes in tobacco BY-2 suspension culture cells (Tse et al., 2004).

10 Microscopic Methods

Microscopic methods including epifluorescence, CLSM and EM allowed visualization of endocytosis and endocytic compartments in both fixed and living plant cells. Additionally, in combination with other methods they contributed substantially to our understanding of endocytic internalization, compartmentalization of endocytic pathways on the subcellular level, the dynamic behaviour of endocytic molecules and their spatio-temporal interactions which in turn regulate the endocytic process.

10.1 Electron Microscopy

EM was used to visualize clathrin-coated pits, coated vesicles and endocytic compartments in chemically fixed and freeze-fixed plant cells (e.g. Derksen et al., 2002; Tse et al., 2004). Additionally, EM was often used to monitor the internalization at the subcellular level of endocytic markers such as cationized ferritin or gold-labelled lectin into isolated protoplasts (Hubner et al., 1985; Hillmer et al., 1986; Horn et al., 1990). Several endocytic compartments, such as partially coated reticulum, pre-vacuolar compartment and multivesicular bodies, were discovered and described using EM methods.

The pharmacological effect of specific inhibitors of vesicular trafficking, such as brefeldin A (BFA), on the morphology and ultrastructure of ER, Golgi and other endomembranes was also monitored using EM methods (Ritzenthaler et al., 2002; Grebe et al., 2003). Moreover, EM in combination with immunogold labelling and/or biochemical fractionation was recently used to monitor endocytic internalization of the fluid-phase marker LY (Baluška et al., 2004) and for precise localization of molecules that specifically associate with the late endosomal compartment, e.g. localization of BP80 to multivesicular bodies (Tse et al., 2004). Additionally, arabinogalactan proteins (AGPs) associated with the plasma membrane through their GPI anchoring were reported to undergo internalization into plant cells and endocytic trafficking via pre-vacuolar compartments with subsequent delivery to vacuoles for degradation (Šamaj et al., 2000). Interestingly, EM clearly revealed tight association of AGPs both with plasma membrane and with vacuolar tonoplast, suggesting that they play a role in the integrity of tonoplast and transvacuolar strands (Šamaj et al., 2000). Additionally, multivesicular bodies enriched with the AGP epitopes LM2 and JIM13 can be found in cells of intact maize roots and in glandular cells of *Drosera* tentacles (Fig. 1d). These secretory cells are supposed to balance their higher rates of exocytosis with compensatory endocytosis.

10.2
Epifluorescence and Confocal Laser Scanning Microscopy (CLSM)

Epifluorescence and CLSM in combination with immunolabelling with antibodies were used to study endocytic internalization and recycling of plasma membrane proteins as well as cell wall pectins and xyloglucans (e.g. Geldner et al., 2001, 2003; Baluška et al., 2002, 2005; Grebe et al., 2003; Tse et al., 2004). Additionally, these methods were extensively used to monitor internalization of fluorescent endosomal markers such as FM dyes and LY as well as structural sterols labelled with filipin (Emans et al., 2002; Grebe et al., 2003; Baluška et al., 2004; Meckel et al., 2004; Tse et al., 2004). Importantly, CLSM was used to study the subcellular localization and dynamic behaviour of endocytic marker proteins such as Rab GTPases, SNAREs and GNOM, as well as FYVE-domain fusion peptide along within actin (Voigt et al., 2005).

10.3
Fluorescence Recovery after Photobleaching (FRAP)

FRAP allows the study of the reappearance of fluorescently labelled proteins in membrane domains after a specific region in the membranes has been photobleached. This method was used to study cycling and the reappearance of protein LTI6a (LTI=low-temperature induced) tagged with GFP on the plasma membrane and endosomal BFA compartments (Grebe et al., 2003).

10.4
Fluorescence Resonance Energy Transfer (FRET)

FRET is an advanced CLSM method which is used to monitor physical interactions between proteins. In order to do this, both reaction partners or interacting protein domains need to be fused to variants of the green fluorescent protein, which are designed such that radiation-free energy can be transferred between them as soon as they come into close proximity. This method was applied in the detection of physical interaction between receptor kinase SERK1 and PP2C-type phosphatase KAPP (kinase-associated phosphatase) (Shah et al., 2002). It was revealed that SERK1 tagged with CFP colocalizes with KAPP tagged with YFP on both the plasma membrane and endosomes; however, only on the endosomes do they come into close enough proximity to interact with one another, which suggests a role of KAPP in SERK1 endocytic internalization.

11 Conclusions and Future Prospects

Diverse biophysical, biochemical, molecular biological, genetic and microscopic methods and techniques were used in the field of plant endocytosis. Especially, the combination of these various methods has recently started to provide useful information about the structural and functional organization of endocytic pathways in plants. One of the reasons why the field of plant endocytosis research seems to be lagging behind that of animal endocytosis is the slow adaptation of methods to the particular nature of plant cells and tissues. This problem has been encountered in most areas of plant cell biology. However, as shown in this chapter, the repertoire of experimental approaches and specific methods redesigned from animal research for use on plant systems has surpassed a critical mass, so that the pace of progress is steadily rising.

The molecular aspects of endosomal interactions are clearly coming into focus and in a short-term perspective, the combination of genetic, biochemical and cell biological approaches should be most fruitful in answering specific questions related to the functions of the diverse endosome associated and/or resident proteins and other cargoes. Advanced microscopic techniques such as FRAP, FLIM and FRET should be instrumental for further progress in this field.

Acknowledgements I thank Diedrik Menzel for critical reading of the manuscript and Ursulla Mettbach and Claudia Heym for excellent technical assistance, as well as to Mary Preuss and Erik Nielsen for providing Fig. 1a. This work was supported by a grant from the Slovak Grant Agency APVT (grant no. APVT-51-002302), Bratislava, Slovakia.

References

Bahaji A, Cornejo MJ, Ortiz-Zapater E, Contreras I, Aniento F (2001) Uptake of endocytic markers by rice cells: variations related to the growth phase. Eur J Cell Biol 80:178–186

Bahaji A, Aniento F, Cornejo MJ (2003) Uptake of endocytic marker by rice cells: variations related to osmotic and saline stress. Plant Cell Physiol 44:1100–1111

Baluška F, Hlavacka A, Šamaj J, Palme K, Robinson DG, Matoh T, McCurdy DW, Menzel D, Volkmann D (2002) F-actin-dependent endocytosis of cell wall pectins in meristematic root cells: insights from brefeldin A-induced compartments. Plant Physiol 130:422–431

Baluška F, Šamaj J, Hlavacka A, Kendrick-Jones J, Volkmann D (2004) Actin-dependent fluid-phase endocytosis in inner cortex cells of maize root apices. J Exp Bot 55:463–473

Baluška F, Baroja-Fernandez E, Pozueta-Romero J, Hlavacka A, Etxeberria E, Šamaj J (2005) Endocytic uptake of nutrients, cell wall molecules, and fluidized cell wall portions into heterotrophic plant cells (in this volume). Springer, Berlin Heidelberg New York

Barth M, Holstein SHE (2004) Identification and functional characterization of *Arabidopsis* AP180, a binding partner of plant αC-adaptin. J Cell Sci 117:2051–2062

Berczi A, Horvath G (2003) Lipid rafts in the plant plasma membrane? Acta Biol Szeged 47:7–10

Blackbourn HD, Jackson AP (1996) Plant clathrin heavy chain: sequence analysis and restricted localization in growing pollen tubes. J Cell Sci 109:777–786

Bloch D, Lavy M, Efrat Y, Efroni I, Bracha-Drori K, Abu-Abied M, Sadot E, Yalovsky S (2005) Ectopic expression of an activated RAC in *Arabidopsis* disrupts membrane cycling. Mol Biol Cell 16:1913–1927

Bolte S, Brown S, Satiat-Jeunemaitre B (2004) The N-myristoylated Rab-GTPase m-Rabmc is involved in post-Golgi trafficking events to the lytic vacuole in plant cells. J Cell Sci 117:943–954

Boonsirichai K, Sedbrook JC, Chen R, Gilroy S, Masson P (2003) ALTERED RESPONSE TO GRAVITY is a peripheral membrane protein that modulates gravity-induced cytoplasmic alkalinization and lateral auxin transport in plant statocytes. Plant Cell 15:2612–2625

Borner GHH, Sherier DJ, Weimar T, Michaelson LV, Hawkins ND, Macaskill A, Napier JA, Beale MH, Lilley KS, Dupree P (2005) Analysis of detergent-resistant membranes in *Arabidopsis*. Evidence for plasma membrane lipid rafts. Plant Physiol 137:104–116

Camacho L, Malho R (2003) Endo/exocytosis in the pollen tube apex is differentially regulated by Ca^{2+} and GTPase. J Exp Bot 54:83–92

Carroll AD, Moyen C, van Kesteren WJP, Tooke F, Battey NH, Brownlee C (1998) Ca^{2+}, annexins, and GTP modulate exocytosis from maize root cap protoplasts. Plant Cell 10:1267–1276

Da Silva LLP, Taylor JP, Hadlington JL, Hanton SL, Snowden CJ, Fox SJ, Foresti O, Brandizzi F, Denecke J (2005) Receptor salvage from the prevacuolar compartment is essential for efficient vacuolar protein targeting. Plant Cell 17:132–148

Depta H, Robinson DG, Holstein SEH, Lützelschwab M, Michalke W (1991) Membrane markers in highly purified clathrin-coated vesicles from Cucurbita hypocotyls. Planta 183:434–442

Derksen J et al. (2002) Growth and cellular organization of *Arabidopsis* pollen tubes in vitro. Sex Plant Reprod 15:133–139

Emans N, Zimmermann S, Fischer R (2002) Uptake of a fluorescent marker in plant cells is sensitive to brefeldin A and wortmannin. Plant Cell 14:71–86

Geldner N, Friml J, Stierhof Y-D, Jürgens G, Palme K (2001) Auxin transport inhibitors block PIN1 cycling and vesicle trafficking. Nature 413:425–428

Geldner N, Anders N, Wolters H, Keicher J, Kornberger W, Muller P, Delbarre A, Ueda T, Nakano A, Jürgens G (2003) The *Arabidopsis* GNOM ARF-GEF mediates endosomal recycling, auxin transport, and auxin-dependent plant growth. Cell 112:219–230

Geldner N (2004) The plant endosomal system – its structure and role in signal transduction and plant development. Planta 219:547–560

Gifford ML, Robertson FC, Soares DC, Ingram GC (2005) *Arabidopsis* CRINKLY4 function, internalization, and turnover are dependent on the extracellular crinky repeat domain. Plant Cell 17:1154–1166

Grebe M, Xu J, Möbius W, Ueda T, Nakano T, Geuze HJ, Rook MB, Scheres B (2003) *Arabidopsis* sterol endocytosis involves actin-mediated trafficking via ARA6-positive early endosomes. Curr Biol 13:1378–1387

Gross A, Kapp D, Nielsen T, Niehaus K (2004) Endocytosis of Xanthomonas campestris pathovar campestris lipopolysaccharides in non-host plant cells of Nicotiana tabacum. New Phytol 165:215–226

Hillmer S, Depta H, Robinson DG (1986) Confirmation of endocytosis in higher plant protoplasts using lectin-gold conjugates. Eur J Cell Biol 41:142–149

Holstein SHE, Drucker M, Robinson DG (1994) Identification of a β-type adaptin in plant clathrin-coated vesicles. J Cell Sci 107:945–953

Holstein SE (2002) Clathrin and plant endocytosis. Traffic 3:614–620

Homann U (1998) Osmotically induced excursions in the surface area of guard cell protoplasts. Planta 206:329–333

Homann U, Thiel G (1999) Unitary exocytotic and endocytotic events in guard-cell protoplasts during osmotically driven volume changes. FEBS Lett 460:495–499

Homann U (2005) Endocytosis in guard cells (in this volume). Springer, Berlin Heidelberg New York

Horn MA, Heinstein PF, Low PS (1990) Biotin-mediated delivery of exogenous macromolecules into soybean cells. Plant Physiol 93:1492–1496

Horn MA, Heinstein PF, Low PS (1992) Characterization of parameters influencing receptor-mediated endocytosis in cultured soybean cells. Plant Physiol 98:673–679

Hubner R, Depta H, Robinson DG (1985) Endocytosis in maize root cap cells: evidence obtained using heavy metal salt solutions. Protoplasma 129:214–222

Hurst AC, Meckel T, Tayefeh S, Thiel G, Homann U (2004) Trafficking of the plant potassium inward rectifier KAT1 in guard cell protoplasts of Vicia faba. Plant J 37:391–397

Kubitscheck U, Homann U, Thiel G (2000) Osmotically evoked shrinking of guard-cell protoplasts causes vesicular retrieval of plasma membrane into the cytoplasm. Planta 210:423–431

Lazzaro MD, Thomson WW (1992) Endocytosis of lanthanum nitrate in the organic acid-secreting trichomes of chickpea (Cicer arietinum). Am J Bot 79:1113–1118

Meckel T, Hurst AC, Thiel G, Homann U (2004) Endocytosis against high turgor: intact guard cells of Vicia faba constitutively endocytose fluorescently labelled plasma membrane and GFP-tagged K^+-channel KAT1. Plant J 39:182–193

Mongrand S, Morel J, Laroche J, Claverol S, Carde JP, Hartmann MA, Bonneu M, Simon-Plas F, Lessire R, Bessoule JJ (2004) Lipid rafts in higher plant cells: purification and characterization of Triton X-100-insoluble microdomains from tobacco plasma membrane. J Biol Chem 279:36277–36286

Murphy AS, Bandyopadhyay A, Holstein SE, Peer WA (2005) Endocytotic cycling of PM proteins. Annu Rev Plant Biol 56:221–251

Ovecka M, Lang I, Baluška F, Ismail A, Illeš P, Lichtscheidl IK (2005a) Endocytosis and vesicle trafficking during tip growth of root hairs. Protoplasma (in press)

Ovecka M, Lichtscheidl IK (2005) Sterol endocytosis and trafficking in plant cells (in this volume). Springer, Berlin Heidelberg New York

Parton RM, Fischer-Parton S, Watahiki MK, Trewavas AJ (2001) Dynamics of the apical vesicle accumulation and the rate of growth are related in individual pollen tubes. J Cell Sci 114:2685–2695

Preuss ML, Serna J, Falbel TG, Bednarek SY, Nielsen E (2004) The *Arabidopsis* Rab GTPase RabA4b localizes to the tips of growing root hairs. Plant Cell 16:1589–1603

Ritzenthaler C, Nebenfuhr A, Movafeghi A, Stussi-Garaud C, Behnia L, Pimpl P, Staehelin LA, Robinson DG (2002) Reevaluation of the effects of brefeldin A on plant cells using tobacco Bright Yellow 2 cells expressing Golgi-targeted green fluorescent protein and COPI antisera. Plant Cell 14:237–261

Russinova E, Borst J-W, Kwaaitaal M, Cano-Delgado A, Yin Y, Chory J, de Vries SC (2004) Heterodimerization and endocytosis of *Arabidopsis* brassinosteroid receptors BRI1 and AtSERK3 (BAK1). Plant Cell 16:3216–3229

Ryan TA, Reuter H, Smith SJ (1997) Optical detection of a quantal presynaptic membrane turnover. Nature 388:478–482

Šamaj J, Šamajová O, Peters M, Baluška F, Lichtscheidl I, Knox JP, Volkmann D (2000) Immunolocalization of LM2 arabinogalactan-protein epitope associated with endomembranes of plant cells. Protoplasma 212:186–196

Šamaj J, Ovecka M, Hlavacka A, Lecourieux F, Meskiene I, Lichtscheidl I, Lenart P, Salaj J, Volkmann D, Bogre L, Baluška F, Hirt H (2002) Involvement of the mitogen-activated protein kinase SIMK in regulation of root hair tip growth. EMBO J 21:3296–3306

Šamaj J, Baluška F, Voigt B, Schlicht M, Volkmann D, Menzel D (2004) Endocytosis, actin cytoskeleton and signaling. Plant Physiol 135:1150–1161

Šamaj J, Read N, Baluška F (2005) Endocytosis in plants and filamentous fungi. Trends Cell Biol (in press)

Scheele U, Holstein SHE (2002) Functional evidence for the identification of an *Arabidopsis* clathrin light chain polypeptide. FEBS Lett 514:355–360

Shah K, Russinova E, Gadella TW Jr, Willemse J, De Vries SC (2002) The *Arabidopsis* kinase-associated protein phosphatase controls internalization of the somatic embryogenesis receptor kinase. Genes Dev 16:1707–1720

Shope JC, DeWald DB, Mott KA (2003) Changes in surface area of intact guard cells are correlated with membrane internalization. Plant Physiol 133:1314–1321

Sohn EJ, Kim ES, Zhao M, Kim SJ, Kim H, Kim Y-W, Lee YJ, Hillmer S, Sohn U, Jiang L, Hwang I (2003) Rha1, an *Arabidopsis* Rab5 homolog, plays a critical role in the vacuolar trafficking of soluble cargo proteins. Plant Cell 15:1057–1070

Timmers AC, Auriac MC, de Billy F, Truchet G (1998) Nod factor internalization and microtubular cytoskeleton changes occur concomitantly during nodule differentiation in alfalfa. Development 125:339–349

Tse YC, Mo B, Hillmer S, Zhao M, Lo SW, Robinson DG, Jiang L (2004) Identification of multivesicular bodies as prevacuolar compartments in Nicotiana tabacum BY-2 cells. Plant Cell 16:672–693

Ueda T, Yamaguchi M, Uchimiya H, Nakano A (2001) Ara6, a plant-unique novel type Rab GTPase, functions in the endocytic pathway of *Arabidopsis* thaliana. EMBO J 20:4730–4741

Ueda T, Uemura T, Sato MH, Nakano A (2004) Functional differentiation of endosomes in *Arabidopsis* cells. Plant J 40:783–789

Uemura T, Ueda T, Ohniwa RL, Nakano A, Takeyasu K, Sato MH (2004) Systematic analysis of SNARE molecules in *Arabidopsis*: dissection of the post-Golgi network in plant cells. Cell Struct Funct 29:49–65

Villanueva MA, Taylor J, Sui X, Griffing LR (1993) Endocytosis in plant protoplasts: visualization and quantification of fluid-phase endocytosis using silver-enhanced bovine serum albumin-gold. J Exp Bot 44:275–281

Voigt B, Timmers ACJ, Šamaj J, Hlavacka A, Ueda T, Preuss M, Nielsen E, Mathur J, Emans N, Stenmark H, Nakano A, Baluška F, Menzel D (2005) Actin-propelled motility of endosomes is tightly linked to polar tip-growth of root hairs. Eur J Cell Biol 84:609–621

Voigt B, Timmers T, Šamaj J, Müller J, Baluška F, Menzel D (2005) GFP-FABD2 fusion construct allows in vivo visualization of the dynamic actin cytoskeleton in all cells of *Arabidopsis* seedlings. Eur J Cell Biol 84:595–608

Walther A, Wendland J (2004) Apical localization of actin patches and vacuolar dynamics in Ashbya gossypii depend on the WASP homolog Wal1p. J Cell Sci 117:4947–4958

Xu J, Scheres B (2005) Dissection of *Arabidopsis* ADP-Ribosylation Factor 1 function in epidermal cell polarity. Plant Cell 17:525–536

Yano K, Matsui S, Tsuchiya T, Maeshima M, Kutsuna N, Hasezawa S, Moriyasu Y (2004) Contribution of the plasma membrane and central vacuole in the formation of autolysosomes in cultured tobacco cells. Plant Cell Physiol 45:951–957

Plant Cell Monogr (1)
J. Šamaj · F. Baluška · D. Menzel: Plant Endocytosis
DOI 10.1007/7089_003/Published online: 12 October 2005

Endocytic Uptake of Nutrients, Cell Wall Molecules and Fluidized Cell Wall Portions into Heterotrophic Plant Cells

František Baluška[1,4] (✉) · Edurne Baroja-Fernandez[2] · Javier Pozueta-Romero[2] · Andrej Hlavacka[1] · Ed Etxeberria[2,3] · Jozef Šamaj[1,5]

[1]Institute of Cellular and Molecular Botany, Rheinische Friedrich-Wilhelms-University Bonn, Department of Plant Cell Biology, Kirschallee 1, 53115 Bonn, Germany
baluska@uni-bonn.de

[2]Agrobioteknologia eta Natura Baliabideetako Instituta Nafarroako Unibertsitate Publikoa and Consejo Superior de Investigaciones Científicas Mutiloako etorbidea zenbaki gabe, 31192 Mutiloabeti, Nafarroa, Spain

[3]University of Florida, IFAS, Citrus Research and Education Center, 700 Experiment Station Road, Lake Alfred, FL, 33850, USA

[4]Institute of Botany, Slovak Academy of Sciences, Dúbravská cesta 14, 84223 Bratislava, Slovakia
baluska@uni-bonn.de

[5]Institute of Plant Genetics and Biotechnology, Slovak Academy of Sciences, Akademicka 2, 95007 Nitra, Slovakia

Abstract After arrival at the surface of heterotrophic cells, nutrients are taken up by these cells via endocytosis to sustain metabolic processes. Recent advances in plant endocytosis reveal that this is true for their heterotrophic cells, either cultivated in suspension cultures or for intact root apices. Importantly, sucrose appears to act as a specific stimulus for fluid-phase endocytosis. Uptake of extracellular nutrients by endocytosis is not in direct conflict with transport through membrane-bound carriers given that cell homeostasis can be better maintained if both these mechanisms operate in parallel. Besides nutrients, plant cells also accomplish internalization of cell wall molecules, such as xyloglucans and boron/calcium cross-linked pectins. Even large portions of apparently fluidized cell wall together with symbiotic bacteria can be internalized into some plant cells, suggesting that they can perform phagocytosis-like tasks despite their robust cell walls. Internalized cell wall molecules allow effective adaptation to osmotic stress, and also may serve for nutritive purposes. Plant endosomes enriched with the internalized cell wall molecules are used for new cell wall formation during plant cytokinesis. Moreover, rapid remodeling of cell walls through endosomal recycling is likely involved in opening/closing movements of stomata, and perhaps also in the formation of wall papillae during pathogen attacks and in recovery of cells from plasmolysis.

1 Introduction

Endocytosis is an inherent feature of all eukaryotic cells. The most notable role of endocytosis, elaborated especially in amebae and *Dictyostelium* cells, is cell nutrition via internalization of extracellular nutritive molecules and solutes (Marsch 2002). While vesicle-mediated nutrient uptake had been demonstrated in other organisms, corresponding studies in plants were derailed by: (i) studies suggesting the possible involvement of ion channels in the uptake of Lucifer Yellow when this fluorochome was actually intended to serve as a fluid phase marker (Cole et al. 1991); and (ii) by the demonstration of sugar transporters at both the plasma membrane (Williams et al. 2000; Lemoine 2000) and the tonoplast (Getz 1991).

Early reports on the engulfment of multilamellar and multivesicular compartments, now known to represent the plant late endosomes (Tanchak and Fowke 1987; Tse et al. 2004), by the central vacuole (Herman and Lamb 1991), as well as on their fusion with the plasma membrane resulting in so-called paramural bodies (Roland 1972), were dismissed as fixation artifacts. Early indications, that endocytosis may participate directly in the trapping, distribution, and sorting of extracellular components, were inherent in several papers published from the seventies up to the early nineties. Unfortunately, these early studies were not accepted by the mainstream plant cell biology community, since the general view was, that the high turgor pressure makes endocytosis in plant cells unfeasible (reviewed by Šamaj et al. 2004, 2005). As a result, the role of endocytosis as an inherent part of the overall mechanism of nutrient uptake into heterotrophic plant cells remained a controversial issue until recently (Echeverría 2000; Baluška et al. 2004; Etxeberria et al. 2005a, 2005b, 2005c).

The concept that dissolved nutrients in the extracellular milieu are potentially carried to the vacuole by an endocytic-related network was revived using a variety of membrane impermeable soluble dyes which eventually appeared in the vacuole, for instance in tobacco cultured cells (Emans et al. 2002; Yamada et al. 2005). Moreover, new studies reported internalization of fluid-phase endocytosis markers into cells of onion and maize root apices (Cholewa and Peterson 2001; Baluška et al. 2004), as well as into tobacco suspension culture cells (Yano et al. 2004). These studies using the fluorescent membrane impermeable dyes Alexa-568, 8-hydroxy-1,3,6-pyrenetrisulphonate, and Lucifer Yellow (LY), helped to overcome previous doubts and put to rest criticisms expressed on early experiments performed with these endocytic tracers (see the chapter by Šamaj, this volume).

2
Endocytic Uptake of Solutes and Sucrose into Suspension Plant Cells

That a portion of the nutrients stored in the vacuole are taken up by endocytosis was recently established using sycamore cell cultures in conjunction with the endocytic inhibitors wortmannin and LY294002, and Lucifer Yellow as the fluid-phase endocytosis marker (Etxeberria et al. 2005a). When transferred into a sucrose-rich medium, cells accumulated sucrose rapidly for approximately 60 min. Sucrose uptake during this period proved to be wortmannin and LY294002 insensitive. After 90 min incubation, the rate of sucrose uptake increased rapidly in a linear manner for an additional 6 h. This second phase was strongly suppressed by the endocytic inhibitors wortmannin and LY294002, which would be in conformity with the existence of an endocytic transport of sucrose into the cells. Complete cessation of sucrose uptake by wortmannin occurred at a time when sucrose had already commenced to accumulate rapidly, this strongly substantiates these observations.

Possible involvement of fluid-phase endocytosis in sucrose uptake was further investigated in experiments where LY was added together with sucrose. LY accumulation followed a pattern very similar to that of sucrose after the initial 90 min of culture, and inclusion of either wortmannin or LY294002 greatly inhibited LY uptake. If both sucrose and the membrane impermeable LY were transported together into the vacuole by the same non-selective endocytic mechanism, the fluorescent marker would be expected in the vacuole of cultured cells. Incubation of starved cultured cells with sucrose and LY confirmed this scenario (Etxeberria et al. 2005a). A strong fluorescence appeared within the entire vacuolar space, after starved cells were supplemented with sucrose. Wortmannin completely abolished accumulation of LY within the vacuoles, as was the case in control samples incubated in LY without added sucrose. Shorter incubation times allowed the visualization of early uptake events including formation of endocytic vesicles that progressed towards larger compartments of various sizes and configurations with the eventual appearance in the central vacuole (Etxeberria et al. 2005a).

A peculiarity noticed during the studies described above was the tight dependence of endocytosis to the presence of sucrose. Although low levels of endocytosis were observed in the presence of other simple sugars (i.e., trehalose, glucose and fructose, or a combination of both), uptake of LY at equimolar concentrations of sucrose was approximately 10 times higher than that for hexoses. At this point we can only speculate that, although is likely that heterotrophic cells come into contact with various sugars in the apoplastic milieu, sucrose has evolved as a favored regulatory molecule (Etxeberria et al. 2005a).

Any claim that endocytosis as a mechanism of nutrient uptake in sycamore cultured cells may be a unique feature of this artificial cell system and not applicable to *in planta* conditions was dismissed by a series of succeeding experiments using *Citrus* juice cells (Etxeberria et al. 2005a, 2005b) and turnip storage roots (unpublished data). *Citrus* juice cells are enclosed in sac-like structures (juice sacs) that can be easily excised and experimentally manipulated. When samples of juice sacs were incubated with two membrane impermeable fluorescent endocytic markers differing in size and ionic properties (Alexa-488 and 3000 mw dextran-Texas red), a similar sequence of events as those described above for sycamore was observed. Early endocytic vesicles contained both endocytic markers, and their co-localization demonstrated the non-specific nature of the uptake system characteristic of fluid phase endocytosis.

Importantly, uptake of extracellular nutrients by endocytosis is not in direct conflict with transport through membrane-bound carriers given that cell homeostasis can be better maintained if both these mechanisms operate in parallel. For example, "reserve" sucrose to be accumulated in the vacuole is transported in bulk flow through a mechanism that bypasses the cytosol, whereas "transitory" sucrose immediately needed by the cytosolic metabolism is transported by plasma membrane-bound carriers and funneled directly towards sites of catalytic activities (Etxeberria et al. 2005c). In this manner, the highly regulated cytosol is not disrupted by the constantly changing flow of metabolites arriving from source cells. The elusive tonoplast associated sucrose carrier (Lalonde et al. 1999) likely operates in the fine regulation of cytosolic sucrose concentration and in the export of vacuolar sucrose at times of high demands (Etxeberria and Gonzalez 2003).

A dual system for extracellular nutrient uptake is highly compatible and may well explain inconsistencies observed in numerous studies of sugar uptake into plant cells, where biphasic kinetic uptake curves have been obtained (Felker and Goodwin 1988; Getz et al. 1987; Saftner et al. 1983). Common to all these studies is a concentration uptake curve in which a hyperbolic phase at low external sugar concentrations is followed by a linear phase at increasingly higher concentrations. We can only speculate at this point, but a highly regulated uptake system at low external sugar concentrations does not appear compatible with a sudden non-regulated, "open flow gates" diffusion-like uptake at high external concentrations. This second linear phase likely corresponds to an endocytic system triggered, when the external sugar concentration exceeds minimum nutrient requirements and becomes sufficient to support vacuolar storage, and/or when osmotic conditions trigger uptake changes for intracellular osmotic adjustments. A linear increase in the uptake, which is proportional with external concentrations, is a characteristic feature of an endocytic transport system.

3
Fluid-Phase Endocytosis is Accomplished Preferentially by the Inner Cortex Cells Located near the Unloading Phloem Elements

Heterotrophic plant cells, such as root and suspension culture cells, as well as dark-grown plant cells are dependent on external nutrient supply. Sucrose starvation induces autophagy and formation of autolysosomes in plant cells (Yano et al. 2004). Within the plant body, phloem elements redistribute assimilates synthesized in leaves and transport them towards sink tissues. One of the best studied sink tissues is that of root apices. In root apices, unloading phloem elements release large amounts of sucrose, literally flooding the neighboring cells. Sucrose is transported from cell-to-cell symplastically via plasmodesmata (Oparka and Cruz 2000; Baluška et al. 2001c, Sadler et al. 2005). However, calculations made for maize root apices revealed, that the number of plasmodesmata can not satisfy the high demand for sucrose established by their large meristems and by the root caps (Bret-Harte and Silk 1995). Another popular scenario is that sucrose is enzymatically cleaved by cell wall invertase and the products are then loaded into cells via plasma membrane sugar transporters (Williams et al. 2000).

Detailed analysis of maize root apices submerged into LY solution revealed that endocytic LY uptake was accomplished preferentially in the inner cortex cells located in the transition zone interpolated between the meristem and elongation region (Baluška et al. 2001a, 2004). As these cells are exposed to a large amount of sucrose released from the unloading phloem, it is not surprising to find that they internalize LY into endosomes and subsequently into vacuoles via the fluid-phase endocytosis to fulfill nutritive function for actively growing root apices. Interestingly in this respect, mycorrhizal arbuscules develop specifically in cells of the inner root cortex via invagination of the plasma membrane and intracellular ramification of fungal hyphae (for recent review see Harrison 2005). Additionally, root nodules possessing internalized symbiotic bacteria develop preferentially from inner cortex cells in leguminous plants (Goormachtig et al. 2004).

4
Exogenous Sucrose Regulates Growth of Roots Both in Culture Cells and in Intact Seedlings

If supplied with the adequate nutrition and oxygen, excised roots grow efficiently and can be maintained almost infinitely in culture conditions. This suggests that the symplastic pathway, although a major route within the intact plant body (Oparka and Cruz 2000; Baluška et al. 2001c, Sadler et al. 2005), is not essential and that either the plasma membrane sugar transporters (Williams et al. 2000) or endocytic processes (Echeverría 2000; Baluška et al.

2004; Etxeberria et al. 2005a, 2005b) can fully satisfy all nutritive requirements of growing roots. *Arabidopsis* roots are extremely sensitive to sucrose and, in fact, slow their growth if supplies of external sucrose drop down. For instance, addition of 4.5% sucrose into the medium increased the number of dividing cells and enlarged the size of the apical meristem of *Arabidopsis* roots (Hauser and Bauer 2000). In particular, the basal size limit of the apical meristem was clearly shifted up from about 162 to about 300 μm measured from the root cap junction upwards (Hauser and Bauer 2000).

Similar, but less striking is the size dependence of the maize root apical meristem on the exogenous supply of sucrose (Muller et al. 1998). Exogenous sucrose also induces formation of adventitious roots in *Arabidopsis*, regulates cell cycle (Riou-Khamlichi et al. 2000), cytosolic calcium levels (Furuichi et al. 2001) and diverse signalling cascades interacting with those induced by plant hormones. Obviously, sucrose has evolved as a major regulatory molecule not only for the fluid-phase endocytosis but also for a myriad of other processes (Gibson et al. 2004).

5
Endocytic Internalization of Cell Wall Molecules

Topologically, the endosomal interior belongs to the extracellular space. Therefore, it should not be surprising to find cell wall molecules within endosomes. The importance of endocytosis and endocytic membrane networks for cell wall assembly and remodeling is evident in the mutant *emb30/gnom* and the double mutant of ADL1A and ADL1E dynamins which have aberrantly organized thickened cell walls (Shevell et al. 2000; Kang et al. 2003). Particularly, JIM5- but not JIM7-reactive pectins are affected in *emb30/gnom* mutant cells (Shevell et al. 2000). This corresponds well with the finding that JIM5- but not JIM7-reactive pectins are internalized into cells of maize root apices (Baluška et al. 2002).

JIM5-reactive pectins accumulate in BFA compartments and within cell plates together with boron and calcium crossed-linked RGII pectins (Baluška et al. 2002, 2005; Šamaj et al. 2004). In addition, they were reported to localize also to plasma membrane invaginations and adjacent multivesicular bodies in stylar transmitting tissue of *Datura* (Hudák et al. 1993). In contrast, Golgi-derived JIM7-reactive pectins did not show this endocytic localization. Hudák et al. (1993) showed that plasma membrane invaginations as well as multivesicular bodies contain carbohydrates and are filled with fibrillar material resembling cell wall components. Similar fibrillar material of cell wall origin, identified as arabinogalactan-type pectins, was reported in multilamellar compartments invaginating into vacuoles of bean root cells and accumulating within cell plates (Northcote et al. 1989). Besides cross-linked cell wall pectins, arabinogalactan proteins (AGPs) were also reported to be

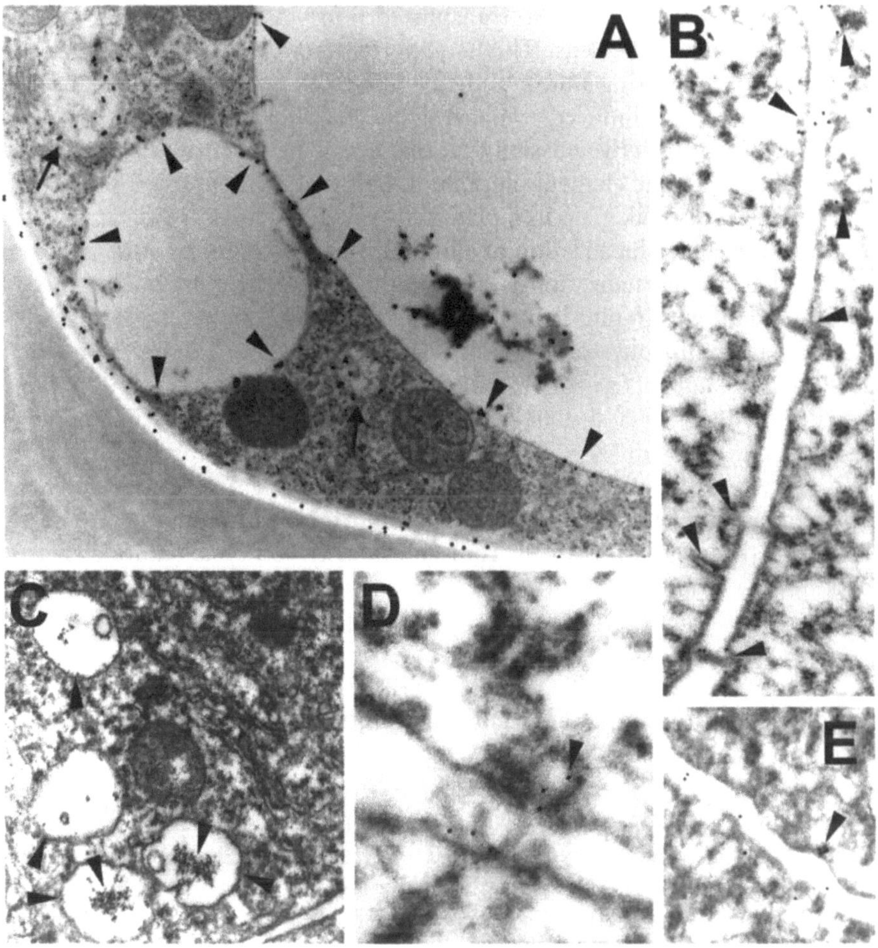

Fig. 1 Endocytosis of arabinogalactan-proteins (**A**) and fluid-phase marker LY (**B–E**) into plant cells. (**A**) Arabinogalactan-protein (AGP) epitope JIM13 is internalized from the plasma membrane into small and bigger vacuoles via pre-vacuolar compartments (indicated by *arrows*) as revealed by immunogold labeling with the JIM13 antibody in the *Drosera* glandular cell. Note, that this AGP epitope is associated with tonoplast in both small and big vacuoles (indicated by *arrowheads*). (**B–E**) Endocytic internalization of LY as revealed by immunogold EM labeling with LY-specific antibody (**B**, **D** and **E**) and LY precipitation via $BaCl_2$ (**C**) in maize root cortex cells. Note, that LY is preferably internalized from plasmodesmata domains via tubulo-vesicular protrusions (*arrowheads* in **B** and **D**) into vesicles (*arrowhead* in **E**) and small vacuoles (*arrowheads* in **C**)

internalized via multivesicular bodies and pre-vacuolar compartments (Herman and Lamb 1991; Šamaj et al. 2005). Interestingly in this respect, AGPs coat, not only plasma membrane, but also vacuolar membranes (Šamaj et al. 2000, Fig. 1A). Fusion of AGP-enriched endosomes with vacuoles is one possible mechanism how AGPs reach the tonoplast.

Boron and calcium are also transported into plant cells together with internalized cell wall pectins. This is apparent from the fact that antibodies specifically recognizing boron and calcium cross-linkages label endosomes and endocytic BFA-induced compartments (Baluška et al. 2002, 2005; Šamaj et al. 2004). An attractive possibility is that heavy metals such as lead and cadmium, and the toxic element aluminum, being often complexed with pectic networks, are also taken up into plant cells via endocytosis. In support of this notion, subcellular localization of aluminum in cells of maize root apices revealed that its internalization was accomplished at the cross-walls, where it was abundant within multilamellar compartments (Vázquez 2002). As pectic matrix has unique physical properties (Ridley et al. 2001), it might be speculated that it even acts as some sort of "smart matrix" (for recycling synaptic vesicles see Reigada et al. 2003) exerting essential functions within endocytic vesicles and endosomes, which are then important for endosomal functions. It is therefore not surprising that aluminum affects processes dependent on endosomes and endosomal recycling. Importantly, aluminum inhibits the basipetal transport of auxin in root apices of *Arabidopsis* (Kollmeier et al. 2000).

6
Endocytic Internalization of Whole Portions of Fluidized Cell Walls and Bacteria

One of the most spectacular examples of internalization into plant cells is the endocytic uptake of symbiotic bacteria, embedded within fluidized cell wall portions (Brewin 2004), into the newly divided cells of nodule primordia (Verma 1992). Besides this, bacteria can be internalized also by plant protoplasts (Davey and Cocking 1972). Endocytosis of bacteria is dependent on the action of endosomal Rab GTPases and on the generation of endosomal PI(3)P (Cheon et al. 1993; Hong and Verma 1994). Bacteria seem to be participating in this cell wall fluidization (van Spronsen et al. 1994). However, there must be also some plant-specific mechanism for the cell wall fluidization (Brewin 2004) allowing cell wall endocytosis as evidenced by the internalization of so-called "infection-thread wall degradation vesicles" (IWDV), which lack bacteria and are apparently filled only with the fluidized cell wall portions (Basset et al. 1977; Roth and Stacey 1989a). In soybean mutants which fail to internalize bacteria, the IWDVs massively internalize large portions of fluidized cell walls into cells of nodule primordia (Roth and Stacey 1989b). Internalized cell wall complexes are presumably degraded within endosomes as can be inferred from their very loose arrangement (Roth and Stacey 1989a,b) as well as from the fact that cysteine proteases were localized both to these vacuolar bodies (Vincent and Brewin 2000) and to endosomes (Yamada et al. 2005). They can be used for regulation of osmotic balance and serve also for nutritional purposes as sucrose starvation induces autophagy and formation of autolysosomes (Yano et al. 2004).

Large-scale internalization of apparently fluidized cell wall material into small vacuoles is a characteristic feature also for the cell wall thinning during bulge formation in trichoblasts initiating root hairs (Ciamporova et al. 2003). Internalization of polysaccharide-based material and fluids from the extracellular space (apoplast) via both multilamellar and multivesicular carriers was also described for root cells of zucchini (Coulomb and Coulomb 1976) and rice, where this process was proposed to be relevant for the uptake of nutrients (Nishizawa and Mori 1977).

These endocytic processes and structures are especially prominent in osmotically stressed root cells (Ciamporová and Mistrík 1993) and those under chilling stress (Stefanowska et al. 2002), suggesting possible roles of multilamellar and multivesicular endosomes in stress adaptation. Interestingly in this respect, bulge formation during root hair initiation might represent some sort of "physiological wounding" experiencing both osmotic and mechanical stress (Baluška et al. 2002), and stress-activated MAP kinases are recruited to these subcellular domains in large amounts (Šamaj et al. 2002; Ovecka et al. 2005). Intriguingly, boron deficiency inhibits both internalization of cell wall pectins into root cells (Yu et al. 2002) as well as uptake of bacteria into host nodule cells (Bolanos et al. 1996).

7 Plasmodesmata/Pit-fields as Subcellular Domains Specialized for Endocytosis of Cell Wall Molecules and Fluidized Cell Wall Portions?

Plasmodesmata and pit-fields are known to be enriched with pectins and depleted of cellulose microfibrils (reviewed by Baluška et al. 2001c). Recently, we established a link between fluid-phase endocytosis and plasmodesmata/pit-fields (Baluška et al. 2004, Fig. 1B–E). This link gets further support from the recent studies of plant-viral movement proteins which target plasmodesmata and interact with endosomal KNOLLE (Laporte et al. 2003; Uemura et al. 2004). Moreover, these proteins co-localize with Ara7 endosomal Rab GTPase within endosomes (Haupt et al. 2005) and endosomal Rab11 was reported in the plasmodesmata (Escobar et al. 2003).

Obviously, callose- and pectin-enriched plasmodesmata not only recruit vesicle trafficking pathways but also act as effective platforms for rapid endocytosis (Baluška et al. 2004, 2005a; Oparka et al. 2004). In this scenario, endocytosis and recycling of cross-linked pectins would allow flexible remodelling of the cell wall at plasmodesmata (Baluška et al. 2001c). Unique cell walls around plasmodesmata sleeves must be capable of large-scale movements in order to allow architectural re-arrangements of the inner plasmodesmal structures. This scenario is inevitable for performing active gating of these cell-cell channels which are embedded within the cell walls. It is of great interest in this respect, that pectin methylesterase interacts physically with viral movement proteins which gate the plasmodesmata for cell-cell transport of macromolecules (Chen et al. 2000). Moreover, this enzyme which modifies cell wall pectins *de muro* (within cell walls) is also essential for the systemic spread of tobacco mosaic virus (Chen and Citovsky 2003).

8 Cytokinesis, Guard Cell Movements, Papilla Formation, and Re-Plasmolysis: Processes Relying on Endosomes Enriched with Cell Walls Molecules?

For over four decades, plant cell cytokinesis has been considered to be driven via fusion of Golgi-derived vesicles to form the cell plate, a primordial cell wall. It is astonishing that this popular concept is based solely on the similarity between phragmoplast vesicles and vesicles seen in the vicinity of Golgi stacks. However, there are several problems with this popular concept. First, plant cytokinesis is completed in a matter of minutes. During this time almost one third of the original cell surface is rebuilt, in a time window, when Golgi stacks have to fulfill another task, namely to divide and partition into the daughter cells. Secondly, vesicles initiating and driving cell plate formation are performing homotypic fusions via finger-like tubular protrusions (Samuels et al. 1995). This feature is characteristic for endosomes but not for Golgi-derived vesicles, which can fuse only with the parent plasma membrane. Our detailed analysis of dividing maize root cells revealed that all cell wall pectin epitopes, which accomplish endocytic internalization, are also abundant in both early and late cell plates, whereas Golgi-derived JIM7-reactive pectins do not accumulate within cell plates (Baluška et al. 2005b). Moreover, growing cell plates also accumulate the endocytic tracer FM4-64 (Belanger and Quatrano, 2000) as well as LY (Baluška F., unpublished data).

Stomatal guard cells are performing dramatic changes in their surface areas within a short time period in order to open or close stomata. Their cell walls are well-known to be very rich in pectins (Majewska-Sawka et al. 2002). Indeed, analysis using FM4-64 confirmed that endocytosis was responsible for decreases of their surface area (Shope et al. 2003), while secretory endosomes filled with cell wall pectins would be ideally suited to allow rapid recovery of the original surface areas, if opening of stomata would be needed. Furthermore, rhamnogalacturonan 1 (RG-1) pectins decorated with galactan and arabinan side chains were reported to be essential for proper guard cell movements (Jones et al. 2003), when both RG-1 pectins and pectins enriched with arabinan side chains are among those undergoing endocytic internalization in root apex cells (Baluška et al. 2005b).

Another situation in which plant cells require extremely rapid secretion of large amounts of preformed cell wall molecules is encountered at the sites of pathogen attack, which are effectively sealed off by so-called papillae (Schulze-Lefert 2004). In barley, this polarized secretion is accomplished via unusually large secretory compartments, having up to 1 μm in diameter, that are enriched with reactive oxygen species (Hückelhoven et al. 1999; Collins et al. 2003). Moreover, these secretory vesicles are also associated with PEN1 (Assaad et al. 2004) which is a close homologue of the SNARE SYP122 (VAMP721) and a component of the plasma membrane and endosomes in *Arabidopsis* (Uemura et al. 2004; see the chapter by Sato et al., this volume). Our own preliminary data have revealed that GFP-PEN1 is also localized to endosomes, BFA-induced compartments, and cell plates of *Arabidopsis* root cells (B. Voigt, Thordal-Christensen H., and Baluška F., unpublished data). Last but not least, endosomes represent a ready source of plasma membrane supply to satisfy the need for membrane replenishment during rapid replasmolysis of plant cells (Oparka et al. 1994).

9
Conclusions and Future Prospects

Indisputably, plant endocytosis is presently undergoing explosive development (Geldner 2004; Šamaj et al. 2004, 2005). Although studies devoted to the endocytic internalization of plasma membrane proteins and their recycling are more advanced (reviewed by Murphy et al. 2005), endocytic internalization of external fluids and nutrients also emerge to be inherent both to suspension plant cells as well as to intact organs such as growing root apices and storage roots. Additionally, plant cells can internalize several cell wall molecules such as pectins, xyloglucans, and AGPs, as well as whole portions of apparently fluidized cell wall, and use endosomes filled with these molecules for secretion in situations where extremely rapid cell wall assembly is needed, such as cell plate formation in cytokinetic cells, stomata movements, and perhaps papilla formation during pathogen attack.

Electron microscopy studies in the 1960s and 1970s reported fusions of multivesicular compartments with the plasma membrane forming so-called paramural bodies and on engulfment of multilamellar and multivesicular compartments by the central vacuole (Roland 1972). Unfortunately, all these observations were considered to be classical examples of fixation artifacts. Today, both engulfments of late endosomes known as multivesicular bodies by the lysosomes, as well as their fusions with the plasma membrane, releasing exosomes, is well-known for animal cells.

Internalized cell wall pectins co-localize with recycling plasma membrane proteins within endocytic BFA compartments (Šamaj et al. 2004), suggesting that they accomplish rapid recycling too. This would implicate that plant cells can actively remodel existing cell walls using the endocytic machinery. Such reports do not exist in animal, *Dictyostelium*, or yeast literature yet. Obviously, despite lagging considerably behind these more developed model objects, endocytic plant research can obtain pioneering achievements in some specialized areas of the field.

Acknowledgements We thank Irene Lichtscheidl for providing us with high-pressure freezed *Drosera* samples and Ursulla Mettbach for excellent technical assistance. This work was supported by a grant from the Slovak Grant Agency APVT (grant no. APVT-51-002302) and Vega (Grant Nr. 2/5085/25), Bratislava, Slovakia.

References

Assaad FF, Qiu JL, Youngs H, Ehrhardt D, Zimmerli L, Kalde M, Wanner G, Peck SC, Edwards H, Ramonell K, Somerville CR, Thordal-Christensen H (2004) The PEN1 syntaxin defines a novel cellular compartment upon fungal attack and is required for the timely assembly of papillae. Mol Biol Cell 11:5118–5129

Baluška F, Volkmann D, Barlow PW (2001a) A polarity crossroad in the transition growth zone of maize root apices: cytoskeletal and developmental implications. J Plant Growth Regul 20:170–181

Baluška F, Jásik J, Edelmann HG, Salajová T, Volkmann D (2001b) Latrunculin B induced plant dwarfism: plant cell elongation is F-actin dependent. Dev Biol 231:113–124

Baluška F, Cvrcková F, Kendrick-Jones J, Volkmann D (2001c) Sink plasmodesmata as gateways for phloem unloading. Myosin VIII and calreticulin as molecular determinants of sink strength? Plant Physiol 126:39–46

Baluška F, Hlavacka A, Šamaj J, Palme K, Robinson DG, Matoh T, McCurdy DW, Menzel D, Volkmann D (2002) F-actin-dependent endocytosis of cell wall pectins in meristematic root cells: insights from brefeldin A-induced compartments. Plant Physiol 130:422–431

Baluška F, Šamaj J, Hlavacka A, Kendrick-Jones J, Volkmann D (2004) Actin-dependent fluid phase endocytosis in inner cortex cells of maize root apices. J Exp Bot 396:463–473

Baluška F, Volkmann D, Menzel D (2005a) Plant synapses: actin-based adhesion domains for cell-to-cell communication. Trends Plant Sci 10:106–111

Baluška F, Hlavacka A, Liners F, Schlicht M, Van Cutsem P, McCurdy D, Menzel D (2005b) Cell wall pectins and xyloglucans are internalized into dividing root cells and accumulate within cell plates during cytokinesis. Protoplasma 225:141–155

Basset B, Goodman RN, Novacky A (1977) Ultrastructure of soybean nodules. I. Release of rhizobia from infection thread. Can J Microbiol 23:573–582

Belanger KD, Quatrano RS (2000) Membrane recycling occurs during asymmetric tip growth and cell plate formation in Fucus distichus zygotes. Protoplasma 212:24–37

Bolanos L, Brewin NJ, Bonilla I (1996) Effects of boron on Rhizobium-legume cell-surface interactions and nodule development. Plant Physiol 110:1249–1256

Bret-Harte MS, Silk WK (1995) Nonvascular, symplasmic diffusion of sucrose cannot satisfy the carbon demands of growth in the primary root tip of Zea mays L. Plant Physiol 105:19–33

Brewin NJ (2004) Plant cell wall remodelling in the Rhizobium-legume symbiosis. Crit Rev Plant Sci 23:293–316
Chen MH, Citovsky V (2003) Systemic movement of a tobamovirus requires host cell pectin methylesterase. Plant J 35:386–392
Chen MH, Sheng J, Hind G, Handa AK, Citovsky V (2000) Interaction between the tobacco mosaic virus movement protein and host cell pectin methylesterases is required for viral cell-to-cell movement. EMBO J 19:913–920
Cheon CI, Lee NG, Siddique AB, Bal AK, Verma DP (1993) Roles of plant homologs of Rab1p and Rab7p in the biogenesis of the peribacteroid membrane, a subcellular compartment formed de novo during root nodule symbiosis. EMBO J 12:4125–4133
Cholewa E, Peterson CA (2001) Detecting exodermal Casparian bands in vivo and fluid phase endocytosis in onion (Allium cepa L.) roots. Can J Bot 79:30–37
Ciamporová M, Mistrík I (1993) The ultrastructural response of root cells to stressful conditions. Environm Exp Bot 33:11–26
Ciamporová M, Dekánková K, Hanácková Z, Ovecka M, Baluška F (2003) Structural aspects of root hair initiation in Vicia sativa roots treated with F-actin polymerisation inhibitor latrunculin B. Plant and Soil 255:1–7
Cole L, Coleman J, Kearns A, Morgan G, Hawes C (1991) The organic anion transport inhibitor, probenecid, inhibits the transport of Lucifer Yellow at the plasma membrane and the tonoplast in suspension cultured plant cells. J Cell Sci 99:545–555
Collins NC, Thordal-Christensen H, Lipka V, Bau S, Kombrink E, Qiu JL, Huckelhoven R, Stein M, Freialdenhoven A, Somerville SC, Schulze-Lefert P (2003) SNARE-protein-mediated disease resistance at the plant cell wall. Nature 425:973–977
Coulomb S, Coulomb S (1976) Endocytosis in Cucurbita pepo root meristems: coated vesicles, multivesicular bodies and vacuole relationship. CR Acad Sci Paris III 319:377–383
Davey MR, Cocking EC (1972) Uptake of bacteria by isolated higher plant protoplasts. Nature 239:455–456
Echeverría E (2000) Vesicle-mediated solute transport between the vacuole and the plasma membrane. Plant Physiol 123:1217–1226
Emans N, Zimmermann S, Fischer R (2002) Uptake of a fluorescent marker in plant cells sensitive to brefeldin A and wortmannin. Plant Cell 14:71–86
Escobar NM, Haupt S, Thow G, Boevink P, Chapman S, Oparka K (2003) High-throughput viral expression of cDNA-green fluorescent protein fusions reveals novel subcellular addresses and identifies unique proteins that interact with plasmodesmata. Plant Cell 15:1507–1523
Etxeberria E, González PC (2003) Evidence for a tonoplast associated form of sucrose synthase and its potential involvement in sucrose mobilization from the vacuole. J Exp Bot 54:1407–1414
Etxeberria E, Baroja-Fernández E, Muñoz FJ, Pozueta-Romero J (2005a) Sucrose inducible endocytosis as a mechanism for nutrient uptake in heterotrophic plant cells. Plant Cell Physiol 46:474–481
Etxeberria E, González PC, Pozueta-Romero J (2005b) Sucrose transport into Citrus juice cells: evidence for an endocytic transport system. J Am Soc Hort Sci 130:269–274
Etxeberria E, González PC, Tomlinson P, Pozueta-Romero J (2005c) Existence of two parallel mechanisms for glucose uptake in heterotrophic plant cells. J Exp Bot 56:1905–1912
Felker FC, Goodwin J (1988) Sugar uptake by maize endosperm suspension cultures. Plant Physiol 88:1235–1239

Furuichi T, Mori IC, Takahashi K, Muto S (2001) Sugar-induced increase in cytosolic Ca^{2+} in *Arabidopsis* thaliana whole plants. Plant Cell Physiol 42:1149–1155
Geldner N (2004) The plant endosomal system—its structure and role in signal transduction and plant development. Planta 219:547–560
Geldner N, Friml J, Stierhof Y-D, Jürgens G, Palme K (2001) Auxin-transport inhibitors block PIN1 cycling and vesicle trafficking. Nature 413:425–428
Geldner N, Anders N, Wolters H, Keicher J, Kornberger W, Muller P, Delbarre A, Ueda T, Nakano A, Jürgens G (2003) The Arabidopsis GNOM ARF-GEF mediates endosomal recycling, auxin transport, and auxin-dependent plant growth. Cell 112:219–230
Getz HP (1991) Sucrose transport in tonoplast vesicles of red beet roots is linked to ATP hydrolysis. Planta 185:261–268
Getz HP, Knawer D, Willenbrink J (1987) Transport of sugars across the plasma membrane of beet root protoplasts. Planta 171:185–196
Gibson SI (2004) Sugar and phytohormone response pathways: navigating a signalling network. J Exp Bot 55:253–264
Goormachtig S, Capoen W, Holsters M (2004) Rhizobium infection: lessons from the versatile nodulation behaviour of water-tolerant legumes. Trends Plant Sci 9:518–522
Harrison MJ (2005) Signaling in the arbuscular mycorrhizal symbioosis. Ann Rev Microbiol 59:19–42
Haupt S, Cowan GH, Ziegler A, Roberts AG, Oparka KJ, Torrance L (2005) Two plant-viral movement proteins traffic in the endocytic recycling pathway. Plant Cell 17:164–181
Hauser M-T, Bauer E (2000) Histochemical analysis of root meristem activity in Arabidopsis thaliana using a cyclin: GUS (β-glucuronidase) marker line. Plant and Soil 226:1–10
Hawes C, Crooks K, Coleman J, Satiat-Jeunematrie B (1995) Endocytosis in plants: fact or artefact. Plant Cell Environm 18:1245–1252
Herman EM, Lamb CJ (1991) Arabinogalactan-rich glycoproteins are localized on the cell surface and in intravacuolar multivesicular bodies. Plant Physiol 98:264–272
Hillmer S, Deptam H, Robinson DG (1986) Confirmation of endocytosis in higher plant protoplasts using lectin-gold conjugates. Eur J Cell Biol 41:142–149
Hong Z, Verma DP (1994) A phosphatidylinositol 3-kinase is induced during soybean nodule organogenesis and is associated with membrane proliferation. Proc Natl Acad Sci USA 91:9617–9621
Hubner R, Depta H, Robinson DG (1985) Endocytosis in maize root cap cells: evidence obtained using heavy metal salt solutions. Protoplasma 129:214–222
Hückelhoven R, Fodor J, Preis C, Kogel KH (1999) Hypersensitive cell death and papilla formation in barley attacked by the powdery mildew fungus are associated with hydrogen peroxide but not with salicylic acid accumulation. Plant Physiol 119:1251–1260
Hudák J, Walles B, Vennigerholz F (1993) The transmitting tissue in Brugmansia suaveolens L.: ultrastructure of the stylar transmitting tissue. Ann Bot 71:177–186
Jones L, Milne JL, Ashford D, McQueen-Mason SJ (2003) Cell wall arabinan is essential for guard cell function. Proc Natl Acad Sci USA 100:11783–11788
Kang B-H, Busse JS, Bednarek SY (2003) Members of the Arabidopsis dynamin-like gene family, ADL1A, are essential for plant cytokinesis and polarized cell growth. Plant Cell 15:899–913
Keller F (1992) Transport of stachyose and sucrose by vacuoles of Japanese artichoke (Stachys sieboldii) tubers. Plant Physiol 98:442–445
Kollmeier M, Dietrich P, Bauer CS, Horst WJ, Hedrich R (2000) Aluminum activates a citrate-permeable anion channel in the aluminum-sensitive zone of the maize root

apex. A comparison between an aluminum-sensitive and an aluminum-resistant cultivar. Plant Physiol 126:397–410

Lalonde S, Boles E, Hellmann H, Barker L, Patrick JW, Frommer WB, Ward JM (1999) The dual function of sugar carriers: transport and sugar sensing. Plant Cell 11:707–726

Laporte C, Vetter G, Loudes A-M, Robinson DG, Hillmer S, Stussi-Garaud C, Ritzenthaler C (2003) Involvement of the secretory pathway and the cytoskeleton in intracellular targeting and tubule assembly of Grapevine fanleaf virus movement protein in tobacco BY-2 cells. Plant Cell 15:2058–2075

Lazzaro MD, Thompson WW (1992) Endocytosis of lanthanum nitrate in the organic acid-secreting trichomes of chickpea (Cicer arietinum). Am J Bot 79:1113–1118

Lemoine R (2000) Sucrose transporters in plants: update on function and structure. Biochem Biophys Acta 1465:246–262

Majewska-Sawka A, Münster A, Rodríguez-García MI (2002) Guard cell wall: immunocytochemical detection of polysaccharide components. J Exp Bot 53:1067–1079

Muller B, Stosser M, Tardieu F (1998) Spatial distributions of tissue expansion and cell division rates are related to irradiance and to sugar content in the growing zone of maize roots. Plant Cell Environm 21:149–158

Murphy AS, Bandyopadhyay A, Holstein SE, Peer WA (2005) Endocytic cycling of PM proteins. Annu Rev Plant Biol 56:221–251

Nebenführ A, Ritzenthaler C, Robinson DG (2002) Brefeldin A: deciphering an enigmatic inhibitor of secretion. Plant Physiol 130:1102–1108

Nishizawa N, Mori S (1977) Invagination of plasmalemma: its role in the absorption of macromolecules in rice roots. Plant Cell Physiol 18:767–782

Northcote DH, Davey R, Lay J (1989) Use of antisera to localize callose, xylan and arabinogalactan in the cell-plate, primary and secondary walls of plant cells. Planta 178:353–366

Oparka KJ (2003) Getting the message across: how do plant cells exchange macromolecular complexes? Trends Plant Sci 9:33–41

Oparka KJ, Cruz SS (2000) The great escape: phloem transport and unloading of macromolecules. Annu Rev Plant Physiol Plant Mol Biol 51:323–347

Oparka KJ, Murant EA, Wright KM, Prior DAM (1991) The drug probenecid inhibits the vacuolar accumulation of fluorescent anions in onion epidermal cells. J Cell Sci 99:557–563

Oparka KJ, Wright KM, Murant EA, Allan EJ (1993) Fluid-phase endocytosis: do plants need it? J Exp Bot 44:247–255

Oparka KJ, Prior DAM, Crawford JW (1994) Behaviour of plasma membrane, cortical ER and plasmodesmata during plasmolysis of onion epidermal cells. Plant Cell Environm 17:163–171

Ovecka M, Lichtscheidl I, Baluška F, Šamaj J, Volkmann D, Hirt H (2005) Regulation of root hair tip growth: can mitogen-activated protein kinases be taken into account? NATO Series (in press)

Reigada D, Diez-Perez I, Gorostiza P, Verdaguer A, Gomez de Aranda I, Pineda O, Vilarrasa J, Marsal J, Blasi J, Aleu J, Solsona C (2003) Control of neurotransmitter release by an internal gel matrix in synaptic vesicles. Proc Natl Acad Sci USA 100:3485–3490

Ridley BL, O'Neill MA, Mohnen D (2001) Pectins: structure, biosynthesis, and oligogalacturonide-related signaling. Phytochemistry 57:929–967

Riou-Khamlichi C, Menges M, Healy JMS, Murray JAH (2000) Sugar control of the plant cell cycle: differential regulation of Arabidopsis D-type cyclin gene expression. Mol Cell Biol 20:4513–4521

Roland J-C (1973) The relationship between the plasmalemma and plant cell wall. Int Rev Cytol 36:45–92

Roth LE, Stacey G (1989a) Bacterium release into host cells of nitrogen-fixing soybean nodules: the symbiosome membrane comes from three sources. Eur J Cell Biol 49:13–23

Roth LE, Stacey G (1989b) Cytoplasmic membrane systems involved in bacterium release into soybean nodule cells as studied with two Bradyrhizobium japonicum strains. Eur J Cell Biol 49:13–23

Russinova E, Borst J-W, Kwaaitaal M, Cano-Delgado A, Yin Y, Chory J, de Vries SC (2004) Heterodimerization and endocytosis of Arabidospis brassinosteroid receptors BRI1 and AtSERK3 (BAK1). Plant Cell 16:3216–3229

Šamaj J, Šamajová O, Peters M, Baluška F, Lichtscheidl IK, Knox JP, Volkmann D (2000) Immunolocalization of LM2 arabinogalactan-protein epitope associated with endomembranes of plant cells. Protoplasma 212:186–196

Šamaj J, Ovecka M, Hlavacka A, Lecourieux F, Meskiene I, Lichtscheidl I, Lenart P, Salaj J, Volkmann D, Bögre L, Baluška F, Hirt H (2002) Involvement of the mitogen-activated protein kinase SIMK in regulation of root hair tip-growth. EMBO J 21:3296–3306

Šamaj J, Baluška F, Voigt B, Schlicht M, Volkmann D, Menzel D (2004) Endocytosis, actin cytoskeleton and signalling. Plant Physiol 135:1150–1161

Šamaj J, Read ND, Volkmann D, Menzel D, Baluška F (2005) The endocytic network in plants. Trends Cell Biol 15:425–433

Samuels AL, Giddings TH, Staehelin LA (1995) Cytokinesis in tobacco BY-2 and root tip cells: a new model of cell plate formation in higher plants. J Cell Biol 130:1–13

Satiat-Jeunemaitre B, Cole L, Bourett T, Howard R, Hawes C (1996) Brefeldin A effects in plant and fungal cells: something new about vesicle trafficking? J Microsc 181:162–177

Schulze-Lefert P (2004) Knocking on the heaven's wall: pathogenesis of and resistance to biotrophic fungi at the cell wall. Curr Opin Plant Biol 7:1–7

Stadler R, Wright KM, Lauterbach C, Amon G, Gahrtz M, Feuerstein A, Oparka KJ, Sauer N (2005) Expression of GFP-fusions in Arabidopsis companion cells reveals non-specific protein trafficking into sieve elements and identifies a novel post-phloem domain in roots. Plant J 41:319–331

Saftner RA, Daie J, Wyse R (1983) Sucrose uptake and compartmentation in sugar beet taproot tissue. Plant Physiol 72:1–6

Shevell DE, Kunkel T, Chua N-H (2000) Cell wall alteration in the Arabidopsis emb30 mutant. Plant Cell 12:2047–2059

Shope JC, DeWald DB, Mott KA (2003) Changes in surface area of intact guard cells are correlated with membrane internalization. Plant Physiol 133:1314–1321

Stefanowska M, Kuras M, Kacperska A (2002) Low temperature-induced modification in cell ultrastructure and localization of phenolics in winter oilseed rape (Brassica napus L. var. oleifera L.) leaves. Ann Bot 90:637–645

Takahashi F, Sato-Nara K, Kobayshi K, Suzuki M, Suzuki H (2003) Sugar-induced adventitious roots in Arabidopsis seedlings. J Plant Res 116:83–91

Tanchak MA, Fowke LC (1987) The morphology of multivesicular bodies in soybean protoplasts and their role in endocytosis. Protoplasma 134:173–182

Tse YC, Mo B, Hillmer S, Zhao M, Lo SW, Robinson DG, Jiang L (2004) Identification of multivesicular bodies as prevacuolar compartments in Nicotiana tabacum BY-2 cells. Plant Cell 16:672–693

Ueda T, Uemura T, Sato MH, Nakano A (2004) Functional differentiation of endosomes in Arabidopsis cells. Plant J 40:783–789

Uemura T, Ueda T, Ohniwa RL, Nakano A, Takeyasu K, Sato MH (2004) Systematic analysis of SNARE molecules in Arabidopsis: dissection of the post-Golgi network in plant cells. Cell Struct Funct 29:49–65

van Spronsen PC, Bakhuizen R, van Brussel AAN, Kijne JW (1994) Cell wall degradation during infection thread formation by the root nodule bacterium Rhizobium leguminosarum is a two-step process. Eur J Cell Biol 64:88–94

Vázquez MD (2002) Aluminum exclusion mechanism in root tips of maize (Zea mays L.): lysigeny of aluminum hyperaccumulator cells. Plant Biol 4:234–249

Verma DPS (1992) Signals in root nodule organogenesis and endocytosis of Rhizobium. Plant Cell 4:373–382

Villanueva MA, Taylor J, Sui X, Griffing LR (1993) Endocytosis in plant protoplasts: visualization and quantitation of fluid-phase endocytosis using silver-enhanced bovine serum albumin-gold. J Exp Bot 44:275–281

Vincent JL, Brewin NJ (2000) Immunolocalization of a cysteine protease in vacuoles, vesicles, and symbiosomes of pea nodule cells. Plant Physiol 123:521–530

Voigt B, Timmers A, Šamaj J, Hlavacka A, Ueda T, Preuss M, Nielsen E, Mathur J, Emans N, Stenmark H, Nakano A, Baluška F, Menzel D (2005) Actin-based motility of endosomes is linked to the polar tip-growth of root hairs. Eur J Cell Biol 84:609–621

Williams LE, Lemoine R, Sauer N (2000) Sugar transporters in higher plants—a diversity of roles and complex regulation. Trends Plant Sci 5:283–290

Yamada K, Fuji K, Shimada T, Nishimura M, Hara-Nishimura I (2005) Endosomal proteases facilitate the fusion of endosomes with vacuoles at the final step of the endocytotic patway. Plant J 41:888–898

Yano K, Matsui S, Tsuchiya T, Maeshima M, Kutsuna N, Hasezawa S, Moriyasu Y (2004) Contribution of the plasma membrane and central vacuole in the formation of autolysosomes in cultured tobacco cells. Plant Cell Physiol 45:951–957

Yu Q, Hlavacka A, Matoh T, Volkmann D, Menzel D, Goldbach HE, Baluška F (2002) Short-term boron deprivation inhibits endocytosis of cell wall pectins in meristematic cells of maize and wheat root apices. Plant Physiol 130:415–421

Zuo J, Niu Q-W, Nishizawa N, Wu Y, Kost B, Chua N-H (2000) KORRIGAN, an Arabidopsis endo-1,4-β-glucanase, localizes to the cell plate by polarized targeting and is essential for cytokinesis. Plant Cell 12:1137–1152

Plant Cell Monogr (1)
J. Šamaj · F. Baluška · D. Menzel: Plant Endocytosis
DOI 10.1007/7089_004/Published online: 12 October 2005

Plant Prevacuolar Compartments and Endocytosis

Sheung Kwan Lam[1] · Yu Chung Tse[1] · Liwen Jiang[1] (✉) · Peter Oliviusson[2] · Oliver Heinzerling[2] · David G. Robinson[2]

[1]Department of Biology and Molecular Biotechnology Program, The Chinese University of Hong Kong, Shatin, N.T. Hong Kong, P.R. China
ljiang@cuhk.edu.hk

[2]Department of Cell Biology, Heidelberg Institute for Plant Sciences, University of Heidelberg, 69120 Heidelberg, Germany

Abstract Prevacuolar compartments (PVCs) are membrane-bound organelles mediating protein traffic from both Golgi and plasma membrane to vacuoles in eukaryotic cells. Recent studies demonstrate that PVCs in plant cells are multivesicular bodies (MVBs) that merge secretory and endocytic pathways leading to the lytic vacuole, a compartment thought to be equivalent to the mammalian lysosome or the yeast vacuole. In this review, we discuss recent studies on the identity, molecular components and functional roles of plant PVCs and examine whether the plant PVC can also be claimed to be equivalent to the endosome/MVB of mammalian and yeast cells.

1 Introduction

Eukaryotic cells have a secretory pathway which is composed of several functionally distinct membrane compartments. At the same time, eukaryotes have the ability to internalize a variety of macromolecules by endocytosis, a process also involving membrane-bound organelles each with characteristic proteins. Prevacuolar compartments/late endosomes are an organelle where secretory and endocytic traffic to the lytic/vacuolar compartment merge. On the basis of precedents from mammalian and yeast cells, prevacuolar compartments (PVCs) are intermediate organelles on the biosynthetic route to the vacuole and receive cargo delivered by the *trans-Golgi* network (TGN)-derived transport vesicles (Lemmon and Traub 2000; Maxfield and McGraw 2004). Due to the lower pH in the PVC the cargo ligands dissociate from their receptors, and the receptors and missorted proteins are then returned to the Golgi apparatus for another round of cycling (Robinson et al. 2000).

Receptor-ligand complexes internalized at the plasma membrane travel through several endosomal compartments before being deposited in the lysosome/vacuole. These compartments characteristically have internal vesicles, hence the term "multivesicular endosomes or bodies" (MVB). These microvesicles appear to originate in the early or recycling endosomes (Parton et al. 1992), and have a different composition to the limiting membrane

(Griffiths et al. 1990; Kobayashi et al. 1998). Their formation is related to receptor down-regulation (Katzmann et al. 2002), and involves ubiquitinylation as a means of tagging those membrane proteins destined for degradation (Reggiori and Pelham 2001). The microvesicles and the soluble content of the MVBs are most likely delivered into the interior of the lytic compartment via direct fusion (Luzio et al. 2000; Katzmann et al. 2002).

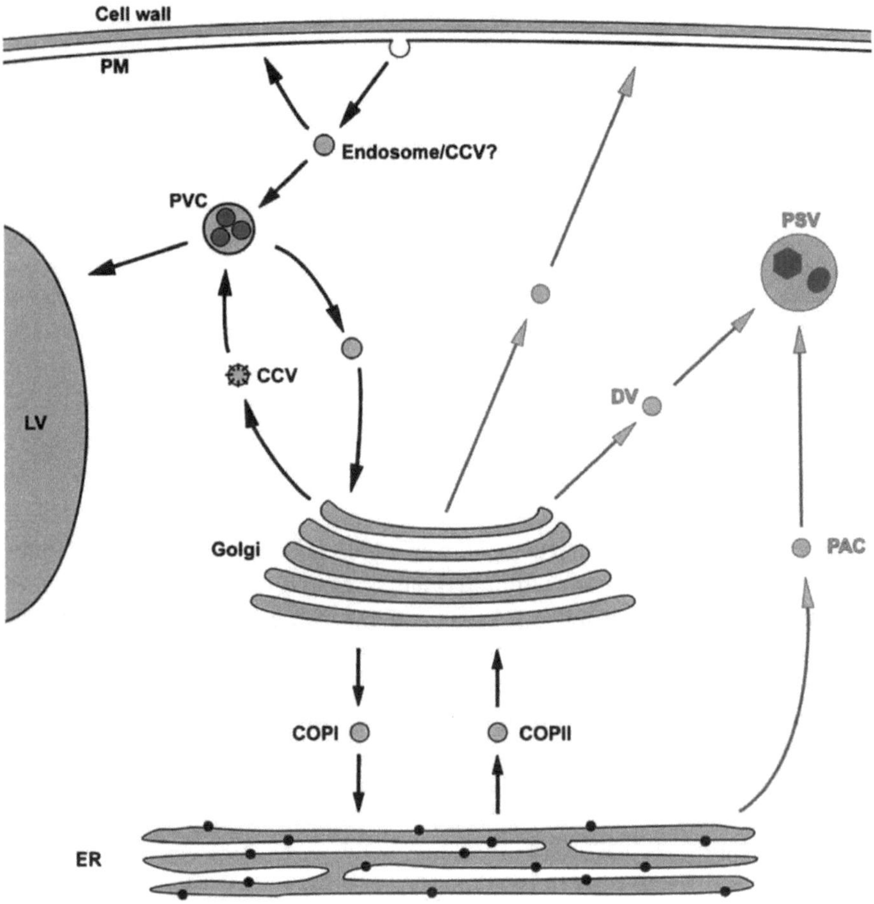

Fig. 1 Working model of protein trafficking in the plant secretory and endocytic pathways. Pathways leading to the lytic vacuole (LV) from either the Golgi or plasma membrane (PM) via a prevacuolar compartment (PVC) are thought to be mediated by clathrin-coated vesicles (CCVs) and are similar to those in mammalian and yeast cells (Jiang and Rogers 2003). In seeds, storage proteins reach protein storage vacuole (PSV) either via a Golgi-dependent pathway that is mediated by dense vesicle (DV) in pea cotyledon (Robinson et al. 1998), or via a Golgi-independent route that is mediated by precursor accumulating (PAC) vesicle in pumpkin seeds (Hara-Nishimura et al. 1998)

In spite of the obviously important role of PVCs in mediating protein traffic to vacuoles in the secretory and endocytic pathways, the identification and characterization of plant PVCs, both functionally and morphologically, has been challenging due to the complexity of the plant vacuolar system and the existence of multiple pathways of vacuolar targeting, especially to the protein storage vacuole (Jiang and Rogers 1998; Bethke and Jones 2000, see also Fig. 1). For the sake of simplicity, the object of this review will be to discuss recent studies on the identity, molecular components, and functional roles of the plant PVC in the secretory and endocytic pathways leading to the lytic vacuole, a compartment thought to be equivalent to the mammalian lysosome or the yeast vacuole. As such, we will examine whether the plant PVC can also be claimed to be equivalent to the endosome/MVB of mammalian and yeast cells.

2 Receptor-Mediated Transport from the Golgi Apparatus to the PVC/Lytic Vacuole

2.1 Current Working Model for Receptor-Mediated Sorting at the TGN

Soluble proteins reach vacuoles because they contain vacuolar sorting determinants (VSDs) that are recognized by integral transmembrane receptor proteins (Neuhaus and Rogers 1998). Such cargo-receptor interaction occurs at the TGN before they are packed into transport vesicles (in particular CCVs—clathrin coated vesicles) for subsequent delivery to PVCs and lytic vacuoles. The best-studied case in plants both in vitro as well as in vivo has been the pea vacuolar sorting receptor (VSR) protein BP-80 and its ligand proaleurain, a barley cysteine protease (see below). Figure 2 summarizes the salient features of this receptor-ligand interaction at TGN, and its relation to the coat proteins of a CCV.

2.2 Isolation and Structure of BP-80

Pea BP-80, the first identified vacuolar sorting receptor protein, was isolated using an affinity column to which a synthetic peptide containing the VSD of proaleurain (NPIR) was coupled. CCVs were purified from developing pea cotyledons and, after detergent solubilization, the CCV proteins were added to the columns (Kirsch et al. 1994). The protein which bound to the column had a molecular mass of 80 kDa and is a type I integral membrane protein that contains a single transmembrane domain (TMD) and a short cytoplasmic tail (CT). It is now recognized as belonging to a gene family encoding

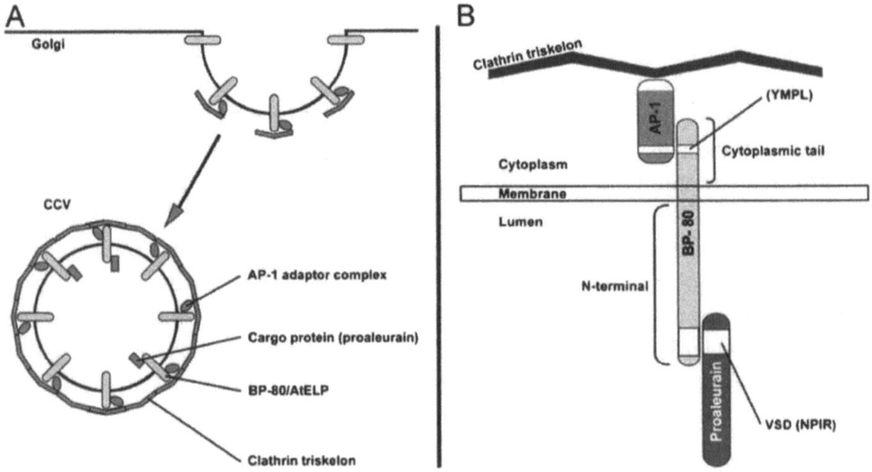

Fig. 2 Working model of receptor-mediated sorting in plants. Depicted are the possible interactions between the receptor BP-80 and the cargo protein proaleurain in the trans-Golgi network (TGN). **A** Formation and budding of clathrin-coated vesicles (CCVs) at the TGN. **B** Specific protein components and their interactions within a CCV. AP-1, adaptor protein complex AP-1; NPIR, vacuolar sorting determinant (VSD) from proaleurain; YMPL, tyrosine motif from the cytoplasmic tail of BP-80

for proteins termed Vacuolar Sorting Receptors (VSRs) (Kirsch et al. 1994; Paris et al. 1996; Paris et al. 1997, see also Chap. IV). A number of VSR homologs have been cloned from other plant species, including *Arabidopsis* and pumpkin (Ahmed et al. 1997b, Shimada et al. 1997).

The short CT of BP-80 and all of its homologs contain a conserved tyrosine motif, YMPL that is thought to be involved in the traffic of the proteins carried by the CCVs (Ahmed et al. 1997b, Paris et al. 1997; Shimada et al. 1997). Pea BP-80 is abundant in highly purified CCVs (Hinz et al. 1999). The CT sequence of an *Arabidopsis* homolog of BP-80 (AtELP) was found to interact in vitro with the mammalian AP-1 clathrin adaptor protein complex (Sanderfoot et al. 1998). A recent study further demonstrated that an *Arabidopsis* TGN-localized μ-adaptin binds to the YMPL motif in the CT of BP-80 (Happel et al. 2004).

2.3 BP-80 and Homologs are Sorting Receptors for Lytic Vacuole Proteins

Several in vitro binding studies have been carried out to demonstrate the functional roles of BP-80 and its homologs as plant vacuolar sorting receptor proteins. First, a synthetic peptide, SRFNPIRLPT, representing the VS of prosporamin, was shown to specifically interact with BP-80 in a pH-dependent manner (Kirsch et al. 1994). The interaction between BP-80 and proaleurain VSD had a narrow pH optimum of 6.0–6.5 and the binding was abolished at pH 4.0. This pattern is consistent with what would be expected for a sorting receptor that would bind proteins at the neutral pH of the TGN and release them into the acidic environment of a PVC (Kirsch et al. 1994; Jiang and Rogers 2003). A second binding study that utilized affinity columns coupled with different synthetic peptides confirmed that BP-80 was able to bind specifically to the VSD of prosporamin, and to a peptide representing the C-terminus of Brazil nut 2S albumin (Kirsch et al. 1996). Similar results were obtained using *Arabidopsis* lysates: binding at a neutral pH was observed for peptides representing the native proaleurain sequences, but no binding was observed for barley lectin and tobacco chitinase C-terminal propeptides (Ahmed et al. 2000).

The ligand binding characteristics of BP-80 was further studied using recombinant truncated BP-80 lacking the TMD/CT and expressed in and purified from *Drosophila* S2 cells. A fluorescent tag was coupled to the proaleurain peptide and used in a binding assay where bound peptide was separated by gel filtration chromatography (Cao et al. 2000). The obtained results indicated that the pH-dependent, NPIR-specific binding site appears to be defined exclusively by the structure of a "unique" domain lacking EGF repeats, whereas a non-NPIR-specific binding site required the presence of the EGF repeats.

Functional proof that BP-80 acts as a VSR in vivo has been given in both yeast (Humair et al. 2001) and in plant cells (daSilva et al. 2005). In the first named study, a petunia aleurain VSD was used to generate a signal peptide-AleurainVSD-GFP fusion (sp-Aleu-GFP). When expressed alone in a mutant yeast strain having a deletion of the yeast vacuolar sorting receptor Vps10p, sp-Aleu-GFP was secreted from the cells. However, when BP-80 and sp-Aleu-GFP were co-expressed in the mutant yeast cells, the receptor directed the sp-Aleu-GFP fusion to the yeast vacuole in a specific manner because the receptor did not direct other control fusions to the yeast vacuole.

A truncated sp-GFP-BP-80 fusion construct containing the BP-80 TMD/CT (Jiang and Rogers 1998) has been used to study ligand-receptor interaction in plant cells (daSilva et al. 2005). This construct was previously shown to localize to PVCs in transgenic tobacco BY-2 cells (Tse et al. 2004). The hypothesis being tested was that the product of the truncated sp-GFP-BP-80 fusion construct would compete with the endogenous BP-80 when transiently expressed in tobacco protoplasts and thus lead to the mis-sorting of cargo proteins. Indeed, when the truncated sp-GFP-BP-80 fusion was coexpressed with a vacuole-directed NPIR-α-amylase fusion construct, the latter was secreted into the culture media (daSilva et al. 2005). These results clearly demonstrate that the transport to the cell surface of a vacuolar cargo protein can be achieved through competition with a non-functional (truncated) BP-80 receptor.

2.4 Subcellular Localization of BP-80 and its Homologues

The distribution of BP-80 and storage proteins in isolated transport vesicles has also been determined in pea cotyledons. Highly purified CCVs contained abundant BP-80 but little of the pea storage proteins vicilin and legumin (Hohl et al. 1996). In contrast, the storage proteins were enriched in so-called dense vesicles which had little or no BP-80 (Hinz et al. 1999). Thus, BP-80 was closely associated with CCV protein trafficking. Further studies using quantitative immunogold electron microscopy to analyze the distribution of the proteins in Golgi complexes demonstrated that BP-80 and storage proteins distributed in Golgi complexes differently (Hillmer et al. 2001). The storage proteins were observed to form aggregates in the *cis*-Golgi where intense labelling with anti-storage protein antibodies was measured, with little labelling in the *trans*-Golgi. In contrast, the distribution of BP-80 was just the reverse, where maximal labelling occurred in the *trans*-Golgi and little labelling was observed in the *cis*-Golgi (Hillmer et al. 2001). These results have recently been confirmed in an investigation on germinating mung bean seeds which clearly shows that VSR antibodies label CCVs but not dense vesicles (Wang, Li, Hillmer, Robinson and Jiang, manuscript in preparation). Taken together, these data are consistent with the notion that BP-80 is a sorting receptor that selects its cargo at the TGN.

Investigations on the location of BP-80 and its homologues have also been performed on non-storage protein producing cells. Immunogold electron microscopy and subcellular fractionation approaches have shown that VSR proteins are not present in the vacuolar membrane (Ahmed et al. 1997; Paris et al. 1997; Sanderfoot et al. 1998). BP-80 has been localized to the Golgi apparatus and to a putative lytic PVC in pea root tip cells (Paris et al. 1997). AtELP, a BP-80 homolog from *Arabidopsis* equivalent to VSR_{At1} (Ahmed et al. 1997) has been localized to the Golgi apparatus and to a putative PVC characterized by ~ 100 nm diameter tubules in *Arabidopsis* root tip cells (Sanderfoot et al. 1998). However, these tubular structures were also consistent with the appearance of TGN (Robinson et al. 2000). Subcellular fractionation demonstrated that AtELP in both *Arabidopsis* (Ahmed et al. 1997; Sanderfoot et al. 1998) and transgenic *Arabidopsis* plants expressing a mammalian Golgi enzyme alpha-2,6-sialyltransferase (Wee et al. 1998) cofractionates with Golgi membranes. Additionally, AtELP co-fractionated in sucrose density gradients of *Arabidopsis* root membrane with the PVC-SNARE AtPep12p (Sanderfoot et al. 1998). Taken together, these results indicate that VSR proteins are localized at both the Golgi apparatus and a putative PVC in vegetative cells.

More recently, the relative distribution of BP-80 type receptors in pea and tobacco cells have been determined using confocal immunofluorescence with antibodies specific for proteins resident in either the Golgi apparatus or the PVC (Li et al. 2002). Five different VSR antibodies were tested in various combinations in double-labelling experiments in pea root tip cells and all colocalized (Li et al. 2002). The results obtained show that VSR-labelled organelles are largely separate from the ER, Golgi apparatus and vacuolar compartments. Labelling with VSR antibodies largely colocalized with anti-AtPep12 (AtSYP21) labelling, and labelling with anti-AtPep12 was largely separate from that for the Golgi markers. Thus, VSR proteins must be concentrated on PVCs under steady state conditions (Li et al. 2002). It also means that VSR proteins only recycle back to Golgi briefly for selection of transit cargo molecules and then return to the PVCs for cargo delivery.

2.5
Chimeric BP-80 Reporter Proteins as PVC Markers

Chimeric fusion proteins employing the TMD and CT sequences of BP-80 have also been used to identify the plant PVC and determine its role in mediating protein trafficking in the plant secretory pathway (Jiang and Rogers 1998). The hypothesis was that the TMD and CT sequences of BP-80 were specific and sufficient for correct targeting of the receptor. A chimeric integral membrane reporter protein was therefore designed and tested by transient expression in tobacco suspension culture cells (Jiang and Rogers 1998). The chimeric reporter protein had a lumenal domain from barley proaleurain with a mutated vacuolar sorting sequence connected to linker sequences that included a FLAG epitope tag and a short proteolytically sensitive Ser/Thr-rich sequence from BP-80. The linker was, in turn, connected to the BP-80 TMD/CT sequences (Jiang and Rogers 1998). The strategy behind the experiments was that the chimeric BP-80 reporter would be processed into a mature form only if it trafficked to its correct destination, which in turn would demonstrate the role of TMD/CT sequences in correct targeting. Indeed, upon expression this reporter protein reached small post Golgi, non-vacuolar compartments where the proaleurain moiety was proteolytically processed into mature aleurain and released from the membrane. Proteolytic processing of proaleurain therefore provided a precise functional assay for the arrival of the reporter in the lytic PVC. As a control, the reporter protein was modified by the insertion of three tandemly repeated Kex2p substrate sequences in the linker region. It then became cleaved by a Golgi-localized Kex2p like protease, but in this case the released proaleurain was secreted (Jiang and Rogers 1998; Jiang and Rogers 1999).

Both subcellular fractionation and confocal immunofluorescence approaches were also used to identify the organelle receiving the reporter protein. When double-labelling experiments using antibodies specific for the reporter protein and for the endogenous tobacco VSR proteins were performed, the reporter protein was localized to organelles of $\leq 1\ \mu m$ in size (Jiang and Rogers 1998). In addition, these organelles floated on Ficoll step gradients together with a fraction rich in vacuoles (Jiang and Rogers 1998).

These results have similarities to studies in yeast, where the TMD alone of some single pass proteins is sufficient to direct the protein to the PVC (Roberts et al. 1992). However, subsequent studies have emphasized that both the length of the TMD and its amino acid composition greatly affect localization of a protein within the secretory pathway (Lewis et al. 2000; Reggiori et al. 2000; Brandizzi et al. 2002).

Because VSR proteins are concentrated on PVCs (Li et al. 2002), and because chimeric BP-80 reporters containing the BP-80 TMD/CT colocalize with endogenous VSR proteins when the reporter is expressed in transgenic tobacco culture cells (Jiang and Rogers, 1998), it is a good assumption that BP-80 reporters can be used as markers for the PVC. This has been confirmed with another fusion approach using yellow fluorescent protein (YFP; Tse et al. 2004). A signal peptide (sp)-YFP fusion construct was made with the TMD/CT of BP-80. In addition, and as a control, a Golgi marker construct GONST1-YFP (Baldwin et al. 2001) was also employed. In both cases typical punctate signals were detected upon expression in tobacco BY-2 cells. The organelles labelled by the sp-YFP-BP-80 reporter proteins colocalized with endogenous VSR proteins, but were separate from the Golgi marker protein mannosidase I. Furthermore, the YFP-labelled PVCs were also distinguished from YFP-tagged Golgi apparatus based on their different sensitivities to the drugs brefeldin A (BFA) and wortmannin. Whereas BFA at low concentrations (5–10 μg/ml) led to dramatic morphological changes of the Golgi apparatus, it had little effect on the PVC. In contrast, wortmannin at concentrations of 8–16 μM caused PVCs to form small vacuoles and had no effect on the Golgi apparatus (Fig. 3). This again points to the positive identification of the PVC through the use of chimeric BP-80 reporters (Tse et al. 2004).

However, two different GFP fusion studies seem to give rise to different conclusions. Brandizzi et al. (2002) fused the TMD and first five residues of the CT of BP-80 to signal peptide-GFP and expressed the chimeric protein in tobacco leaf epidermal cells. This GFP reporter protein colocalized completely with a YFP-tagged Golgi marker. The authors therefore concluded that the BP-80 TMD targeted their protein to the Golgi complex, and that positive information is needed to transit out of the Golgi complex to the lytic PVC. The contradiction between these two studies (Jiang and Rogers 1998; Brandizzi et al. 2002) might possibly be due to the different cell types used or may lie in the nature of the constructs employed. In the proaleurain or YFP reporter constructs of Jiang and Rogers (1998), the spacer amino acids in front of the TMD came from BP-80 and defined a Ser/Thr-rich motif, SKTASQAKST. In contrast, in the GFP reporter construct of Brandizzi et al. (2002), the corresponding amino acids were THGMDELYKST. It is well established that the stalk regions of Golgi glycosyltransferases adjacent to the transmembrane sequences can play important roles in protein-protein interactions that, in turn, participate in retention

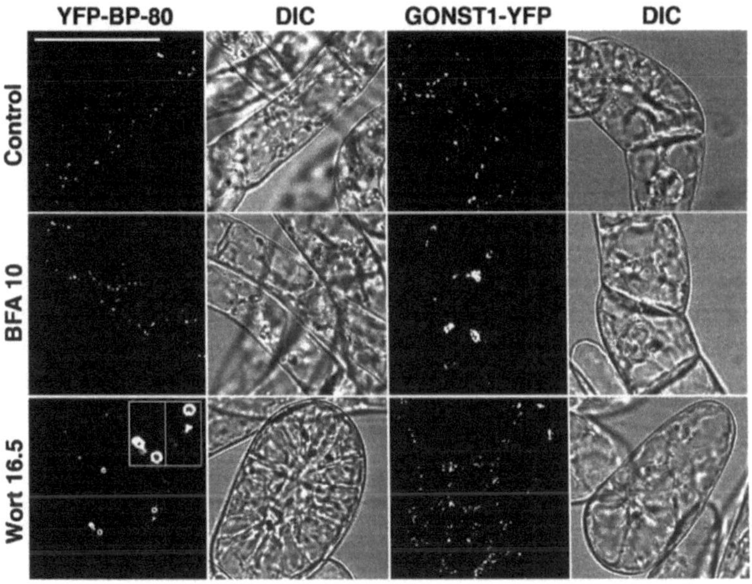

Fig. 3 Dynamics of YFP-tagged prevacuolar compartments in transgenic tobacco BY-2 cells. YFP-BP-80-tagged PVCs in transgenic tobacco BY-2 cells showed typical punctate patterns that were not affected by brefeldin A (BFA) treatment at 10 μg/ml, but were induced to form small vacuoles by wortmannin treatment at 16.5 μg/ml (*left panel*). GONST1-YFP-tagged Golgi stacks showed similar punctate patterns that did not respond to wortmannin treatment but were induced to form aggregates in the presence of BFA (*right panel*). Scale bar = 50 μm

of the proteins to specific Golgi cisternae (Nilsson et al. 1993). This consideration raises the possibility that the native BP-80 sequence could have interacted with other proteins, e.g. endogenous tobacco VSR proteins, and caused the proaleurain reporter (Jiang and Rogers, 1998) or the sp-YFP-BP-80 reporter (Tse et al. 2004) to be passively drawn into vesicles destined for the lytic PVC.

In another study, signal peptide-GFP was fused to the TMD and CT regions of pumpkin PV72, a receptor-like protein isolated from precursor accumulating vesicles to yield a construct that included the PV72 Ser/Thr-rich lumenal sequence (Mitsuhashi et al. 2000). When this sp-GFP-PV72 fusion protein was expressed in tobacco BY-2 suspension culture cells, its localization was dependent upon the state of growth and differentiation of the cells. In three day-old suspension culture cells the GFP localized to small, ≤ 1 μm punctate structures whose appearance was consistent with either Golgi or PVCs. The authors assumed the protein was localized to the Golgi apparatus. However, their conclusion was not verified by either immunogold electron microscopy (EM) or immunofluorescence with established markers for these organelles.

3
Multivesicular Bodies Identified as the Lytic PVC

Multivesicular bodies (MVBs) have been recorded on numerous occasions in the plant literature over the last 25 years (for references see Robinson et al. 2000). The presence of peroxidase and acid phosphatase is an indication that they lie on the biosynthetic pathway to the lytic vacuole (Record and Griffing 1988). Typically, MVBs in plant cells are somewhat smaller than a Golgi stack and have an osmiophilic plaque somewhere on their surface (see Fig. 4A). This is a feature shared by multivesicular endosomes in mammalian cells (Raposo et al. 2001, van Dam and Stoorvogel 2002). Generally MVBs are not seen as frequently as Golgi stacks in thin sections, but definite data on this has

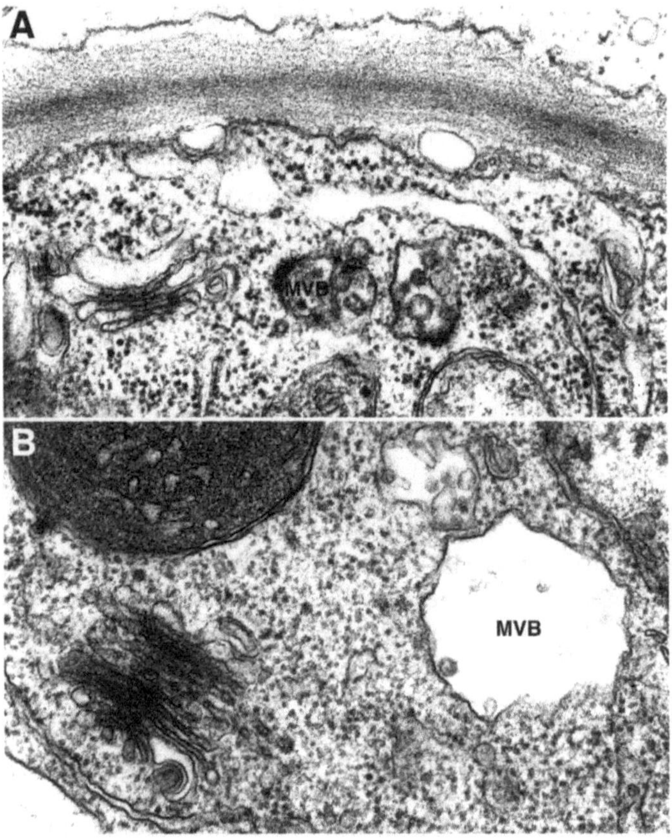

Fig. 4 Multivesicular bodies in BY-2 cells. **A** Morphology of multivesicular bodies in normal grown BY-2 cells; **B** Morphology of MVBs in wortmannin-treated BY-2 cells. Note that the MVBs are dilated and have fewer internal vesicles. Scale bar = 500 nm

not been available until quite recently. Using electron tomography on serial sections prepared from high pressure frozen/freeze-substituted *Arabidopsis* shoot meristem cells, Segui-Simarro and Staehelin (2005) have calculated that roughly one MVB is present for every two Golgi stacks. Moreover, MVBs and Golgi stacks are always in close proximity to one another.

Conclusive proof that MVBs are PVCs was provided by Tse et al. (2004) who demonstrated the presence of VSRs by immunogold EM labelling of sections prepared from high pressure frozen-freeze substituted tobacco BY-2 cells. Interestingly, VSR labelling was restricted to the boundary membrane rather than the internal vesicles of the MVBs suggesting that only a small proportion of the VSRs that reach the PVC at any one time are destined for degradation. By using VSR antibodies to probe subcellular fractions for organelles in which VSRs are concentrated, isolated VSR-labelled membraneous organelles were also identified as MVBs by immunogold EM negative staining. As might be expected, MVBs like Golgi stacks are motile organelles, as demonstrated by live cell imaging of BY-2 cells expressing YFP-tagged PVCs (Tse et al. 2004).

Confirmation that MVBs are PVCs comes from examining the differential effects of the drugs BFA and wortmannin (see above). Whereas BFA at low concentration (5–10 μg/ml) leads to rapid and dramatic morphological changes in the morphology of the Golgi apparatus of BY-2 cells (see Ritzenthaler et al. 2002), this drug does not affect MVBs. Conversely, wortmannin causes MVBs to swell (Fig. 4B; see also Tse et al. 2004), to lose their osmiophilic plaques, and leads to a reduction in the number of internal vesicles. These changes have also been recorded for multivesicular endosomes in mammalian cells (Bright et al. 2001; Sachse et al. 2002). Thus, wortmannin appears to be a useful MVB/PVC-specific drug, and is becoming a most useful tool for perturbing transport to the lytic vacuole in plants (daSilva et al. 2005).

4
Recycling Receptors from the PVC

Sending back receptors to the TGN for another round of sorting is crucial to maintaining the anterograde transport of hydrolases to the lytic compartment of the cell. It is therefore a characteristic feature of the PVC. Interestingly, in comparison to the wealth of information on the sorting events at the TGN, there is relatively less information on the recycling process, especially in mammalian cells. Whilst it is certain that clathrin is not involved, there is still some uncertainty as to the exact roles of the molecules identified as participating in this event. This situation is exacerbated by the fact that two distinct endosome-Golgi retrieval pathways have been identified in both mammalian (Mallet and Maxfield 1999), and yeast (Hettema et al. 2003) cells. This suggests that retrograde traffic out of the PVC, even in plants, may also entail a sorting process.

Several cytosolic proteins have been implicated in recycling from the endosome/PVC: TIP47, the adaptor complex AP-1, and retromer. TIP47, which is only present in mammalian cells, has been shown to bind in vitro to the cytosolic tails of both the large (CI-) and small (CD-) mannose 6-phosphate receptors (MPRs) (Diaz and Pfeffer 1998). However, more recent data suggests that TIP47 in vivo acts instead as a lipid-droplet-binding protein (Hickenbottom et al. 2004). AP-1 is well known to interact with GGAs (Golgi-localizing, -adaptin ear homology domain, ARF-binding proteins) in sorting MPRs at the TGN (Doray et al. 2002), and there is circumstantial evidence that AP-1 is somehow involved in retrograde traffic from endosomes in mammalian cells (Mallard et al. 1998; Meyer et al. 2000), but since clathrin is not required for this process it is unclear how exactly AP-1 can mediate receptor recycling. Moreover, a number of point mutations in yeast AP-1 are without effect on the traffic of carboxypeptidase A to the vacuole, suggesting normal recycling of Vps10p to the Golgi apparatus (Yeung et al. 1999).

4.1 Retromer

Early studies on yeast revealed that two vacuolar protein sorting mutants *vps35* and *vps29* had similar phenotypes to that of *vps10* (Seaman et al. 1997). Vps10p is the receptor equivalent to MPR in yeast, and is known to cycle between the TGN and PVC (Cooper and Stevens 1996), but in the *vps35* and *vps29* mutants Vps10p is relocated to the vacuole and becomes depleted in the Golgi (Seaman et al. 1997). Independent investigations carried out at the same time established the importance of two other *VPS* gene products, Vps17p and Vps5p, in maintaining the correct distribution of Vps10p (Horazdovsky et al. 1997; Nothwehr and Hindes 1997). Together, these results pointed to the importance of Vps35p, Vps29p, Vps17p, and Vps5p in the recycling of Vps10p to the Golgi. Subsequent studies involving cross-linkers demonstrated that these four proteins, together with Vps26p formed a pentameric complex, which bound to the cytosolic tail of Vps10p at the surface of the PVC (Seaman et al. 1998).

Retromer interacts with Vps10p directly via Vps35p, whereas Vps26p enhances the linkage between Vps35p and the membrane (Reddy and Seaman 2001). On the other hand, binding of Vps26p to Vps35p enables attachment of VPs17p and Vps5p to the trimeric retromer subcomplex (reviewed by Pfeffer 2001 and Seaman 2005; see also Fig. 5). Close homologues to four of these proteins have been identified in mammalian cells (Haft et al. 2000). A homologue for Vps17p does not exist, and its place in the mammalian retromer complex appears to be taken by the sorting nexin SNX2, although this notion has recently been challenged (Gullapalli et al. 2004). Interestingly, the mammalian homologue for Vps5p is the sorting nexin SNX1. Typical for both Vps5p and Vps17p, and for SNX1 and SNX2 is the presence of so-called BAR

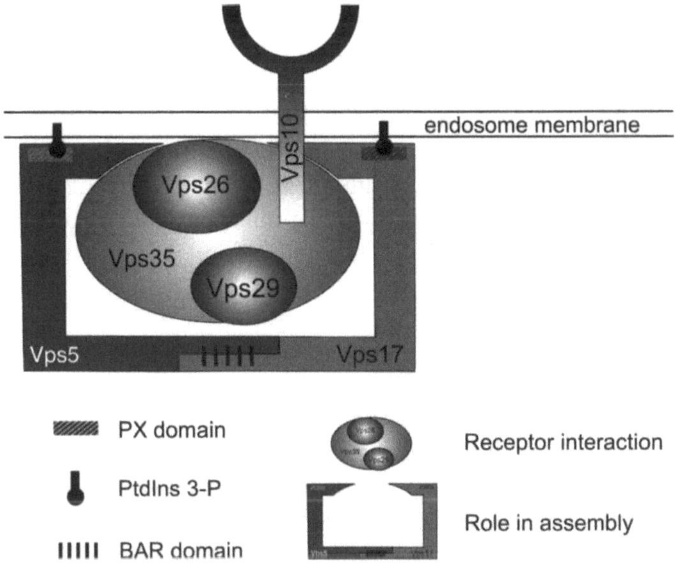

Fig. 5 The molecular components of the retromer complex. Shown are the relationships between the yeast vacuolar sorting receptor Vpsl0p and the components of the two retromer subunits

(coiled-coil) domains in their C-terminal regions through which the dimerization of these two polypeptides is achieved (Seaman and Williams 2002).

PX domains are also characteristic of SNX proteins, and these allow for binding to phosphatidylinositol 3-phosphate (PI-3P). It is therefore not surprising that wortmannin, which is an inhibitor of phosphatidyl-3 kinase (Vanhaesebroek et al. 1997), should influence the association of Vps5p and Vps26 to the endosomal membrane (Cozier et al. 2002; Arighi et al. 2004). In yeast, *VPS34* is the gene for PI-3 kinase and this, together with a Vps30p/Vps38 dimer regulates retromer function in this organism (Burda et al. 2002). As might be expected, *vps34* and *vps38* mutants have the same phenotype as *vps35* and *vps 29* (Seaman et al. 1997).

In agreement with studies on the binding of Vps35p to Vpsl0p (Nothwehr et al. 1999) it has recently been shown that mammalian Vps35 interacts directly with the cytosolic domain of the CI-MPR (Arighi et al. 2004). However, the residues of the CI-MPR which are involved (48–100 and 500y-693) do not have the tyrosine-containing (YSKV, residues 153–162) nor acidic cluster-dileucine (residues 153–162) motifs which are responsible for sorting at the PM and the TGN respectively (Ghosh et al. 2003).

Although the evidence for the participation of retromer in receptor recycling from the PVC/endosome in yeast and mammalian cells is overwhelming, retromer-coated vesicles have neither been demonstrated in situ nor have they been isolated or induced in vitro. Moreover, small GTPases like

Sar1 and/or Arf1, which are required for the coat recruitment in COP- and clathrin-coated vesicles, also do not appear to be necessary. Instead, fusion tubules have been discussed as an alternative transport form (Seaman 2005). The basis for this suggestion not only lies in the detection of Vps35 and SNX1 in tubular structures emanating from multivesiculate endosomes in mammalian cells (Zhong et al. 2002; Arighi et al. 2004), but in the self-assembly facility of SNX1 which can induce tubulation when added to liposomes in vitro (Peter et al. 2004).

4.2 Retromer in Plants

The close homology between yeast and mammalian retromer components suggests that retromer is highly conserved and should also be present in plants. This seems to be borne out in a preliminary data bank search, where putative homologues to Vps35 and Vps 26 have been identified (Dacks et al. 2003). Most significantly, the functionally critical Asp 123 residue in yeast Vps35p (Nothwehr et al. 1999) is also present at this position in plant Vps35. Our group has made a number of observations on plant retromer. Firstly, we have identified and cloned *Arabidopsis* homologues to Vps35 (At3g51310), Vps29 (At3g47810), and Vps26(At5g53530), expressed these as recombinant proteins and generated polyclonal antisera (Oliviusson et al., manuscript in preparation). The antisera recognize specifically proteins in homogenates from *Arabidopsis* and tobacco BY-2 cells of the expected molecular mass (89 kDa, AtVps35; 21 kDa, AtVps29; 35 kDa, AtVps26). The proteins recognized exist together in a 145 kDa complex which associates with microsomal membranes.

Immunogold EM labelling of cryosections reveals the presence of retromer proteins mainly on multivesicular bodies, and also to a lesser extent on the Golgi apparatus. This distribution is confirmed by immunofluorescence (see Fig. 6) where a high degree of colocalization between signals for Vps35, YFP-BP-80 (VSR-At1), and Pep12 was obtained. In agreement with observations on yeast (see above) treatment of cells with wortmannin leads to a separation of the signals for Vps35 and VST-At1 in BY-2 cells. Thus, although the plant PVC does not have tubular outgrowths, retromer is associated with its surface. There may be a direct interaction between plant Vps35 and VSR-At1, and there may be retromer coated vesicles in plants, but this at the moment remains pure speculation.

5 Molecular Markers for Plant PVCs

Two approaches have been used to study molecular components of plant PVCs. The first approach has taken advantage of the known *Arabidopsis*

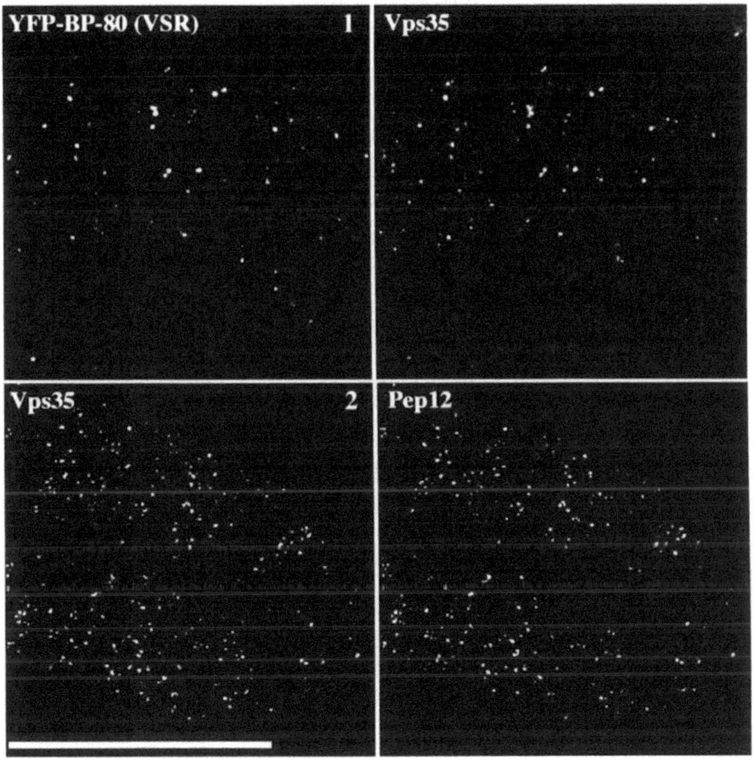

Fig. 6 Vps35 colocalizes with PVC markers in BY-2 cells. Transgenic tobacco BY-2 cells expressing the PVC reporter YFP-BP-80 were fixed and double labelled with anti-GFP (to detect the reporter) and *Arabidopsis* Vps35 antibodies (*panel 1*). Similarly, fixed BY-2 cells were double labelled with AtVps35 and Pep12 antibodies (*panel 2*). Confocal images were collected from the same optical section. A high degree of colocalization is demonstrated between Vps35 and the VSR-reporter on the one hand, and Vps35 and the PVC marker Pep12 on the other hand. Scale bar = 50 μm

genomic sequences and our knowledge on the possible roles of Rab GTPases and SNAREs (soluble N-ethyl-maleimide sensitive factor attachment protein receptor) molecules in protein sorting and vesicle fusion. For example, in yeast, the syntaxin Pep12p is specifically associated with the PVC/endosome (Pelham 2000). In *Arabidopsis*, its ortholog is AtPep12 (AtSYP22), which can functionally complement the yeast *pep12* mutant (Sanderfoot et al. 2000). AtPep12 has been localized to a post-Golgi compartment in *Arabidopsis* root tip cells by immunogold EM and subcellular fractionation (Bassham et al. 1995; Conceição et al. 1997). However, the structure of the PVC was not well defined in these studies. Immunofluorescence colocalization studies performed on pea root tip cells and tobacco BY-2 cells with VSR and Pep12 antibodies has clearly shown a high degree of VSR colocalization with AtPep12, and also that labelling with anti-

AtPep12p was largely separate from that for Golgi markers (Li et al. 2002; Tse et al. 2004).

The *Arabidopsis* genome contains a total of 54 SNARE genes and encodes 57 Rab GTPases (Sanderfoot et al. 2000; Uemura et al. 2004). A strategy has therefore recently been developed whereby various GFP-tagged Rab GTPases and SNARE fusion proteins were constructed and transiently expressed in *Arabidopsis* cells for identification of their subcellular localization. Even though overexpression of fusion proteins might cause mislocalization and loss of function, such an approach has been, in general, successful in studying Rab GTPases and SNAREs. When various GFP-tagged SNARE proteins were transiently expressed in *Arabidopsis* cells, a total of 12 SNARE proteins were found to be resident in the endosome/PVC. Similarly, several Rab GTPases including the subgroup Rab5 members Ara5, Ara6, Ara7 were also found to localize in endosome/PVC of *Arabidopsis* cells (Ueda et al. 2004). When transiently expressed in *Arabidopsis* protoplasts, both Ara6-GFP and GFP-AtSyp22 colocalized with the corresponding endogenous Ara6 and AtSyp22, while GFP-AtSyp22 was even functional in vivo (Ueda et al. 2001; Uemura et al. 2002). Interestingly, distinct endosomes for the endocytic pathway in *Arabidopsis* cells can also be defined: the early endosome, a possible site for GNOM-dependent recycling of plasma membrane protein such as AtPin1 (Geldner et al. 2003), is displayed by the presence of Ara7, Rha1, and AtVam727; while the late endosome/PVC, is characterized by the presence of Ara6, AtSyp21, and AtSyp22 (Ueda et al. 2004).

The second approach is an organelle proteomic study. The establishment of transgenic BY-2 cell lines expressing the PVC marker sp-YFP-BP-80 reporter and the identification of PVC as MVB in these cells have provided useful tools for further isolation of PVCs and their subsequent proteomic analysis (see above and Mo et al. 2003; Tse et al. 2004). Thus far, reliable protocols have been successfully developed for large-scale isolation of VSR-enriched PVCs fractions from both tobacco BY-2 cells and *Arabidopsis* culture cells (Mo, Lo, and Jiang, manuscript in preparation). Further MS/MS analysis on these PVC protein extractions will allow identification of both soluble (cargo) proteins and membrane (receptor) proteins of PVCs. However, such an organelle proteomic analysis might generate unexpected results as recently demonstrated from the proteomic analysis of isolated vacuoles of *Arabidopsis* due to the complexity of protein trafficking in plant cells. For example, several SNARE proteins (AtPep12/AtSYP21, AtSYP22/AtVam3, SYP51, SYP52, AtVAMP711, AtVAMP712, AtVAMP713, AtVTI11, AtVTI13) are found in both PVC and vacuolar membrane and are possibly responsible for triggering membrane fusion for cargo delivery (Carter et al. 2004).

Table 1 summarizes known protein/marker components of endosomes/PVCs in plant cells. We expect that the number of proteins will increase dramatically upon the completion of the PVC proteomic analysis on isolated PVC organelles. Some of the newly identified PVC proteins may provide us with

Table 1 Characteristics of PVC proteins in the lytic vacuolar pathway

Protein	Protein type	Function	Organism	Ref.
BP-80 (VSR_{At-1})	VSR	Receptor for lytic PVC	*PS*	Paris et al. 1997; Li et al. 2002
BP-80 reporter	Reporter fusion	PVC marker	*NT*	Jiang and Rogers 1998; Tse et al. 2004
AtELP	VSR	Receptor for lytic PVC	*AT*	Sanderfoot et al. 2004
μA-adaptin	API-adaptor	CCV formation	*AT*	Happel et al. 2004
Vps35	Retromer subunit	Receptor recycling	*NT/AT*	Oliviusson et al. 2005
AtPep12p AtSYP21 AtSYP22 SYP51 SYP52	SNAREs/ Syntaxins	Vesicle fusion	*AT*	Conceição et al. 1997; Uemura et al. 2004
AtVAMP711 AtVAMP712 AtVAMP713 AtVTI11 AtVTI13	Other SNAREs	Vesicle fusion	*AT*	Zheng et al. 1999; Uemura et al. 2004
Ara7 Rha1	Rab GTPase (Rab5)	Vesicle fusion	*AT*	Sohn et al. 2003; Lee et al. 2004
Pra3	Rab GTPase (Rab11)	Vesicle fusion	*PS*	Inaba et al. 2002

Organism: PS, *Pisum sativum*; NT, *Nicotiana tabacum*; AT, *Arabidopsis thaliana*

tools for future studies on molecular mechanisms of protein sorting and biogenesis of plant endosomes/PVCs in both secretory and endocytic pathways.

6
PVC and Endocytosis Cross Talk

Endocytosis is well-established in mammalian cells, both as clathrin-dependent, receptor-mediated, and lipid-raft pathways (LeRoy and Wrana et al. 2005). Considerable evidence for endocytosis in plants (see Marcote et al. 2000 and Šamaj et al. 2004, and the articles in this book), has accrued over the years, even though unequivocal proof for clathrin-mediated receptor-ligand internalization at the PM and subsequent receptor recycling remains to be published (see Chap. VI). Nevertheless, a number of papers in which electron-dense tracers were used to chart the endocytic pathway provide evidence for the participation of MVBs in this process (Hillmer et al. 1986; Galway et al. 1993). What has been missing until recently, has been the proof that endocytic MVBs and multivesicular PVCs are one and the same organelle. Of considerable value in achieving this goal has been the use of the fluorescent styryl dyes FM4-64 and FM1-43 (see Chap. I). After their insertion in the PM, these fluorescent dyes pass through an endosomal/prevacuolar compartment on their way to the tonoplast (Vida and Emr 1995; Kim et al. 2001; Ueda et al. 2001; Emans et al. 2002). In addition, endocytosis of these dyes from the plasma membrane to the vacuole in tobacco BY-2 cells is temperature dependent, but is inhibited in the presence of wortmannin (Emans et al. 2002). Thus colocalization of proteins with the internalized FM4-64 can be used to determine whether a particular compartment labelled by a specific protein is a potential endosome in plant cells. Proteins localized to endosomes in this way include the *Arabidopsis* Rab GTPases Ara6 and Ara7 (Ueda et al. 2001), the *Arabidopsis* Pra2 (SYP111) (a homolog to the mammalian Rab11) (Inaba et al. 2002), a small GTPase ARF1 in maize (Baluška et al. 2002, 2004), and the *Arabidopsis* GNOM protein (an ARF-guanosine exchange factor that controls the endosomal recycling of a putative auxin-efflux carrier PIN1 in *Arabidopsis* (Geldner et al. 2003).

The coidentity between FM4-64-labelled endosomes and the VSR-labelled PVC was recently demonstrated in tobacco BY-2 cells by Tse et al. (2004). Transgenic tobacco BY-2 cells expressing the PVC-localized YFP-BP-80 reporter was used in an uptake study with the endosomal FM4-64 marker. Consistent with previous results, the dye remained at the cell surface of BY-2 cells during early stages of incubation, but was rapidly taken up into the cells and exhibited first a punctate endosome-like pattern and later labelled the tonoplast. Most of the internalized punctate structures labelled by FM4-64 colocalized with the PVC YFP-BP-80 reporter signal. In contrast, in cells expressing the YFP-Golgi reporter, the YFP-bearing organelles were largely separated from the FM4-64-labelled endosomal structures. Since the YFP-BP-80 reporter colocalizes with endogenous VSR proteins in multivesicular PVCs, these results clearly demonstrate that the YFP-labelled PVC also lies on the endocytic pathway in tobacco BY-2 cells (Tse et al. 2004).

Several other studies support the notion that the secretory pathway and endocytic pathways converge at the same MVB/PVC in plant cells. Sohn et al. (2003) carried out a study to determine the functional role of the *Arabidopsis* Rab5 homolog Rha1 (Sohn et al. 2003). When a dominant-negative mutant of Rha1 was expressed in *Arabidopsis* protoplasts, the vacuolar targeting of GFP-tagged sporamin and aleurain was inhibited and resulted in the secretion of these two soluble vacuolar cargo proteins into the culture medium. The inhibitory effect was specific to Rha1 because overexpression of a similar dominant negative mutant of Ara6 did not affect the vacuolar targeting of these two cargo proteins, although the overexpression of wild type Rha1 protein did partially rescue vacuolar targeting of these cargo proteins. Furthermore, GFP-tagged Rha1 and endogenous Rha1, as well as Ara7 colocalized with VSR or Pep12p proteins in PVCs in *Arabidopsis* protoplasts (Sohn et al. 2003; Lee et al. 2004; Tse et al. 2004). Thus, Rha1 plays an important role in mediating protein traffic from PVC to vacuoles in *Arabidopsis* protoplasts. Moreover, both Ara6 and Ara7 are thought to localize to early endosomes and play a role in endocytosis in *Arabidopsis* cells because they largely colocalized with the internalized FM4-64 (Ueda et al. 2001; Grebe et al. 2003). Taken together, since Rha1 colocalized with VSR in multivesicular PVC, while internalized FM4-64 colocalized with VSR and Ara6 during endocytosis (Sohn et al. 2003; Tse et al. 2004; Ueda et al. 2001), it is thus likely that the same PVC mediates protein traffic for secretory and endocytic pathways.

Finally, a recent study has demonstrated that *Arabidopsis* cells might contain functionally distinct populations of endosomes (Ueda et al. 2004). Both populations colocalized with internalized FM4-64 and displayed punctate patterns, but the early endosome was marked by transiently expressed GFP-tagged Ara7, Rha1, and AtVam727, while the late endosome was marked by GFP-tagged Ara6, AtSYP21 (Pep12), and AtSYP22 (AtVam3) (Ueda et al. 2004, see Chap. X and XI). At this time, we do not know if these two sets of protein markers will define similar distinct endosomal populations in other cell types or plants. It would therefore be interesting to find out if Ara6 or Ara7 colocalize with VSRs in BY-2 cells (Tse et al. 2004) when they can be compared directly via confocal immunofluorescence microscopy in the same cell. It is also possible that both early endosome and late endosome (PVC/MVB) in plant cells have similar multivesicular internal vesicles, which can only be distinguished by immunogold studies using antibodies or protein markers specific to either early or late endosomes.

7
Concluding Remarks

The data presented indicated that PVCs in the secretory and endocytic pathways leading to lytic vacuole in plant cells share a great similarity with the corresponding compartments in animal and yeast cells. However, we are just at the beginning of understanding the functional roles and molecular components of plant PVCs. We also know little about the molecular mechanisms of plant endocytosis and the intermediate compartments in the endocytic pathway. Many questions need to be addressed in future experiments. For example, do all plant cells contain two or multiple endosome/PVC populations? Are there two distinct PVCs for lytic vacuole and protein storage vacuoles? Our current molecular and proteomic analysis on isolated PVCs may well help us in answering some of these questions.

Acknowledgements Our own research reported here was partially supported by grants from the Research Grants Council of Hong Kong (project CUHK4156/01M, CUHK4260/02M and CUHK4307/03M), Area of Excellence, Germany/Hong Kong and France/Hong Kong Joint Research Scheme to L. Jiang, and by the State of Baden-Württemberg (Landesschwerpunktprogramm) to D.G. Robinson.

References

Ahmed SU, Bar-Peled M, Raikhel NV (1997) Cloning and subcellular location of an *Arabidopsis* receptor-like protein that shares common features with protein-sorting receptors of eukaryotic cells. Plant Physiol 114:325–336

Ahmed SU, Rojo E, Kovaleva V, Venkataraman S, Dombrowski JE, Matsuoka K, Raikhel NV (2000) The plant vacuolar sorting receptor AtELP is involved in transport of NH_2-terminal propeptide-containing vacuolar proteins in *Arabidopsis* thaliana. J Cell Biol 149:1335–1344

Arighi CN, Hartnell LM, Aguilar RC, Haft CR, Bonifacino JS (2004) Role of the mammalian retromer in sorting of the cation-independent mannose 6-phosphate receptor. J Cell Biol 165:123–133

Baldwin TC, Handford MG, Yuseff MI, Orellana A, Dupree P (2001) Identification and characterization of GONST1, a golgi-localized GDP-mannose transporter in *Arabidopsis*. Plant Cell 13:2283–2295

Baluška F, Hlavacka A, Šamaj J, Palme K, Robinson DG, Matoh T, McCurdy DW, Menzel D, Volkmann D (2002) F-actin-dependent endocytosis of cell wall pectins in meristematic root cells. Insights from brefeldin A-induced compartments. Plant Physiol 130:422–431

Baluška F, Šamaj J, Hlavacka A, Kendrick-Jones J, Volkmann D (2004) Actin-dependent fluid-phase endocytosis in inner cortex cells of maize root apices. J Exp Bot 55:463–473

Bassham DC, Gal S, Conceicao AS, Raikhel NV (1995) An *Arabidopsis* syntaxin homolog isolated by functional complementation of a yeast pep12 mutant. Proc Natl Acad Sci USA 92:7262–7266

Bethke PC, Jones RL (2000) Vacuoles and prevacuolar compartments. Curr Opin Plant Biol 3:469–475

Brandizzi F, Snapp EL, Roberts AG, Lippincott-Schwartz J, Hawes C (2002) Membrane protein transport between the endoplasmic reticulum and the Golgi in tobacco leaves is energy dependent but cytoskeleton independent: evidence from selective photobleaching. Plant Cell 14:1293–1309

Bright NA, Lindsay MR, Stewart A, Luzio JP (2001) The relationship between lumenal and limiting membranes in swollen late endocytic compartments formed after wortmannin treatment or sucrose accumulation. Traffic 2:631–642

Burda P, Padilla SM, Sarkar S, Emr SD (2002) Retromer function in endosome-to-Golgi retrograde transport is regulated by the yeast Vps34 PtdIns 3-kinase. J Cell Sci 115:3889–3900

Cao X, Rogers SW, Butler J, Beevers L, Rogers JC (2000) Structural requirements for ligand binding by a plant vacuolar sorting receptor. Plant Cell 12:493–506

Carter C, Pan S, Zouhar J, Avila EL, Girke T, Raikhel NV (2004) The vegetative vacuole proteome of *Arabidopsis* thaliana reveals predicted and unexpected proteins. Plant Cell 16:3285–3303

Conceição AS, Marty-Mazars D, Bassham DC, Sanderfoot AA, Marty F, Raikhel NV (1997) The syntaxin homolog AtPEP 12p resides on a late post-Golgi compartment in plants. Plant Cell 9:571–582

Cooper AA, Stevens TH (1996) Vps10p cycles between the late-Golgi and prevacuolar compartments in its function as the sorting receptor for multiple yeast vacuolar hydrolases. J Cell Biol 133:529–541

Cozier GE, Carlton J, McGregor AH, Gleeson PA, Teasdale RD, Mellor H, Cullen PJ (2002) The phox homology (PX) domain-dependent, 3-phosphoinositide-mediated association of sorting nexin-1 with an early sorting endosomal compartment is required for its ability to regulate epidermal growth factor receptor degradation. J Biol Chem 277:48730–48736

Dacks JB, Davis LA, Sjogren AM, Andersson JO, Roger AJ, Doolittle WF (2003) Evidence for Golgi bodies in proposed "Golgi-lacking" lineages. Proc R Soc Lond B Biol Sci 270 Suppl 2:S168–171

daSilva LL, Taylor JP, Hadlington JL, Hanton SL, Snowden CJ, Fox SJ, Foresti O, Brandizzi F, Denecke J (2005) Receptor salvage from the prevacuolar compartment is essential for efficient vacuolar protein targeting. Plant Cell 17:132–148

Diaz E, Pfeffer SR (1998) TIP47: a cargo selection device for mannose 6-phosphate receptor trafficking. Cell 93:433–443

Doray B, Ghosh P, Griffith J, Geuze HJ, Kornfeld S (2002) Cooperation of GGAs and AP-1 in packaging MPRs at the trans-Golgi network. Science 297:1700–1703

Emans N, Zimmermann S, Fischer R (2002) Uptake of a fluorescent marker in plant cells is sensitive to brefeldin A and wortmannin. Plant Cell 14:71–86

Galway ME, Rennie PJ, Fowke LC (1993) Ultrastructure of the endocytotic pathway in glutaraldehyde-fixed and high-pressure frozen/freeze-substituted protoplasts of white spruce (Picea glauca). J Cell Sci 106 (Pt 3):847–858

Geldner N, Anders N, Wolters H, Keicher J, Kornberger W, Muller P, Delbarre A, Ueda T, Nakano A, Jurgens G (2003) The *Arabidopsis* GNOM ARF-GEF mediates endosomal recycling, auxin transport, and auxin-dependent plant growth. Cell 112:219–230

Ghosh P, Dahms NM, Kornfeld S (2003) Mannose 6-phosphate receptors: new twists in the tale. Nat Rev Mol Cell Biol 4:202–212

Grebe M, Xu J, Mobius W, Ueda T, Nakano A, Geuze HJ, Rook MB, Scheres B (2003) *Arabidopsis* sterol endocytosis involves actin-mediated trafficking via ARA6-positive early endosomes. Curr Biol 13:1378–1387

Griffiths G, Matteoni R, Back R, Hoflack B (1990) Characterization of the cation-independent mannose 6-phosphate receptor-enriched prelysosomal compartment in NRK cells. J Cell Sci 95 (Pt 3):441–461

Gullapalli A, Garrett TA, Paing MM, Griffin CT, Yang Y, Trejo J (2004) A role for sorting nexin 2 in epidermal growth factor receptor down-regulation: evidence for distinct functions of sorting nexin 1 and 2 in protein trafficking. Mol Biol Cell 15:2143–2155

Haft CR, de la Luz Sierra M, Bafford R, Lesniak MA, Barr VA, Taylor SI (2000) Human orthologs of yeast vacuolar protein sorting proteins Vps26, 29, and 35: assembly into multimeric complexes. Mol Biol Cell 11:4105–4116

Happel N, Honing S, Neuhaus JM, Paris N, Robinson DG, Holstein SE (2004) *Arabidopsis* mu A-adaptin interacts with the tyrosine motif of the vacuolar sorting receptor VSR-PS1. Plant J 37:678–693

Hara-Nishimura I, Shimada T, Hatano K, Takeuchi Y, Nishimura M (1998) Transport of storage proteins to protein storage vacuoles is mediated by large precursor-accumulating vesicles. Plant Cell 10:825–836

Hettema EH, Lewis MJ, Black MW, Pelham HR (2003) Retromer and the sorting nexins Snx4/41/42 mediate distinct retrieval pathways from yeast endosomes. EMBO J 22:548–557

Hickenbottom SJ, Kimmel AR, Londos C, Hurley JH (2004) Structure of a lipid droplet protein; the PAT family member TIP47. Structure (Camb) 12:1199–1207
Hillmer S, Depta H, Robinson DG (1986) Confirmation of endocytosis in higher plant protoplast using lectin-gold conjugates. Eur J Cell Biol 42:142–149
Hillmer S, Movafeghi A, Robinson DG, Hinz G (2001) Vacuolar storage proteins are sorted in the cis-cisternae of the pea cotyledon Golgi apparatus. J Cell Biol 152:41–50
Hinz G, Hillmer S, Bäumer M, Hohl I (1999) Vacuolar storage proteins and the putative sorting receptor BP-80 exit the Golgi apparatus of developing pea cotyledons in different transport vesicles. Plant Cell 11:1509–1524
Hohl I, Robinson DG, Chrispeels MC, Hinz G (1996) Transport of storage proteins to the vacuole is mediated by vesicles without a clathrin coat. J Cell Sci 109:2539–2550
Horazdovsky BF, Davies BA, Seaman MN, McLaughlin SA, Yoon S, Emr SD (1997) A sorting nexin-1 homologue, Vps5p, forms a complex with Vps17p and is required for recycling the vacuolar protein-sorting receptor. Mol Biol Cell 8:1529–1541
Humair D, Hernández Felipe D, Neuhaus JM, Paris N (2001) Demonstration in yeast of the function of BP-80, a putative plant vacuolar sorting receptor. Plant Cell 13:781–792
Inaba T, Nagano Y, Nagasaki T, Sasaki Y (2002) Distinct localization of two closely related Ypt3/Rab11 proteins on the trafficking pathway in higher plants. J Biol Chem 277:9183–9188
Jiang L, Rogers JC (1998) Integral membrane protein sorting to vacuoles in plant cells: evidence for two pathways. J Cell Biol 143:1183–1199
Jiang L, Rogers JC (1999) Functional analysis of a Golgi-localized Kex2p-like protease in tobacco suspension culture cells. Plant J 18:23–32
Jiang L, Rogers JC (2003) Sorting of lytic enzymes in the plant Golgi apparatus. Annu Plant Rev 9:114–140
Katzmann DJ, Odorizzi G, Emr SD (2002) Receptor downregulation and multivesicular-body sorting. Nat Rev Mol Cell Biol 3:893–905
Kim DH, Eu YJ, Yoo CM, Kim YW, Pih KT, Jin JB, Kim SJ, Stenmark H, Hwang I (2001) Trafficking of phosphatidylinositol 3-phosphate from the trans-Golgi network to the lumen of the central vacuole in plant cells. Plant Cell 13:287–301
Kirsch T, Paris N, Butler JM, Beevers L, Rogers JC (1994) Purification and initial characterization of a potential plant vacuolar targeting receptor. Proc Natl Acad Sci USA 91:3403–3407
Kirsch T, Saalbach G, Raikhel NV, Beevers L (1996) Interaction of a potential vacuolar targeting receptor with amino- and carboxyl-terminal targeting determinants. Plant Physiol 111:469–474
Kobayashi T, Stang E, Fang KS, de Moerloose P, Parton RG, Gruenberg J (1998) A lipid associated with the antiphospholipid syndrome regulates endosome structure and function. Nature 392:193–197
Le Roy C, Wrana JL (2005) Clathrin- and non-clathrin-mediated endocytic regulation of cell signalling. Nat Rev Mol Cell Biol 6:112–126
Lee GJ, Sohn EJ, Lee MH, Hwang I (2004) The *Arabidopsis* rab5 homologs rha1 and ara7 localize to the prevacuolar compartment. Plant Cell Physiol 45:1211–1220
Lemmon SK, Traub LM (2000) Sorting in the endosomal system in yeast and animal cells. Curr Opin Cell Biol 12:457–466
Lewis MJ, Nichols BJ, Prescianotto-Baschong C, Riezman H, Pelham HR (2000) Specific retrieval of the exocytic SNARE snc1p from early yeast endosomes. Mol Biol Cell 11:23–38
Li YB, Rogers SW, Tse YC, Lo SW, Sun SS, Jauh GY, Jiang L (2002) BP-80 and Homologs are Concentrated on Post-Golgi, Probable Lytic Prevacuolar Compartments. Plant Cell Physiol 43:726–742

Luzio JP, Rous BA, Bright NA, Pryor PR, Mullock BM, Piper RC (2000) Lysosome-endosome fusion and lysosome biogenesis. J Cell Sci 113 (Pt 9):1515–1524

Mallard F, Antony C, Tenza D, Salamero J, Goud B, Johannes L (1998) Direct pathway from early/recycling endosomes to the Golgi apparatus revealed through the study of shiga toxin B-fragment transport. J Cell Biol 143:973–990

Mallet WG, Maxfield FR (1999) Chimeric forms of furin and TGN38 are transported with the plasma membrane in the trans-Golgi network via distinct endosomal pathways. J Cell Biol 146:345–359

Marcote MJ, Carbonell J (2000) Transient expression of a pea MAP kinase gene induced by gibberellic acid and 6-benzyladenine in unpollinated pea ovaries. Plant Mol Biol 44:177–186

Maxfield FR, McGraw TE (2004) Endocytic recycling. Nat Rev Mol Cell Biol 5:121–132

Meyer C, Zizioli D, Lausmann S, Eskelinen EL, Hamann J, Saftig P, von Figura K, Schu P (2000) mu1A-adaptin-deficient mice: lethality, loss of AP-1 binding and rerouting of mannose 6-phosphate receptors. EMBO J 19:2193–2203

Mitsuhashi N, Shimada T, Mano S, Nishimura M, Hara-Nishimura I (2000) Characterization of organelles in the vacuolar-sorting pathway by visualization with GFP in tobacco BY-2 cells. Plant Cell Physiol 41:993–1001

Mo B, Tse YC, Jiang L (2003) Organelle identification and proteomics in plant cells. Trends Biotech 21:331–332

Neuhaus JM, Rogers JC (1998) Sorting of proteins to vacuoles in plant cells. Plant Mol Biol 38:127–144

Nilsson T, Slusarewica P, Hoe MH, Warren G (1993) Kin recognition. A model for the retention of Golgi enzymes. FEBS Lett 330:1–4

Nothwehr SF, Hindes AE (1997) The yeast VPS5/GRD2 gene encodes a sorting nexin-1-like protein required for localizing membrane proteins to the late Golgi. J Cell Sci 110 (Pt 9):1063–1072

Nothwehr SF, Bruinsma P, Strawn LA (1999) Distinct domains within Vps35p mediate the retrieval of two different cargo proteins from the yeast prevacuolar/endosomal compartment. Mol Biol Cell 10:875–890

Paris N, Rogers SW, Jiang L, Kirsch T, Beevers L, Phillips TE, Rogers JC (1997) Molecular cloning and further characterization of a probable plant vacuolar sorting receptor. Plant Physiol 115:29–39

Paris N, Stanley CM, Jones RL, Rogers JC (1996) Plant cells contain two functionally distinct vacuolar compartments. Cell 85:563–572

Parton RG, Schrotz P, Bucci C, Gruenberg J (1992) Plasticity of early endosomes. J Cell Sci 103 (Pt 2):335–348

Pelham HR (2000) SNAREs and the secretory pathway—lessons from yeast. Exp Cell Res 247:1–8

Peter BJ, Kent HM, Mills IG, Vallis Y, Butler PJ, Evans PR, McMahon HT (2004) BAR domains as sensors of membrane curvature: the amphiphysin BAR structure. Science 303:495–499

Pfeffer SR (2001) Membrane transport: retromer to the rescue. Curr Biol 11:R109–R111

Raposo G, Tenza D, Murphy DM, Berson JF, Marks MS (2001) Distinct protein sorting and localization to premelanosomes, melanosomes, and lysosomes in pigmented melanocytic cells. J Cell Biol 152:809–824

Record RD, Griffing LR (1988) Convergence of the endocytic and lysosomal pathways in soybean protoplasts. Planta 176:425–432

Reddy JV, Seaman MN (2001) Vps26p, a component of retromer, directs the interactions of Vps35p in endosome-to-Golgi retrieval. Mol Biol Cell 12:3242–3256

Reggiori F, Black MW, Pelham HRB (2000) Polar transmembrane domains target proteins to the interior of the yeast vacuole. Mol Biol Cell 11:3737–3749

Reggiori F, Pelham HR (2001) Sorting of proteins into multivesicular bodies: ubiquitin-dependent and -independent targeting. EMBO J 20:5176–5186

Ritzenthaler C, Nebenfuhr A, Movafeghi A, Stussi-Garaud C, Behnia L, Pimpl P, Staehelin LA, Robinson DG (2002) Reevaluation of the effects of brefeldin A on plant cells using tobacco Bright Yellow 2 cells expressing Golgi-targeted green fluorescent protein and COPI antisera. Plant Cell 14:237–261

Roberts CJ, Nothwehr SF, Stevens TH (1992) Membrane protein sorting in the yeast secretory pathway: evidence that the vacuole may be the default compartment. J Cell Biol 119:63–83

Robinson DG, Hinz G, Holstein SEH (1998) The molecular characterization of transport vesicles. Plant Mol Biol 38:49–76

Robinson DG, Rogers JC, Hinz G (2000) Post-Golgi, prevacuolar compartments. Annu Plant Rev 5:270–298

Sachse M, Urbe S, Oorschot V, Strous GJ, Klumperman J (2002) Bilayered clathrin coats on endosomal vacuoles are involved in protein sorting toward lysosomes. Mol Biol Cell 13:1313–1328

Šamaj J, Baluška F, Voigt B, Schlicht M, Volkmann D, Menzel D (2004) Endocytosis, actin cytoskeleton, and signaling. Plant Physiol 135:1150–1161

Sanderfoot AA, Ahmed SU, Marty-Mazars D, Rapoport I, Kirchhausen T, Marty F, Raikhel NV (1998) A putative vacuolar cargo receptor partially colocalize with AtPEP12p on a prevacuolar compartment in *Arabidopsis* roots. Proc Natl Acad Sci USA 95:9920–9925

Sanderfoot AA, Assaad FF, Raikhel NV (2000) The *Arabidopsis* genome. An abundance of soluble N-ethylmaleimide-sensitive factor adaptor protein receptors. Plant Physiol 124:1558–1569

Seaman MN (2005) Recycle your receptors with retromer. Trends Cell Biol 15:68–75

Seaman MN, Williams HP (2002) Identification of the functional domains of yeast sorting nexins Vps5p and Vps17p. Mol Biol Cell 13:2826–2840

Seaman MN, McCaffery JM, Emr SD (1998) A membrane coat complex essential for endosome-to-Golgi retrograde transport in yeast. J Cell Biol 142:665–681

Seaman MN, Marcusson EG, Cereghino JL, Emr SD (1997) Endosome to Golgi retrieval of the vacuolar protein sorting receptor, Vps10p, requires the function of the VPS29, VPS30, and VPS35 gene products. J Cell Biol 137:79–92

Segui-Simarro JM, Staehelin LA (2005) Cell cycle-dependent changes in Golgi stacks, vacuoles, clathrin-coated vesicles and multivesicular bodies in meristematic cells of *Arabidopsis* thaliana: a quantitative and spatial analysis. Planta (in press)

Shimada T, Kuroyanagi M, Nishimura M, Hara-Nishimura I (1997) A pumpkin 72-kDa membrane protein of precursor-accumulating vesicles has characteristics of a vacuolar sorting receptor. Plant Cell Physiol 38:1414–1420

Sohn EJ, Kim ES, Zhao M, Kim SJ, Kim H, Kim YW, Lee YJ, Hillmer S, Sohn U, Jiang L, Hwang I (2003) Rha1, an *Arabidopsis* Rab5 homolog, plays a critical role in the vacuolar trafficking of soluble cargo proteins. Plant Cell 15:1057–1070

Tse YC, Mo B, Hillmer S, Zhao M, Lo SW, Robinson DG, Jiang L (2004) Identification of multivesicular bodies as prevacuolar compartments in Nicotiana tabacum BY-2 cells. Plant Cell 16:672–693

Ueda T, Yamaguchi M, Uchimiya H, Nakano A (2001) Ara6, a plant-unique novel type Rab GTPase, functions in the endocytic pathway of *Arabidopsis* thaliana. EMBO J 20:4730–4741

Ueda T, Uemura T, Sato MH, Nakano A (2004) Functional differentiation of endosomes in *Arabidopsis* cells. Plant J 40:783–789

Uemura T, Yoshimura SH, Takeyasu K, Sato MH (2002) Vacuolar membrane dynamics revealed by GFP-AtVam3 fusion protein. Genes Cells 7:743–753

Uemura T, Ueda T, Ohniwa RL, Nakano A, Takeyasu K, Sato MH (2004) Systematic analysis of SNARE molecules in *Arabidopsis*: dissection of the post-Golgi network in plant cells. Cell Struct Funct 29:49–65

Van Dam EM, Ten Broeke T, Jansen K, Spijkers P, Stoorvogel W (2002) Endocytosed transferrin receptors recycle via distinct dynamin and phosphatidylinositol 3-kinase-dependent pathways. J Biol Chem 277:48876–48883

Vanhaesebroeck B, Leevers SJ, Panayotou G, Waterfield MD (1997) Phosphoinositide 3-kinases: a conserved family of signal transducers. Trends Biochem Sci 22:267–272

Vida TA, Emr SD (1995) A new vital stain for visualizing vacuolar membrane dynamics and endocytosis in yeast. J Cell Biol 128:779–792

Yeung BG, Phan HL, Payne GS (1999) Adaptor complex-independent clathrin function in yeast. Mol Biol Cell 10:3643–3659

Wee EG, Sherrier DJ, Prime TA, Dupree P (1998) Targeting of active sialytransferase to the plant Golgi apparatus. Plant Cell 10:1759–1768

Zheng H, von Mollard GF, Kovaleva V, Stevens TH, Raikhel NV (1999) The plant vesicle-associated SNARE AtVTI1a likely mediates vesicle transport from the trans-Golgi network to the prevacuolar compartment. Mol Biol Cell 10:2251–2264

Zhong Q, Lazar CS, Tronchere H, Sato T, Meerloo T, Yeo M, Songyang Z, Emr SD, Gill GN (2002) Endosomal localization and function of sorting nexin 1. Proc Natl Acad Sci USA 99:6767–6772

Plant Cell Monogr (1)
J. Šamaj · F. Baluška · D. Menzel: Plant Endocytosis
DOI 10.1007/7089_005/Published online: 22 September 2005

Plant Vacuoles: from Biogenesis to Function

Jean-Marc Neuhaus[1] · Nadine Paris[2] (✉)

[1]Laboratoire de Biochimie, rue E. Argand 9, BP2, 2007 Neuchâtel, Switzerland

[2]CNRS-UMR 6037, IFRMP23, Université de Rouen, 76821 Mont Saint Aignan, France
nadine.paris@univ-rouen.fr

Abstract The plant vacuolar system is far more complex than originally expected and multiple sorting pathways leading to various types of vacuoles can be found depending on the cell type and on the stage of development. In addition, the vacuolar system is highly dynamic and can adjust to environmental signals to meet the changing needs of the plant. Some recent advances have been made in the identification of the molecular mechanisms by which such a complex compartmentation develops and evolves over time. In this review, we present an update of the latest results in this exciting field and propose distinct biogenesis models for the formation of vacuoles in vegetative and seed tissues, taking into account some apparently contradictory results.

1 Introduction

The etymology of vacuole derives from *vacuus*, meaning empty. The term refers in fact to the large fluid-filled, seemingly empty compartment originally identified in plant cells when they were first observed under a microscope. The vacuolar content (vacuolar sap) is separated from the embedding cytoplasm by a single membrane called the tonoplast. As early as the 1960s, biologists observed heterogeneity in both size and content of these vacuoles. Under a light microscope, some vacuoles could clearly be seen to contain pigments while others appeared dense in the electron microscope. Presently, after tremendous progress in plant molecular biology and genetics, it is quite exciting to realise that the cellular machinery leading to these vacuolar systems is indeed highly complex. Actually, we are only starting to uncover the pathways generating the different types of vacuoles with their specific functions. In terms of intracellular trafficking, the vacuole is also often described as a "final destination". This idea is no longer up to date since it is now clear that intracellular trafficking can involve all membranes of the secretory pathway, for example vesicle-mediated solute transport from the vacuole to the apoplast (Echeverría 2000), recycling of lipids and some other key factors such as the SNARE proteins involved in specific fusion events between two types of membranes (Bonifacino and Glick 2004). It also does not fit with the novel idea that the vacuolar system may be modified upon various stimuli leading for example to controlled fusion events (Jauh et al. 1999) or to a change of the function of a vacuole from lytic to storage vacuole and back (Murphy et al. 2005). Within the last 15 years, numerous publications identified various compartments as vacuoles mostly because they appeared much larger than vesicles (i.e. much larger than the limit of optical resolution). The term provacuole is sometimes preferred for a small vacuole when a larger vacuole can be also detected in the same cell, although the term suggests a precursor compartment in the formation of a larger vacuole. This is usually difficult to appreciate given the time that may be necessary before the fusion of provacuoles to a larger central vacuole. Importantly, a prevacuole (or prevacuolar compartment, PVC) is an entirely different organelle (the equivalent of a late endosome in animal cells, which is also a multi-vesicular body), which is an intermediate sorting compartment where vacuolar receptors release their ligands and from where they are believed to recycle (see Lam et al. 2005). A precise distinction between vacuole, pro- and prevacuole would therefore require long-term dynamic studies of single cells that are technically difficult and therefore these organelles are likely to often be confused in the literature. Our goal is to review some of the latest results since the last excellent reviews made in this area a few years ago (Robinson and Rogers 2000; Sanderfoot and Raikhel 2003). We would like especially to urge to distinguish between biogenesis and function of vacuoles and for example not to conclude, that a target vacuole has a lytic character only because of the involvement of VSR/BP 80 protein.

2 Identification of Different Vacuoles

One of the most commonly used dyes to visualise the vacuolar pH is neutral red, which is membrane-permeable in its unprotonated form, and trapped in the protonated form in acidic compartments, colouring acidic vacuoles. The use of various laser lines, usually associated with confocal spectral detection systems now allows us to use a more sophisticated probe, the lysosensor yellow/blue DND-160 with pH-dependent spectral peaks (Swanson et al. 1998; Diwu et al. 1999). It then becomes clear that the vacuolar pH can vary widely within a cell population. Expressed under the same 35S promoter, two different soluble vacuolar GFP markers were targeted by two different types of vacuolar sorting determinants (see below) to label either the acidic or the neutral vacuoles (Di Sansebastiano et al. 1998; Di Sansebastiano et al. 2001). Interestingly, when expressed stably in *Arabidopsis*, these GFP vacuolar markers do not systematically highlight the central vacuole. For example in leaves, the central vacuole of epidermal cells accumulates the acidic vacuolar marker (Aleu-GFP) while in the mesophyll cells the central vacuole accumulates the marker for a neutral compartment (GFP-Chi). In each cell type the other marker absent in the central vacuole is either restricted to small peripheral compartments or is not visible at all. In elongating cells (root hairs, trichomes), only Aleu-GFP is visible in the central vacuole, but after the end of elongation, both markers accumulate in the same vacuole, suggesting a hybrid nature (Flückiger et al. 2003). These results suggest that the ability for a given cell to transport a soluble protein to the central vacuole is cell-specific and is also regulated by developmental stages, indicating thereby that the trafficking pathways are not necessarily equivalent for every cell.

In addition, the vacuoles are highly dynamic, as became recently visible with GFP fused to an orthologue of cauliflower TIP1;1, a tonoplast intrinsic protein (Reisen et al. 2003) or with GFP fused to *Arabidopsis* TIP2;1 in a pollen tube (Cutler et al. 2000; Hicks et al. 2004). As an additional complexity in some plant species, seed cells have compartments (globoids and crystalloids) within larger protein storage vacuoles. The mechanism leading to such a complex vacuole is unknown but is likely to differ from a simple phagocytosis-like uptake of one vacuole by another since they are only separated by a single lipid bilayer (Jiang et al. 2001).

3
Functions of Vacuoles

The main plant-specific function of a vacuole is to produce and maintain the osmotic pressure by the accumulation of solutes, which cause the accumulation of water. This osmotic pressure is countered by the rigid cell wall, contributing to the mechanical stability of plant tissues. Primary metabolites such as carbohydrates, amino acids and organic acids, as well as inorganic ions such as chloride, nitrate, potassium and sodium, are among the solutes usually present in the cell sap (Martinoia et al. 2000). Most of these solutes are only temporarily stored in the vacuole. Plant cells can simply control their internal pressure by adjusting the solute content of their vacuoles. In certain cells, such as guard cells or pulvini in *Mimosa pudica* leaves, the changes can be rapid and lead to movements: stomata close (Zeiger and Zhu 1998), leaflets fold back (Fleurat-Lessard et al. 1997; MacRobbie 1999) etc. Cell growth is controlled by complex mechanisms where the turgor pressure plays an essential role (Cosgrove 1993; Cosgrove 1997; Cosgrove 2000). Elongation of plant cells under the stimulation of hormones such as auxins is mediated by a change in the resistance of the cell wall against the turgor pressure (Smart et al. 1998; Netting 2002).

Many compounds of secondary plant metabolism are thought to be sequestered definitively within the vacuole. Since many of them are toxic and often play a role in plant defence, detoxification of the cytosol and storage of "weapons" against predators can be regarded as further functions of the vacuole. Xenobiotics are also often stored in vacuoles after chemical modifications turned them into substrates for active pumps in the tonoplast (Martinoia et al. 2000).

Most enzymes detected in vacuole extracts are hydrolytic, which explains why vacuoles are considered the lysosomes of plant cells. However, some vacuoles may represent storage compartments for lytic enzymes rather than actively digestive lysosomes, e.g. aleurain-containing vacuoles in aleurone cells, "KDEL-vesicles" or provacuoles which store SH-EP for later degradation of globulins during seed germination. Ricinosomes are believed to be protease-storage vacuoles which function as suicide bombs in the final step of programmed cell death (Gietl et al. 2000). More recently, small senescence-associated vacuoles containing a papain-like cystein protease (SAG12) have been identified in leaves which may play a role similar to ricinosomes (Otegui et al. 2005).

Storage is indeed a major function of many vacuoles. Sucrose is stored in certain tissues and in several plants it is used to synthesise and store fructans. Pigments of the anthocyanin or betalain types are stored in flower or fruit tissues for obvious signalling purposes, but also in epidermal layers of stems or leaves, as part of an adaptive process to cold or to high light intensity. Protein storage vacuoles are most prominent in seeds, sometimes displaying a complex structure with internal compartments (Jiang et al. 2000). In vegetative tissues, protein storage vacuoles of a different kind have been described (Jauh et al. 1999). Defence-related enzymes such as chitinases or glucanases may accumulate to high concentrations, representing an important store of amino acids as well as preparing plants to fight off pathogens. The central vacuole of paraveinal mesophyll cells of soybean can change from a lytic to a storage type depending on the physiological conditions, with concomitant change of the aquaporin type (Murphy et al. 2005).

4 Biogenesis of Different Vacuoles

4.1 Transport of Soluble Proteins

Three different types of vacuolar sorting determinants (VSD) have been described: C-terminal (ctVSD), sequence-specific (ssVSD) and condensation-dependent (conVSD, suggested here as a more adequate term than physical-structure, see also Neuhaus and Rogers 1998).

C-terminal VSDs have been identified first in barley lectin and in tobacco chitinase A. Deletion and mutation studies with these two proteins indicated that accessibility from the C-terminus is critical, as vacuolar sorting can be blocked by terminal glycine(s) or by an N-glycan at the most terminal possible position (Dombrowski et al. 1993; Neuhaus et al. 1994). On the other hand, there is very little sequence conservation and no conserved physical property (such as hydrophobicity, as often erroneously stated). The minimal length is four amino acids. This low sequence specificity makes it difficult to prove that a candidate sequence is indeed a ctVSD, since many random C-terminal peptides are sufficient to target a reporter protein to a vacuole. The main requirement is probably that this peptide is not part of the compact structure of the folded protein, but accessible as a loose end. The sorting receptor for such signals might actually have chaperone-like binding properties. All vacuolar isoforms of pathogenesis-related (PR) proteins seem to have this type of VSD. Storage proteins related to phaseolin have a hydrophobic C-terminal vacuolar sorting determinant (AFVY in phaseolin, Frigerio et al. 2001) that could also belong to this type.

Sequence-specific VSDs have been best studied in barley aleurain and in sweet potato sporamin. Both proteins have an N-terminal propeptide that contains an NPIR motif. Systematic mutagenesis of the sporamin propeptide indicated the central role of the Ile residue, which can only be replaced by Leu. The residues at positions – 2 and + 2 also play a role, suggesting that the VSD is bound in an extended conformation (Matsuoka and Nakamura 1999; Matsuoka and Neuhaus 1999). Deletion analysis of the aleurain propeptide indicated that further residues upstream and downstream may contribute to a tighter binding by interacting with different domains of the receptor (Holwerda et al. 1992; Cao et al. 2000). Further ssVSDs have been identified in other protein precursors, and while they have no NPIR motif, an essential Ile or Leu could be identified. Some ssVSDs have also been found in an internal (ricin, Frigerio et al. 2001) or C-terminal propeptide (castor bean 2S albumin, Brown et al. 2003).

Condensation-dependent VSDs are involved in the formation of dense vesicles either at the ER or at the Golgi, depending on the need of a maturation process to allow condensation. This process has been described only for seed storage proteins so far. Interestingly, it seems that this type of protein sorting mainly occurs early in the secretory pathway since vesicles containing such proteins can be seen apparently forming from the ER up to the trans-Golgi. Recently, it was demonstrated that the C-terminal propeptide of phaseolin contributes to its transient attachment to membranes before reaching the Golgi (Castelli and Vitale 2005), in a process reminiscent of the membrane association of pea prolegumin, another storage protein (Hinz et al. 1997). This suggests that the propeptide of phaseolin acts as a condensation-dependent VSD, rather than as a ct-VSD (Castelli and Vitale 2005). The ten amino acid long C-terminal propeptide of the related 7S globulins of soybean (the α-, α'- and β-subunits of β-conglycinin) ends with a phaseolin-related motif (AFY). However, this motif is not necessary for proper targeting to the protein storage vacuole, as deletion of the terminal six amino acids from α'-conglycinin is not sufficient to lose this targeting, while deletion of the whole propeptide causes secretion. Interestingly, Nishizawa and collaborators performed their localisation studies for soybean β-conglycinin α' subunit in seed tissues that contain a complex type of protein storage vacuole (PSV). When this soybean 7S globulin was expressed in *Arabidopsis*, it accumulated massively in crystalloid-like structures within the PSV, while a form lacking the ten C-terminal aminoacids was secreted. In addition, these ten aminoacids were sufficient to send a reporter to the matrix of the PSV (Nishizawa et al. 2003). The C-terminal propeptide of Brazil nut 2S albumin (IAGF) could also belong to this type of VSD.

4.2 Vacuolar Sorting Receptors (VSRs) Homologous to Pea BP80

BP80 (binding protein of 80 kDa) was first identified in 1994 by its ability to bind in vitro the vacuolar sorting determinant of barley aleurain. The protein was isolated from a pea cotyledon clathrin-coated vesicle preparation and was shown to bind the vacuolar signal in a pH-dependent manner (Kirsch et al. 1994). The cDNA encoding BP80 was later cloned and the protein was further characterised. This pea variant and possible cross-reacting homologues localised in developing cotyledons in the vicinity of the *trans* face of the Golgi apparatus, a possible plant functional equivalent of the animal *trans*-Golgi network (TGN), i.e. the compartment from which clathrin-coated vesicles bud and where some vacuolar soluble proteins are supposed to be sorted out of the secretion route. In cotyledons, the native BP80 was also found in patches on the outer surface of small compartments, next to the vacuole, that are possible provacuoles (Paris et al. 1997). Using fluorescence imaging, it was also shown that tobacco BP80 homologues were predominantly localised in a prevacuolar compartment (defined by its labelling with Pep12 antibodies, Li et al 2002) that appears as a multivesicular body at the electron microscopic level in BY2 cells (Jiang et al. 2002; Lam 2005). The vacuolar sorting function of BP80 was qualitatively demonstrated using a functional assay in yeast (Humair et al. 2001; Hodel Hernández et al. 2005). Several studies also showed in vitro interactions between VSRs and vacuolar signals carried by storage proteins such as 2S albumins (Kirsch et al. 1996). In parallel, one BP80 homologue, AtELP, was identified in *Arabidopsis* by separate approaches (Ahmed et al. 1997; Laval et al. 1999). Finally, a pumpkin homologue, PV72, was also identified in precursor-accumulating compartments (PACs, Shimada et al. 1997) that are proposed as intermediates in the transport of proforms of storage proteins directly from the ER to the protein storage vacuole (Hara-Nishimura et al. 1998). In contrast to BP80, PV72 binding to its supposed ligand is calcium-dependent, as it is released at a low calcium concentration. Most recently, AtELP was shown to play a role in sorting storage proteins in seeds, showing again calcium-dependent binding (Shimada et al. 2003). A proteomic analysis of subcellular fractions from *Arabidopsis* suspension cells identified this isoform in the Golgi fraction, while isoforms more closely related to pea BP80 were found in the prevacuolar fraction (Paul Dupree, personal communication). This result indicates that different isoforms may be involved in different sorting pathways. Immunolocalisation of pea BP80 or AtELP to two different compartments had been interpreted as indication of their shuttling between the Golgi and the prevacuolar compartment, but might instead correspond to the localisations of two different but immunologically cross-reacting isoforms.

The nomenclature of VSRs is confusing, since several names are being used for the same protein (BP80, ELP, VSR) and they are even numbered in several ways in the case of *Arabidopsis*. We suggest a new unifying nomenclature for all higher plants (angiosperms, and possibly also gymnosperms) based on their phylogenetic relationships. Using full length or near full-length sequences from various plants, this analysis indicates the existence of three subfamilies which are conserved in *Arabidopsis*, poplar, grapevine, legumes and rice. At least two of them have been detected as EST sequences from *Pinus*. We propose to name the subfamilies VSR1, VSR2 and VSR3 (see Fig. 1A), with numbers within each subfamily (e.g. *Arabidopsis* AtELP/AtVSR1 is AtVSR1;1, while pea BP80 is PsVSR2;1) as was done for TIPs. In *Arabidopsis*, expression of all variants, except for VSR1;2, was detected in most tissues (Laval et al. 2003). Recently, transcriptome analysis became possible through various web services (e.g. http://bbc.botany.utoronto.ca/), allowing us to assess a fine spectrum of expression patterns for these seven variants (actually, the AtVSR2;1 and 2;2 cannot be distinguished). Interestingly, none of the seven *Arabidopsis* variants is equally expressed in all the conditions recorded in the ATGE database (Fig. 1B) but three expression subgroups also appear to fit the phylogenetic classification. This sequence conservation together with the expression patterns suggest different functions for each subfamily, but a similar sorting mechanism within each of them.

Indeed, both *Arabidopsis* AtVSR1;1 and its pumpkin (*Cucurbita sp.*) orthologue PV72 (CsVSR1;1) have been implicated in the biogenesis of protein storage vacuoles in seeds and both have a calcium-dependent affinity to ssVSDs. We have no functional information yet on AtVSR1;2, but given its very specific expression pattern restricted to flowers and mature pollen (Fig. 1B), it is likely that this particular variant is not present in most extracts used for in vitro binding studies. The VSR2 subfamily is expected to show pH-dependent binding like pea BP80 (PsVSR2;1). Interestingly, these VSRs are induced in senescent leaves and may well participate in providing enzymes for the various degradations associated with cell death and possibly also in the formation of the recently described senescence-associated vacuoles (Otegui et al. 2005). The VSR3 subfamily is composed of three members in *Arabidopsis*.

A

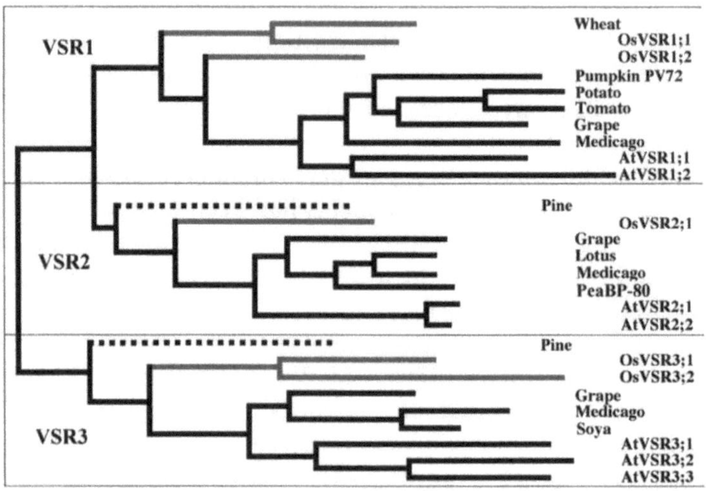

B

	root	leaves	senecent leaves	Stem 2nd internode 1st node	Flower	Mature pollen	Seed
VSR1;1 At3g52850	- /+	- /+	++	- /+	- /+	- -	++++
VSR1;2 At2g30290	0	0	0	0	++++++ (stamen)	+++++ +	0
VSR2;1 and 2;2 At2g14720 and At2g14740	- /+	- /+	++++	- /+	- /+	- -	- /+
VSR3;1 At4g20110	+++	- to 0	++	- to 0	+++ (sepal)	- to 0	- to 0
VSR3;2 At2g34940	+++	++	+	+	+	++++	+
VSR3;3 At1g30900	++	- /+	++	+++	++ (sepal)	- /+	+

Fig. 1 Phylogenetic and transcriptomic analysis of the vacuolar sorting receptor (VSR) family. **A** Phylogenetic analysis of plant VSR, performed using the DNAML program from the PHYLIP package (Felsenstein 1988). Accession numbers available on request. The tree shows three subfamilies named VSR1, VSR2 and VSR3 and each VSR is numbered within the subfamily for Arabidopsis (e.g. AtVSR1;1 corresponds to AtELP1) and rice (OsVSR) similarly to the nomenclature for TIPs (Fig. 3). For the other species the common names are indicated for easier identification. The pea BP-80 would be PsVSR2;1 while pumpkin PV72 would be CsVSR1;1. **B** Transcriptome data were analysed using the Arabidopsis Functional Genomics Tools on the Botany Beowulf Cluster (BBC) from the Department of Botany at the University of Toronto (http://bbc.botany.utoronto.ca/). The expression profiles were obtained using the ExpressionAngler program to search in the AtGenExpress data set. The results were viewed after median centring and normalisation and the most significant inductions or repressions for various tissues were summarised in the table. Zero means no expression, – /+ a basic level of expression below the value of 0.1. Induction or repression factors are indicated in + or – respectively (one + stands roughly for a level of induction equal to 0.1)

Preliminary results indicate that it is the major subfamily involved in sorting ssVSD-containing precursors to acidic vacuoles in leaves (Nguemelieu and Neuhaus, unpublished observations). All together, the AtVSRs are expressed in all tissue types and are likely to play a similar role in the whole plant (Fig. 1B). Interestingly, the moss *Physcomitrella patens* has at least five homologues of VSRs but they form a separate subfamily (not shown in Fig. 1A, but used to root the tree between VSR3 and the other two subfamilies), suggesting that the divergence and proposed specialisation of the three subfamilies have occurred after the divergence of vascular plants from mosses.

4.3 The Receptor-like RMR Proteins

RMR1 (for ReMembR-H2) is a protein that accumulates in the crystalloid, a component of protein storage vacuoles in seeds (Jiang et al. 2000). RMR1 was shown to bind in vitro to the C-terminal VSD from tobacco chitinase (John Rogers, personal communication) and was therefore proposed as a vacuolar sorting receptor for this particular type of soluble vacuolar protein. The molecular processes involved in RMR-mediated trafficking are entirely unknown, but a 1 : 1 stoichiometry is unlikely and the accumulation of RMR inside crystalloids suggest a structural role in building such a complex protein storage vacuole. For these reasons, the term receptor-like for RMR is to be favoured. RMR homologues have been found in many plants, including dicots, monocots and moss, but phylogenetic comparisons show no indication of subfamilies (not shown). Rice has two homologues while *Arabidopsis* has six, one of which (AtRMR2) is more diverged, but has no direct homologue in either rice or poplar. All RMR homologues share the luminal PA domain (Protease-Associated domain, also found in VSRs and in various animal proteases, Jiang et al. 2000), one transmembrane domain and the cytosolic RingH2 domain. Most RMRs also have a Ser-rich cytosolic tail of variable length. RMR homologues also exist in animals, but little is known about their function. The expression profile for the six *Arabidopsis* RMRs was also compared (Fig. 2). While all six isoforms are expressed in seed tissues, all vegetative tissues have at least one expressed RMR. Similarly to AtVSR1;2, the variant AtRMR3 is highly specific for pollen. RMRs have been immunolocalised in crystalloids with antibodies raised against AtRMR1's luminal domain or against a peptide from the cytosolic domain, which is well conserved at least also in AtRMR5 and 6. One or both of these RMRs are highly expressed in seeds and may be directly involved in the biogenesis of the crystalloids. AtRMR1 is more likely to participate in the biogenesis of neutral vacuoles in vegetative tissues.

	Average	root	hypocotyl	leaves	senescent leaves	Stem 2nd internode	Flower	Mature pollen		Seed
RMR1 At1g71980	+/-	+++	+/-	+	+/-	+	- stamen	+/-		+++ Stage 8 to 10
RMR2 At5g66160	+/-	++	+/-	+/-	+	+	+ petale stamen stage 15	+/-		+ Stage 8 to 10
RM3 At1g22670	+/-		+/-	+/-	+/-	+/-	+/-	++++		+ Stages 8 to 10
RMR4 At4g09560	+/-	-	-	+	+	+/-	+ Stamen F12	-		++ Stages 8 to 10
RMR5 At1g35630 RMR6 At1g35625	0	+	+	0	0	0	0	0		++++ to + Stages 4, 5, 6 and 7

Fig. 2 Transcriptomic analysis of the ReMembRingH2 (RMR) family. The transcriptome data was analysed as described in Fig. 1

4.4 Tonoplast Intrinsic Proteins (TIPs)

TIPs belong to the family of aquaporins that are mainly described as water channels, but may also allow diffusion of other small neutral molecules such as glycerol (Maurel and Chrispeels 2001). Some years ago, it was found that the type of isoform of TIP found in the tonoplast of a given vacuole could be correlated with its content. More precisely, the γ-TIP isoform was found in vacuoles accumulating aleurain while α-TIP was delimiting a neutral type of vacuole accumulating barley lectin (Paris et al. 1996). Therefore, it was proposed that the TIP isoform could be a signature for the content of the vacuole and for its function. Later, John Rogers and his group identified δ-TIP vacuoles as vegetative storage vacuoles and further showed that, in root tip cells of young seedlings, the tonoplast could show almost all possible combinations of the three isoforms (Jauh et al. 1999). The mechanisms by which membrane proteins including TIPs are sorted within the secretory pathway and delivered to tonoplast are not known. While α-TIP is present in Golgi membranes of developing pea seeds (Hinz et al. 1999), others find its transport to be Golgi-independent, as indicated by its BFA-insensitivity and lack of glycan modifications (Gomez and Chrispeels 1993; Jiang and Rogers 1998; Park et al. 2004). The conflict may either stem from expression artefacts in a heterologous system and tissue or reflect genuine differences in TIP trafficking between seeds and other tissues. Phylogenetic study of TIPs shows five (or possibly seven) subfamilies (Fig. 3A, see also Johanson et al. 2001). Interestingly, the transcriptome analysis of TIP expression in *Arabidopsis* shows subclasses as well (Fig. 3B). The AtTIP2 subfamily (also called δ-TIP) seems to be expressed in most tissues including seeds and therefore may be associated with a basic and conserved type of vacuole. AtTIP2;1 and 3 were recently shown to facilitate ammonia transport (Loqué et al. 2005). All other subgroups are apparently expressed in a more tissue-specific manner. Surprisingly, the AtTIP1

A

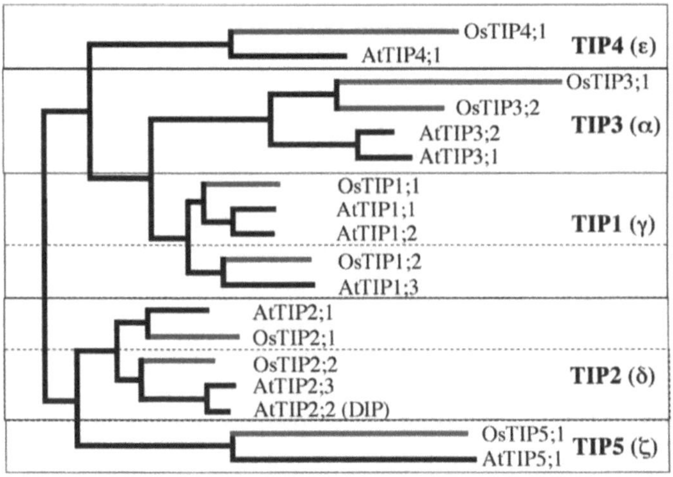

B

	Average	root	leaves	senecent leaves	1st node	Flower	stamen	cauline	Mature pollen	Seed
TIP1;1	0 to +	++	0 to +	-	+	0 to +	0 to +	0 to +	-	- to - - Stage 3 to 7
TIP1;2	+/-	+	+/-	-	+/-	-	- -	+/-	-	- to - - Stage 3 to 6-10
TIP1;3	0	0	0	0	0	+	+++++	0	++	0
TIP2;1	+/-	+	-	- -	average	average	-	-	- -	- - Stage 4 to 10
TIP2;2	+	++	+	+	+	+	+	++	+	+
TIP2;3	0	+++	0	0	0	0	0	0	0	0
TIP3;1	0	0	0	0	0	0	0	0	0	+ to +++ Stage 5 to 10
TIP3;2	0	0	0	0	0	0	0	0	0	+ to +++ Stage 5 to 10
TIP4;1	0	+++	0	0	0	0	0	0	0	0
TIP5;1	0	0	0	0	0	0	++	0	+++++	0

Fig. 3 Phylogenetic and transcriptomic analysis of the tonoplast intrinsic protein (TIP) family. **A** The phylogenetic analysis of plant TIPs was performed as described in Fig. 1. The tree is subdivided into the five subfamilies TIP1 to TIP5 (with corresponding Greek letters, Johanson et al. 2001), but possible functional subdivisions of the subfamilies TIP2 and TIP3 are indicated by the *dotted lines*. **B** The transcriptome data was analysed as described in Fig. 1

subgroup (equivalent to γ-TIP associated to a lytic compartment) is root-specific. Additionally, this subgroup is repressed in seed tissue and contains one specific variant (AtTIP1;3) for flowers and pollen having a close homologue in rice. This suggests a divergence and specialisation within this TIP1 subfamily. The AtTIP3 subfamily (α-TIP) is highly specific for seed tissues, confirming the detection of α-TIP by western blot analysis in seed tissues and in seedlings up to a week old (Johnson et al. 1989). AtTIP4 and AtTIP5 isoforms are specifically expressed in root and flower tissues, respectively. These expression data fit with the idea that a specific isoform may be linked to a specific type of vacuole. However, when GFP was fused to either a cauliflower TIP1 orthologue (Reisen et al. 2003), AtTIP2;1 (Cutler et al. 2000; Avila et al. 2003; Loqué et al. 2005) or AtTIP2;3 (Loqué et al. 2005), the resulting fusion proteins were all found in the tonoplast of the central vacuole in a wide range of cell types. These fusion proteins are all targeted to vacuoles, but seem to lack the specificity of the native proteins. This may be due to the steric hindrance for a putative sorting protein to bind to the cytosolic C-tail of the TIPs, which may contain a specificity determinant, such as the motif identified in α-TIP (Moriyasu et al. 2003). When GFP was inserted into the third luminal loop of a radish TIP1, the resulting hybrid protein labelled small peripheral vacuoles, but not the central vacuole in *Arabidopsis* protoplasts (Matsuoka and Neuhaus 1999). This may indicate that access to the cytosolic face of TIPs is required for the fine sorting to different vacuoles.

5
Biogenesis and Function of Vacuoles

How can we fit all these various elements into one model? It is now clear that we need to distinguish vegetative from seed tissues. For this reason we propose two models, one of which is restricted to seeds (Fig. 4). An accumulation of results on the VSR family is also challenging the original model where a strict relationship was proposed between VSRs, sequence-specific VSDs and the lytic vacuole. In these new models, all vacuolar proteins may end at some point in a hybrid vacuole, independently of their transport pathways. Storage vacuoles can be formed by different pathways in vegetative and seed tissues and storage proteins can be sorted by ssVSD-, ctVSD- and con-VSD-dependent pathways. A vacuolar protein in one tissue may be secreted by another tissue, as described for phytohemagglutinin E in bean roots (Kjemtrup 1995). This indicates that one of the vacuolar sorting pathways may be inactive or absent in some tissues.

In Fig. 4A we distinguish two pathways in vegetative cells. The pathway to acidic, lytic vacuoles is mediated by VSRs of the VSR3 subfamily. After recognition of their ssVSD ligands in the Golgi, these VSRs are transported by clathrin-coated vesicles to the prevacuolar compartment, where a lower pH causes ligand release and from where the receptors are recycled. A matured prevacuolar compartment (from which all receptors have been removed) could become a small acidic provacuole or fuse with a preexisting large acidic or hybrid vacuole. Recently, we obtained evidence that VSRs can transit by the plasma membrane, most likely to retrieve escaped proteases (Paris et al., unpublished observations). In the other pathway, proteins with C-terminal VSDs, such as chitinases, could be recognised by a receptor-like RMR and sent to a neutral type of vacuole. In vegetative cells, the vacuole containing lytic enzymes has a tonoplast with γ-TIP isoforms (TIP1), while the pH-neutral vacuole has a tonoplast with δ-TIP isoforms (TIP2). Since some vacuoles appear labelled with both TIP isoforms, it is likely that the two distinct vacuoles fused to a hybrid vacuole where the hydrolytic enzymes are functional. This model allows the smooth transition from a mostly lytic to a storage type of vacuole (and back) as described in soybean (Murphy et al. 2005).

Senescing tissues appear to turn on another pathway to senescence-associated vacuoles which are highly acidic but lack γ-TIP (Otegui et al. 2005). The gene expression patterns revealed no TIP isoforms that would be senescence-induced, suggesting the possibility of vacuoles without any TIP. On the other hand, the AtVSR2s are highly induced in senescent leaves and may be implicated in the biogenesis of these vacuoles. The only known specific soluble protein of these senescence-associated vacuoles is the cysteine protease SAG12, which is distantly related to aleurains. While it does not have an NPIR motif in the N-terminal propeptide, there are several Leu and Ile that might contribute to an ssVSD. Pollen seems also to be equipped with a specific type of vacuole with the TIP5 subfamily, which is conserved in monocots and dicots. *Arabidopsis* has also a pollen-specific AtVSR1;2, but it is not part of a separate subfamily.

A large subset of results implicates VSRs in sorting storage proteins such as globulins and albumins to the storage vacuole in seed tissues (for the latest results, see Shimada et al. 2003; Jolliffe et al. 2004). The VSR1 subfamily is specifically expressed during seed formation in *Arabidopsis*. Significantly, the TIP3 subgroup (also known as α-TIP) is highly seed specific. Both *Arabidopsis* members (AtTIP3;1 and 3;2) are coexpressed with 99% correlation. β-VPE (vacuolar processing enzymes, At1g62710), a 12S seed storage protein (At1g03890) and 2S albumin (At4g27160) are also coexpressed with these TIPs at more than 95% correlation. All this information lead us to propose a second model for vacuole biogenesis that is highly specific for seed cells (Fig. 4B) where an additional machinery is turned on. Three pathways are proposed in seeds, leading to the complex protein storage vacuole composed of a matrix compartment, which contains for example chitinases, and encloses globoids, which contain for example aleurain and crystalloids, which contain condensed proteins such as globulins. The bulk of the storage proteins are condensing in the ER or Golgi, forming the core of the PACs or DVs, while a membrane-proximal layer may be accumulated with the help of sorting receptors (Wenzel et al. 2005). The involvement of VSR1 isoforms is linked to their calcium-dependent binding of proteins with ssVSDs, distinct from the pH-dependent binding observed with VSR2 isoforms. It is likely that they are linked to the formation of PAC vesicles, which are proposed to

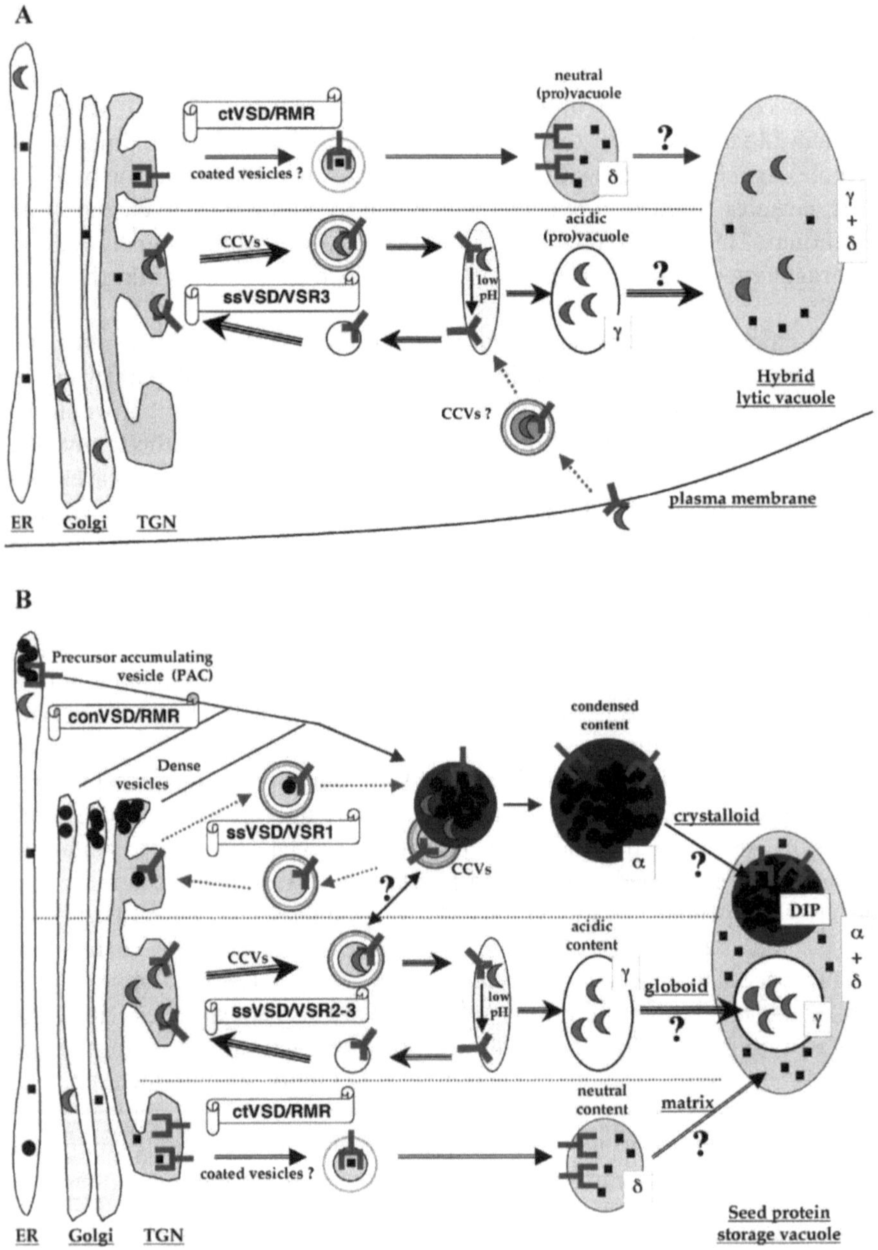

derive directly from the ER. There are much less VSR1s than storage proproteins in PAC vesicles and they are not found in dense vesicles, except possibly when a clathrin-coated vesicle forms on them. The calcium concentration is higher in the ER and Golgi (1 mM, resp. 300 μM in animal cells, Montero

◀ **Fig. 4** Model for vacuolar biogenesis in plant vegetative or seed cells. **A** For vegetative cells the model proposes two pathways: the first transports proteins with a C-terminal VSD, involves RMRs and leads to a (pro)vacuole with a neutral pH and a δ-TIP (TIP2) in its tonoplast. The mechanisms involved in the RMR pathway are unknown and the implication of coated vesicles and of a prevacuole are hypothetical. The second pathway transports soluble proteins with sequence specific VSDs, involves clathrin-coated vesicles and VSRs (most likely of the subfamily 3). This pathway leads to a prevacuole with a low pH which releases the ligand from the receptor. The ligand is further transported (by a VSR-independent mechanism) to an acidic (pro)vacuole with a γ-TIP (TIP1) in its tonoplast. It is possible that some VSR can also cycle via the plasma membrane, possibly to retrieve escaped vacuolar proteins. The (pro)vacuoles labelled with δ-TIP or γ-TIP are able to fuse by a controlled fusion process leading to a hybrid vacuole with mixed sap and tonoplast markers. **B** Model for vacuole biogenesis in seed cells. As for vegetative cells, the two pathways leading to δ- and γ-TIP (pro)vacuoles are conserved but VSR2 is also involved in addition to VSR3 in the ssVSD-dependent transport to the acidic compartment. A seed-specific pathway for proteins with condensation VSDs is likely to involve RMRs. This pathway starts at the ER to generate Precursor Accumulating Compartments (PAC) or at the *cis*-Golgi to generate the dense vesicles that form from the Golgi. The intermediate compartment is likely to have a complex stratified structure where condensation and maturation of the seed storage proteins occurs in order to become a (pro)vacuole with α-TIP (TIP3) in its tonoplast. A possible role for the VSR1 subfamily with calcium dependent binding would be to retrieve proteins that escaped the condensation process and transport them to the dense vesicles or PACs. VSR2s could possibly also transport maturation enzymes to the dense compartment for maturation of storage proteins. In both cases the clathrin-coated pits seen to form on dense vesicles would be recycling the empty receptors to the Golgi. The three provacuolar compartments with α-, γ- or δ-TIP then generate a complex protein storage vacuole with crystalloids (containing RMR and DIP, a TIP2 isoform, in addition to the condensed storage proteins) and globoids both embedded in a matrix surrounded by a tonoplast with both α- and δ-TIP

et al. 1997; Pinton et al. 1998) than in PACs (50 μM, Watanabe et al. 2002), which is not compatible with a PAC-to-Golgi salvage role for the receptors. It is possible that the VSR1s participate in the condensation of storage proteins either directly at the ER for PACs or in the Golgi for dense vesicles, but more likely they transport in a calcium-dependent manner storage pro-proteins that escaped the early condensation process and reached the Golgi by mistake (Fig. 4B upper part). Clathrin-coated pits, which are occasionally visible by electron microscopy at the periphery of PACs or dense vesicles, probably transport free VSR1s back to the Golgi. Interestingly, the internal propeptide of *Ricinus communis* 2S albumin, a proven ssVSD (Brown et al. 2003), binds in vitro VSR proteins from developing castor bean endosperm in a pH-dependent manner (Jolliffe et al. 2004) while an equivalent propeptide in pumpkin 2S albumin is recognised by PV72 (the pumpkin VSR1) in a calcium-dependent process (Watanabe et al. 2002). How can we reconcile these conflicting results? It is possible that calcium-dependent, VSR-mediated transport is restricted to the direct pathway from ER to storage protein vacuoles while transport via the Golgi apparatus would use a pH-dependent VSR.

It is however also possible that the in vitro assay with the castor bean extract also contained other calcium-dependent VSRs that were not recovered by the low pH elution. The strongest evidence for a role of a VSR1 in sorting storage proteins to the vacuole comes from an atvsr1;1 mutant of *Arabidopsis* that accumulates a high level of storage proteins in the extracellular wall space but it is unfortunately not known whether these secreted forms are processed. Strikingly, the profile of total storage proteins extracted from these mutant seeds is almost identical to the profile obtained from *Arabidopsis* knocked-out for various vacuolar processing enzymes (VPE, Shimada et al. 2003; Shimada et al. 2003). Therefore, it is possible that the secretion of storage proteins in the atvsr1;1 mutant is due to an impaired vacuolar transport of VPE, leading to an inefficient maturation of storage proteins and therefore indirectly to secretion of these precursors. The three vacuolar sorting pathways (considering PACs and dense vesicles to be homologous, differing mainly by the initial compartment of condensation) then achieve a complex engulfment or hemifusion process which leads to the embedding of the storage provacuole as a crystalloid and of the lytic provacuole as a globoid inside a matrix that derives from the neutral provacuole. The overall tonoplast now contains both δ-TIP (a TIP2) from this neutral provacuole and α-TIP (TIP3) from the storage while RMRs and DIP (another TIP2) are within the crystalloid. At this stage we can only speculate how this transformation can occur (Jiang et al. 2002).

6 Conclusion and Future Prospects

As often in science, the more we discover in a field, the more questions open up. It is now clear that plants have a tremendous and exciting potential to generate a complex vacuolar system "on demand" and for this purpose they have at least three specific machineries to send soluble vacuolar proteins to these vacuoles. It also clear that pathways leading soluble vacuolar proproteins to compartments from the ER or the Golgi are strongly interconnected, especially in developing seed cells. High-throughput methods will offer valuable information and tools for further studies on plant vacuole biogenesis and function (Avila et al. 2003; Carter et al. 2004; Surpin et al. 2005).

References

Ahmed SU, Bar-Peled M, Raikhel NV (1997) Cloning and subcellular location of an *Arabidopsis* receptor-like protein that shares common features with protein-sorting receptors of eukaryotic cells. Plant Physiol 114:325–336

Avila E, Zouhar J, Agee A, Carter D, Chary S, Raikhel N (2003) Tools to study plant organelle biogenesis. Point mutation lines with disrupted vacuoles and high-speed confocal screening of green fluorescent protein-tagged organelles. Plant Physiol 133:1373–1376

Bonifacino JS, Glick BS (2004) The mechanisms of vesicle budding and fusion. Cell 116:153–166

Brown JC, Jolliffe NA, Frigerio L, Roberts LM (2003) Sequence-specific, Golgi dependent targeting of the castor bean 2S albumin to the vacuole in tobacco protoplasts. Plant Journal 36:711–719

Cao X, Rogers SW, Butler J, Beevers L, Rogers JC (2000) Structural requirements for ligand binding by a probable plant vacuolar sorting receptor. Plant Cell 12:493–506

Carter C, Pan S, Zouhar J, Avila E, Girke T, Raikhel N (2004) The vegetative vacuole proteome of *Arabidopsis thaliana* reveals predicted and unexpected proteins. Plant Cell 16:3285–3303

Castelli S, Vitale A (2005) The phaseolin vacuolar sorting signal promotes transient, strong membrane association and aggregation of the bean storage protein in transgenic tobacco. J Exp Bot 56:1379–1387

Cosgrove DJ (1997) Assembly and enlargement of the primary cell wall in plants. Annu Rev Cell Dev Biol 13:171–201

Cosgrove DJ (1993) How do plant cell walls extend? J Plant Physiol 102:1–6

Cosgrove DJ (2000) Loosening of plant cell walls by expansins. Nature 407:321–326

Cutler SR, Ehrhardt DW, Griffitts JS, Somerville CR (2000) Random GFP: cDNA fusions enable visualization of subcellular structures in cells of *Arabidopsis* at a high frequency. Proc Natl Acad Sci USA 97:3718–3723

Di Sansebastiano G-P, Paris N, Marc-Martin S, Neuhaus J-M (1998) Specific accumulation of GFP in a non-acidic vacuolar compartment via a C-terminal propeptide-mediated sorting pathway. Plant J 15:449–457

Di Sansebastiano G-P, Paris N, Marc-Martin S, Neuhaus J-M (2001) Regeneration of a lytic central vacuole and of neutral peripheral vacuoles can be visualised by GFP targeted to either type of vacuoles. Plant Physiol 126:78–86

Diwu Z, Chen C-S, Zhang C, Klaubert DH, Haugland RP (1999) A novel acidotropic pH indicator and its potential application in labeling acidic organelles of live cells. Chem Biol 6:411–418

Dombrowski JE, Schroeder MR, Bednarek SY, Raikhel NV (1993) Determination of the functional elements within the vacuolar targeting signal of barley lectin. Plant Cell 5:587–596

Echeverría E (2000) Vesicle-Mediated Solute Transport between the Vacuole and the Plasma Membrane. Plant Physiol 123:1217–1226

Felsenstein J (1988) Phylogenies from molecular sequences: inference and reliability. Annu Rev Genet 22:521–565

Fleurat-Lessard P, Frangne N, Maeshima M, Ratajczak R, Bonnemain JL, Martinoia E (1997) Increased expression of vacuolar aquaporin and H+-ATPase related to motor cell function in *Mimosa pudica L.* Plant Physiol 114:827–834

Flückiger R, De Caroli M, Piro G, Dalessandro G, Neuhaus J-M, Di Sansebastiano G-P (2003) Vacuolar system distribution in *Arabidopsis* tissues, visualized using GFP fusion proteins. J Exp Bot 54:1577–1584

Frigerio L, Foresti O, Hernández Felipe D, Neuhaus J-M, Vitale A (2001) The C-terminal tetrapeptide of phaseolin is sufficient to target green fluorescent protein to the vacuole. J Plant Physiol 158:499–503

Frigerio L, Jolliffe NA, Di Cola A, Hernández Felipe D, Paris N, Neuhaus J-M, Lord JM, Ceriotti A, Roberts LM (2001) The internal propeptide of the ricin precursor carries a sequence-specific determinant for vacuolar sorting. Plant Physiol 126:167–175
Gietl C, Schmid M, Simpson DJ (2000) Ricinosomes and aleurain-containing vacuoles (ACVs): proteases-storing organelles. In: Robinson DG and Rogers JC (eds) Vacuolar compartments, vol 5, pp 90–11
Gomez L, Chrispeels MJ (1993) Tonoplast and soluble vacuolar proteins are targeted by different mechanisms. Plant Cell 5:1113–1124
Hara-Nishimura I, Shimada T, Hatano K, Yakeuchi Y, Nishimura M (1998) Transport of storage proteins to protein storage vacuoles is mediated by large precursor-accumulating vesicles. Plant Cell 10:825–836
Hicks G, Rojo E, Hong S, Carter D, Raikhel N (2004) Germinating pollen has tubular vacuoles, displays highly dynamic vacuole biogenesis, and requires VACUOLESS1 for proper function. Plant Physiol 134:1227–1239
Hinz G, Hillmer S, Baumer M, Hohl I (1999) Vacuolar storage proteins and the putative vacuolar sorting receptor BP80 exit the Golgi apparatus of developing pea cotyledons in different transport vesicles. Plant Cell 11:1509–1524
Hinz G, Menze A, Hohl I, Vaux D (1997) Isolation of prolegumin from developing pea seeds: its binding to endomembranes and assembly into prolegumin hexamers in the protein storage vacuole. J Exp Botany 48:139–149
Hodel Hernández D, Paris N, Neuhaus J-M, Deloche O (2005) The yeast *S. cerevisiae* is not an efficient tool for in vivo studies of plant vacuolar sorting receptors. Plant Cell
Holwerda BC, Padgett HS, Rogers JC (1992) Proaleurain vacuolar targeting is mediated by short contiguous peptide interactions. Plant Cell 4:307–318
Humair D, Hernández Felipe D, Neuhaus J-M, Paris N (2001) Demonstration in yeast of the function of BP80, a putative plant vacuolar sorting receptor. Plant Cell 13:781–792
Jauh G-Y, Phillips TE, Rogers JC (1999) Tonoplast intrinsic protein isoforms as markers for vacuolar functions. Plant Cell 11:1867–1882
Jiang L, Erickson A, Rogers JC (2002) Multivesicular bodies: a mechanism to package lytic and storage functions in one organelle? Trends Cell Biol 12:362–367
Jiang L, Phillips TE, Hamm CA, Drozdowicz YM, Rea PA, Maeshima M, Rogers SW, Rogers JC (2001) The protein storage vacuole: a unique compound organelle. J Cell Biol 155:991–1002
Jiang L, Phillips TE, Rogers SW, Rogers JC (2000) Biogenesis of the protein storage vacuole crystalloid. J Cell Biol 150:755–770
Jiang LW, Rogers JC (1998) Integral membrane protein sorting to vacuoles in plant cells: evidence for two pathways. J Cell Biol 143:1183–1199
Johanson U, Karlsson M, Johansson I, Gustavsson S, Sjövall S, Fraysse L, Weig AR, Kjellbom P (2001) The complete set of genes encoding major intrinsic proteins in *Arabidopsis* provides a framework for a new nomenclature for major intrinsic proteins in plants. Plant Physiol 126:1358–1369
Johnson KD, Herman EM, Chrispeels MJ (1989) An abundant, highly conserved tonoplast protein in seeds. Plant Physiol 91:1006–1013
Jolliffe N, Brown J, Neumann U, Vicré M, Bachi A, Hawes C, Ceriotti A, Roberts LM, Frigerio L (2004) Transport of ricin and 2S albumin precursors to the storage vacuoles of *Ricinus communis* endosperm involves the Golgi and VSR-like receptors. Plant Journal 39:821–833
Kirsch T, Paris N, Butler JM, Beevers L, Rogers JC (1994) Purification and initial characterization of a potential plant vacuolar targeting receptor. Proc Natl Acad Sci USA 91:3403–3407

Kirsch T, Saalbach G, Raikhel NV, Beevers L (1996) Interaction of a potential vacuolar targeting receptor with amino- and carboxyl-terminal targeting determinants. Plant Physiol 111:469–474

Laval V, Chabannes M, Carrière M, Canut H, Barre A, Rougé P, Pont-Lezica R, Galaud J-P (1999) A family of *Arabidopsis* plasma membrane receptors presenting animal β-integrin domains. Biochim Biophys Acta 1435:61–70

Laval V, Masclaux F, Serin A, Carrière M, Roldan C, Devic M, Pont-Lezica RF, Galaud J-P (2003) Seed germination is blocked in *Arabidopsis* putative (atbp80) antisense transformants. J Exp Bot 54:213–221

Loqué D, Ludewig U, Yuan L, von Wiren N (2005) Tonoplast Intrinsic Proteins AtTIP2;1 and AtTIP2;3 Facilitate NH3 Transport into the Vacuole. Plant Physiol 137:671–678

MacRobbie E (1999) Vesicle trafficking: a role in trans-tonoplast ion movements. J Exp Bot 50:925–934

Martinoia E, Massonneau A, Frangne N (2000) Transport processes of solutes across the vacuolar membrane of higher plants. Plant Cell Physiol 41:1175–1186

Matsuoka K, Nakamura K (1999) Large alkyl side-chains of isoleucine and leucine in the NPIRL region constitute the core of the vacuolar sorting determinant of sporamin precursor. Plant Mol Biol 41:825–835

Matsuoka K, Neuhaus J-M (1999) Cis-elements of protein transport to the plant vacuoles. J Exp Bot 50:165–174

Maurel C, Chrispeels M (2001) Aquaporins: a molecular entry into plant water relations. Plant Physiol 125:135–138

Montero M, Alvarez J, Scheenen WJJ, Rizzuto R, Meldolesi J, Pozzan T (1997) Ca2+ homeostasis in the endoplasmic reticulum: coexistence of high and low [Ca2+] subcompartments in intact HeLa cells. J Cell Biol 139:601–611

Moriyasu Y, Hattori M, Jauh GY, Rogers JC (2003) Alpha tonoplast intrinsic protein is specifically associated with vacuole membrane involved in an autophagic process. Plant Cell Physiol 44:795–802

Murphy KA, Rachel A, Kuhle RA, Fischer AM, Anterola AM, Grimes HD (2005) The functional status of paraveinal mesophyll vacuoles changes in response to altered metabolic conditions in soybean leaves. Funct Plant Biol 32:335–344

Netting AG (2002) pH, abscisic acid and the integration of metabolism in plants under stressed and non-stressed conditions. II. Modifications in modes of metabolism induced by variation in the tension on the water column and by stress. J Exp Bot 53:151–173

Neuhaus J, Pietrzak M, Boller T (1994) Mutation analysis of the C-terminal vacuolar targeting peptide of tobacco chitinase: low specificity of the sorting system, and gradual transition between intracellular retention and secretion into the extracellular space. Plant J 5:45–54

Neuhaus J-M, Rogers JC (1998) Sorting of proteins to vacuoles in plant cells. Plant Mol Biol 38:127–144

Nishizawa K, Maruyama N, Satoh R, Fuchikami Y, Higasa T, Utsumi S (2003) A C-terminal sequence of soybean β-conglycinin α' subunit acts as a vacuolar sorting determinant in seed cells. Plant J 34:647–659

Otegui MS, Noh YS, Martinez DE, Vila Petroff MG, Staehelin LA, Amasino RM, Guiamet JJ (2005) Senescence-associated vacuoles with intense proteolytic activity develop in leaves of *Arabidopsis* and soybean. Plant J 41:831–844

Paris N, Rogers SW, Jiang L, Kirsch T, Beevers L, Phillips TE, Rogers JC (1997) Molecular cloning and further characterization of a probable plant vacuolar sorting receptor. Plant Physiol 115:29–39

Paris N, Stanley CM, Jones RL, Rogers JC (1996) Plant cells contain two functionally distinct vacuolar compartments. Cell 85:563–572
Park M, Kim S, Vitale A, Hwang I (2004) Identification of the protein storage vacuole and protein targeting to the vacuole in leaf cells of three plant species. Plant Physiol 134:625–639
Pinton P, Pozzan T, Rizzuto R (1998) The Golgi apparatus is an inositol 1,4,5-trisphosphate-sensitive Ca2+ store, with functional properties distinct from those of the endoplasmic reticulum. EMBO J 17:5298–5308
Reisen D, Leborgne-Castel N, Özalp C, Chaumont F, Marty F (2003) Expression of a cauliflower tonoplast aquaporin tagged with GFP in tobacco suspension cells correlates with an increase in cell size. Plant Mol Biol 52:387–400
Robinson DG, Rogers JC (2000) Vacuolar compartments. Sheffield Academic Press and CRC Press, London, Sheffield, p 314
Sanderfoot AA, Raikhel N (2003) The secretory system of *Arabidopsis*. American Society of Plant Biologists, Rockville, MD
Lam SK, Tse YC, Jiang L, Oliviusson P, Heinzerling O, Robinson DG (2005) Plant prevacuolar compartments and endocytosis (in this volume). Springer, Berlin Heidelberg New York
Shimada T, Kuroyanagi M, Nishimura M, Hara-Nishimura I (1997) A pumpkin 72-kDa membrane protein of precursor-accumulating vesicles has characteristics of a vacuolar sorting receptor. Plant Cell Physiol 38:1414–1420
Shimada T, Fuji K, Tamura K, Kondo M, Nishimura M, Hara-Nishimura I (2003) Vacuolar sorting receptor for seed storage proteins in *Arabidopsis thaliana*. Proc Natl Acad Sci USA 100:16 095–16 100
Shimada T, Yamada K, Kataoka M, Nakaune S, Koumoto Y, Kuroyanagi M, Tabata S, Kato T, Shinozaki K, Seki M, Kobayashi M, Kondo M, Nishimura M, Hara-Nishimura I (2003) Vacuolar processing enzymes are essential for proper processing of seed storage proteins in *Arabidopsis thaliana*. J Biol Chem 278:32 292–32 299
Smart LB, Vojdani F, Maeshima M, Wilkins TA (1998) Genes involved in osmoregulation during turgor-driven cell expansion of developing cotton fibers are differentially regulated. Plant Physiol 116:1539–1549
Surpin M, Rojas-Pierce M, Carter C, Hicks GR, Vasquez J, Raikhel NV (2005) The power of chemical genomics to study the link between endomembrane system components and the gravitropic response. Proc Natl Acad Sci USA 102:4902–4907
Swanson S, Bethke P, Jones R (1998) Barley aleurone cells contain two types of vacuoles. Characterization of lytic organelles by use of fluorescent probes. Plant Cell 10:685–698
Watanabe E, Shimada T, Kuroyanagi M, Nishimura M, Hara-Nishimura I (2002) Calcium-mediated association of a putative vacuolar sorting receptor PV72 with a propeptide of 2S albumin. J Biol Chem 277:8708–8715
Wenzel D, Schauermann G, von Lupke A, Hinz G (2005) The cargo in vacuolar storage protein transport vesicles is stratified. Traffic 6:45–55
Zeiger E, Zhu J (1998) Role of zeaxanthin in blue light photoreception and the modulation of light-CO2 interactions in guard cells. J Exp Bot 49:433–442

Plant Cell Monogr (1)
J. Šamaj · F. Baluška · D. Menzel: Plant Endocytosis
DOI 10.1007/7089_006/Published online: 6 September 2005

Molecular Dissection of the Clathrin-Endocytosis Machinery in Plants

Susanne E. H. Holstein

Heidelberg Institute of Plant Sciences, Cell Biology, Im Neuenheimer Feld 230, 69120 Heidelberg, Germany
Susanne_H.Holstein@urz.uni-heidelberg.de

Abstract In the last few years, the endocytic vesicular uptake in plant cells has gained increasing significance in several physiological processes. Therefore, an insight into plant clathrin endocytosis at the molecular level is essential. Plants do contain homologs to several key proteins of the mammalian clathrin-dependent endocytosis machinery, but so far only very few have been functionally characterized. Thus, this chapter deals first with the description of the molecular mechanism of clathrin-dependent endocytosis of non-plant organisms which is followed by the outline of similarities to plant endocytosis with an emphasis on its clathrin-dependency.

1 Molecular Mechanisms of Clathrin-Dependent Endocytosis in Non-Plant Organisms

1.1 Characterization of Clathrin-Coated Vesicle Coat and Accessory Proteins

Endocytosis, the uptake of nutrients and other macromolecules, starts at the outer border of a cell, the plasma membrane (PM), which invaginates at distinct sites in order to engulf the cargo molecules into endocytic clathrin-coated vesicles (CCV). After pinching off the PM the CCV shed off their coats thus enabling the naked vesicles to fuse with the first compartment, the early endosome, in order to release their content into the endocytic pathway (Gruenberg, 2001). At the molecular level, clathrin-mediated endocytosis (CME) can be subdivided into the distinct stages of cargo selection, recruitment of coat components, deformation of the PM, and finally budding/pinching of the CCV from the PM.

CCV are spherical vesicles surrounded by a lattice-like coat which was named after its main structure protein clathrin (kláthron in Greek meaning lattice) and the vesicles were therefore described to be clathrin-coated (Kanaseki and Kadota, 1969). Clathrin is a heterodimer consisting of one clathrin heavy chain (CHC) polypeptide of $\sim$ 190 kDa and one clathrin light chain (CLC) polypeptide of $\sim$ 25 kDa which trimerize into three-legged structures named triskelions, the subunits of the clathrin lattice (Kirchhausen, 2000a). The curvature of a closed polyhedral coat requires the constant number of 12 pentagons, while the number of hexagons may vary thus giving rise to different vesicle sizes (Crowther et al., 1976; Robinson and Depta, 1988; Mousavi et al., 2004). All non-plant organisms investigated so far contain a maximum of two CLC genes and a single gene for CHC, with the exception of some primates who in addition to the ubiquitously expressed and highly conserved CHC17 gene also contain the muscle-specific CHC22 gene (Liu et al., 2001). While the CHC gene from soybean was identified some years ago (Blackbourn and Jackson, 1996), the first plant CLC was identified only recently (Scheele and Holstein, 2002).

In addition to clathrin, the protein coat of endocytic CCV contains another main component, the heterotetrameric Adaptor (AP) complex AP-2 (Keen et al., 1979; Pearse and Robinson, 1984) which consists of the small σ2-adaptin ($\sim$ 20 kDa), the medium μ2-adaptin ($\sim$ 50 kDa) and two large subunits ($\sim$ 100 kDa), the β2- and α-adaptins (Kirchhausen, 1999). Thus, a cross-section through a CCV reveals a three-layered structure with the innermost vesicle membrane harbouring the transmembrane cargo molecules connected via the middle layer of adaptors to the outermost clathrin layer (Pearse et al., 2000).

Of the four mammalian AP complexes, the endocytic AP-2 complex is the best characterized at the molecular level (Collins et al., 2002). With the exception of the small σ2-adaptin, which is obviously a structural component, all other AP-2 adaptins have well-described functions assigned to their specific domains. Accordingly, the μ2-adaptin is the main receptor binding partner and its receptor-binding domain (RBD) serves as the interaction site for trans-membrane proteins, which in turn harbour the tyrosine-based internalization motif YXXϕ within their cytosolic tails (Bonifacino and Traub, 2003). Both large subunits of AP-2 have a tripartite structure consisting of an amino-terminal trunk portion that is connected via a flexible hinge region to the carboxy-terminal ear-domain. While β2-adaptin contains two clathrin interaction sites within its hinge-ear-region (Shih et al., 1995; Owen et al., 2000), the ear-domain of α-adaptin functions as a binding site for numerous accessory or network proteins which are crucial for vesicle budding (Owen et al., 1999; Conner and Schmid, 2002). Thus, α-adaptin performs a special regulatory role in the endocytic process and is considered as an endocytosis-specific protein, since unlike the σ-, μ- and β-adaptins it lacks isoforms in all other AP complexes. Based on their ability to interact with α-adaptin directly the network proteins are divided into two classes. The members of the primary network proteins, like the serine/threonin kinase α-adaptin-associated kinase (AAK1), epsin, EPS15, amphiphysin, AP180 and dynamin all contain the special α-adaptin binding motifs DPW/F and FXDXF that are present in variable numbers and positions (Owen et al., 1999), while the more extensive class of secondary network proteins are lacking these specific peptide motifs

(Jarousse and Kelly, 2000). Since the members of both classes make multiple moderate strength interactions with each other, they create a highly dynamic protein network.

Ever since its discovery, AP-2 was considered as the only adaptor for endocytic cargo but this dogma has recently been refuted. Additional clathrin-dependent uptake processes of specific cargo molecules by monomeric adaptors have been observed in AP-2 depleted cells (Owen et al., 2004). Compared to wildtype cells the number of clathrin-coated pits (CCP) is remarkably reduced in these cells and endocytosis of the transferrin receptor (Tf-R) is blocked. However, the uptake of the low-density-lipoprotein receptor (LDL-R) and the epidermal growth factor receptor (EGF-R) remained unaffected (Hinrichsen et al., 2003; Motley et al., 2003). Two monomeric proteins, Disabled-2 (Dab2) and autosomal recessive hypercholesterolemia (ARH), have been shown to recruit members of the LDL-R family. Both contain a phosphotyrosine binding domain which is able to interact with the FXNPXY motif of the LDL-Receptors (Traub, 2003; Robinson, 2004). Furthermore, there is compelling evidence that the members of the E/ANTH-superfamily function as monomeric adaptors as well (Wendland, 2002). These proteins contain a phosphatidylinositol-(4,5)-bisphosphate [PI(4,5)P2]-binding site within their $\sim$ 200 amino-terminal residues which is termed ENTH (epsin-*N*-terminal-homology domain) in epsin1 and epsinR, or ANTH (AP180-*N*-terminal-homology-domain) in AP180, CALM and Huntingtin-interacting proteins (HIP)1 and HIP12 (Legendre-Guillemin et al., 2004). Downstream of this lipid-binding site mammalian epsin1 and AP180 contain a disordered region with no obvious secondary structure that instead contains various clathrin and AP-2 interaction motifs (Kalthoff et al., 2001). The binding of the EGF-R is mediated by the ubiquitin-interacting motifs (UIMs) of mammalian epsin1 and the Drosophila epsin homolog *liquid facets* is required for the binding of Delta, the membrane-spanning ligand of the Notch-receptor (Traub, 2003). Likewise, in *Saccharomyces cerevisiae*, ubiquitin moieties function as key signals in receptor-mediated endocytosis, and are recognized by the yeast epsin homologs Ent1p and Ent2p (Shih et al., 2002). In contrast to ENTH-proteins, ANTH-proteins lack UIMs but are also reported to function as monomeric cargo-specific adaptors. Like epsinR, which interacts with the v-SNARE Vti1b (Chidambaram et al., 2003), the AP180 homolog of *C. elegans*, UNC-11, internalizes the v-SNARE synaptobrevin (Nonet et al., 1999), while HIP1 is responsible for the interaction with a glutamate-receptor (Metzler et al., 2003). All of the cargo-specific adaptors bind to clathrin as well as to AP-2 and they colocalize with the main CCV coat components at the PM. However, it is unclear whether these monomeric adaptors can also function as clathrin assembly proteins in vivo. This would indicate a sort of independency of AP-2 complexes in CME in multicellular eukaryotes like in the unicellular yeast (see Sect. 1.2).

Both the cargo-specific adaptors and the network proteins are crucial in distinct steps during vesicle biogenesis at the PM with some of them performing multifunctional tasks. Accordingly, AP180 and epsin promote in addition clathrin assembly, while synaptojanin, amphiphysin and endophilin function as lipid-modifying enzymes (Cremona et al., 1999; Slepnev and De Camilli, 2000; Verstreken, et al., 2003), and auxilin and Hsc70 act as uncoating chaperones (Ungewickell et al., 1995; Morgan et al., 2001). AAK1 and Auxilin2/GAK function as endocytic protein kinases, with additional regulatory roles in the budding and uncoating reactions, respectively (Umeda et al., 2000; Conner and Schmid, 2002), and the adaptor-dimer HIP1/HIP1R functions as a linker between clathrin (HIP1) and the actin cytoskeleton (HIP1R) (McPherson, 2002; Robinson, 2004). Some of these network proteins are permanent coat components like AP180, auxilin and HIP1/HIP1R (Kirchhausen, 1999; Scheele et al., 2001; Legendre-Guillemin et al., 2002), while others associate only transiently with the coat proteins or with other network proteins to perform their functions (Kirchhausen, 2000b; Slepnev and de Camilli, 2000; Brodsky et al., 2001). This highly dynamic network of at least 20 proteins in mammals is based on the interactions of specific motifs with their corresponding domains, like Src-homology 3 (SH3)-motif (e.g. in endophilin) interacting with proline-rich domains (PRD) (e.g. in synaptojanin), or the epsin-homology (EH)-domain (e.g. in EPS15) with the NPF-motif (e.g. in epsin1), while the pleckstrin-homology (PH)-domain (e.g. in dynamin) is required for lipid-interactions (Jarousse and Kelly, 2000). Some network proteins like AP180 and auxilin1 are exclusively restricted to neuronal cells, while their non-neuronal isoforms CALM and Auxilin2/GAK are ubiquitously expressed (Tebar et al., 1999; Umeda et al., 2000). Plant homologs to mammalian network proteins have been identified from databases (Holstein, 2002) and a small number have already been functionally characterized (see Sect. 2.3).

1.2 The Current Model of Mammalian Endocytic CCV Biogenesis

CME starts with the assembly of coat components at distinct sites of the PM, the so-called hot spots. These local high concentrations of the lipid PM-marker PI(4,5)P2 are created from phospoinositides (PI) by the sequential interaction of the PI4-kinase type Iia and the PI4-P-kinase type Iγ, whereby the latter enzyme is recruited to the PM by Arf6-GTP (Krauss et al., 2003; Legendre-Guillemin et al., 2004). The PI(4,5)P2-spots are enriched in the inner leaflet of the PM and are the main recruiting sites for AP-2 and the monomeric A/ENTH adaptors which are the first to interact with the lipids in a clathrin-independent manner (Motley et al., 2003). Thus, epsin1 and AP180 might function as tethers in that they loosely connect the clathrin machinery to the PM thereby initiating CCV biogenesis (Kalthoff et al., 2001).

Since this docking step precedes cargo incorporation into CCPs, a conformational change in the AP-2 complex is required allowing the interaction between cargo and the RBD of μ2-adaptin (Owen, 2004). In the closed, inactive conformation of AP-2, the μ2-RBD is located within a cleft built by the trunk-portions of α-and β2-adaptins (Collins et al., 2002). The RBD-μ2 becomes only exposed in the active AP-2 complex after its threonine residue (^{156}Thr), located within its linker-region, is phosphorylated by AAK1 (Conner and Schmid, 2002; Ricotta et al., 2002). Moreover, only in the active AP-2 complex its two PI(4,5)P2 binding sites, one in α-trunk and the other in μ2-adaptin, are located within one layer and therefore able to bind to PIs simultaneously (Gaidarov et al., 1996; Rohde et al., 2002; Collins et al., 2002). This binding step is supported by synaptotagmin, a PM localized calcium sensor, which is capable to bind to PI(4,5)P2, α-trunk and μ2-adaptin simultaneously and might therefore be considered as an AP-2 docking protein (Takei and Haucke, 2001; Bai et al., 2004).

For the assembly of an initially flat clathrin lattice, the recruitment of cytosolic triskelia to the PM by AP180 and β2-adaptin/AP-2 is already sufficient. As maturation of the CCP proceeds, a more progressing curvature of the PM is required, that finally leaves the vesicles stuck to the membrane via a narrow neck. This membrane deformation is primarily the work of amphiphysin, the lysophosphatidylacid-transferase endophilin and the GTPase dynamin (Slepnev and DeCamilli, 2000). Thereby, amphiphysin is considered as a coordinator of vesicle budding and fission, since it is able to sense membrane curvature via its lipid-binding BIN/amphiphysin/RVS (BAR)-domain and is simultaneously able to bind to AP-2 and clathrin, and also to dynamin via its central region and its proline-rich domain, respectively (Yoshida et al., 2004). Since dynamin and amphiphysin coassemble into rings, the oligomerization of dynamin might be facilitated by the presence of the concentrated amphiphysin molecules which have been recruited before to pre-existing high curvature membranes (Peters et al., 2004). For a long time dynamin has been considered as a mechanical enzyme but recent studies describe it rather as a regulatory protein which recruits the uncoating machinery, auxilin and Hsc70, to the neck of the vesicles in order to promote the budding process by exchanging cage-bound with soluble cytosolic triskelia (Wu et al., 2001; Newmyer et al., 2003; Šamaj et al., 2005). However, the involvement of dynamin in the CCV budding process has been convincingly demonstrated in the temperature-sensitive *shibire* mutants from *Drosophila melanogaster* that display an accumulation of CCPs due to budding restriction (Hinshaw, 2000). In contrast, the dynamin-like proteins from baker's yeast are not involved in the endocytosis of the α-factor pheromon receptor Ste2p which rather depends on the amphiphysin homologs Rvs167p/Rvs161p and the actin cytoskeleton (Munn et al., 1995). Like its mammalian counterpart, Rvs167p also functions as a scaffolding protein in that it interacts with numerous endocytotically active proteins. Endocytosis studies in yeast CHC, CLC, AP180 and

AP-2 deletion mutants revealed that the clathrin machinery is to some degree dispensable (Huang et al., 1999; Yeung et al., 1999). Moreover, the most important endocytosis factor in yeast is the actin cytoskeleton which is reflected by the fact that about 30% of the 50 gene products involved in endocytosis are either actin-binding proteins or regulators of actin assembly. Like the CHCΔ- and CLCΔ-mutants most of the other endocytosis mutants display defects in the actin cytoskeleton as well. In yeast cells clathrin is required for the correct localization of Sla2p/HIP at the cell cortex (Engqvist-Goldstein and Drubin, 2003). Thereby, Sla2p/HIP which harbours an actin binding site, binds to Pan1p/EPS15, that in turn activates the Arp2/3 actin assembly complex. Although, the role of the actin cytoskeleton in mammalian endocytosis is not fully established, a myosinVI variant functions as an actin motor protein and is a component of the CCV coat (Bennett et al., 2001). It is conceivable that in mammalian endocytosis the actin cytoskeleton could participate in three stages during CCV biogenesis, namely as a scaffold at the hot spots of the PM during assembly and recruitment of the clathrin machinery, as an energy provider for the scission process of the vesicle either alone or in combination with dynamin, and finally as a promotor for vesicle transport from the PM into the cytoplasm (Engqvist-Goldstein and Drubin, 2003; Šamaj et al., 2005). However, in vivo studies on mammalian cells point out that actin polymerization takes place in the late stages during CCV biogenesis (Merrifield et al., 2002).

After pinching off the PM, the CCV are uncoated within in a very short time-range, which is a prerequisite for the delivery of the vesicle cargo into the endocytic pathway. Compared to the mechanism of triskelion uncoating from CCV (Ungewickell et al., 1995; Holstein et al., 1996; Ungewickell et al., 1997), the knowledge about AP complex uncoating is only marginal. However, the phosphorylation status of the phospholipids seem to play a major role in this process. In the current model, the network protein endophilin recruits its interaction partner, the poly-phosphoinositol-phosphatase synaptojanin to newly formed synaptic vesicles where it acts as a negative regulator on coat protein PM interactions (Cremona et al., 1999; Verstreken et al., 2003). Since PI(4,5)P2 becomes dephosphorylated to PI(4)P, the uncoating of AP-2 complexes is facilitated by lowering their affinity to the PM and finally, an increased content of PI(3)P, the marker lipid of early endosomal compartments, is achieved (Gruenberg, 2001; LeRoy and Wrana, 2005).

Although, an overwhelming mass of detailed information on the molecular level has been obtained especially in the last years, some basic questions, as for example about the time-point of uncoating (CCP vs. CCV) are still unanswered.

2 Clathrin-Mediated Endocytosis in Plants

2.1 Fluid-Phase Endocytosis

Whether endocytosis is an established phenomenon in plant cells has been a matter of controversy for several decades. Thereby turgor pressure in cell wall surrounded plant cells was considered the main obstacle (Cram, 1980). A turning point in this debate was achieved when thermodynamically based considerations took into account the daily fluctuations of the turgor pressure in single cells as well as the size of endocytic vesicles which may not exceed 1 bar and 100 nm, respectively (Saxton and Breidenbach, 1988; Gradmann and Robinson, 1989). Recent endocytosis studies finally dispelled the last doubts, since in intact guard cells the uptake of a PM protein against high turgor pressures was shown (Meckel et al., 2004).

However, fluid-phase endocytosis has been convincingly demonstrated before on protoplasts and even on intact plant cells by the use of various electron-dense markers (Low and Chandra, 1994) and was even shown to be actin dependent (Baluška et al., 2004; Šamaj et al., 2004). Moreover, filipin-labelled PM sterols (Grebe et al., 2003) and the PM-binding styryl dyes FM1-43 and FM4-64 (Ueda et al., 2001; Emans et al., 2002) served as endocytic tracers. During the unspecific uptake of electron-dense markers into plant protoplasts distinct intracellular compartments were successively labelled: starting with the CCP/CCV at the PM, the labelling proceeded to the partially coated reticulum, from there to multivesicular bodies and finally reached the vacuole (Low and Chandra, 1994). Since these plant organelles resemble morphologically their mammalian counterparts, a similar function along the endocytic pathway was assumed (Holstein, 2002). So far, plant endosomal compartments have been characterized by the colocalization of PM-binding FM-dyes or PM-sterols with Ara6 and Ara7, the plant homologs of mammalian Rab5, an early endosomal marker (Ueda et al., 2001; Grebe et al., 2003). Recently, a functional differentiation into early/recycling and late endosomes has been described based on the colocalization of SNAREs and Rab proteins (Ueda et al., 2004). However, a colocalization of an endocytosed plant PM-receptor after ligand binding with an endosomal marker protein on specific organelles is missing, since receptor-mediated endocytosis in plants is still unproven (Russinova and de Vries, 2005). Nevertheless, some indications of specific cargo uptake have been obtained by elicitor and biotin binding studies showing the typical features, such as temperature- and energy-dependency, saturation of the uptake and competition by free ligands (Low et al., 1993; Bahaij et al., 2001, 2003).

Of the two main endocytosis pathways, the clathrin-dependent and the clathrin-independent but lipid-raft-dependent, the former is by far the better investigated in plants. A caveolin homolog is missing in plant databases (Samaij et al., 2004) and lipid-raft microdomains and GPI-anchored proteins associated with them at the plant PM have been discovered only recently (Borner et al., 2003; Lalanne et al., 2004; Mongrand et al., 2004). In contrast, CCV have already been detected in different plant species since the 1960s (reviewed in Newcomb, 1980) and clathrin has been documented to be present in high amounts not only at the cell plate during cytokinesis (Otegui et al., 2001) but also at the tip of fast growing plant cells such as root hairs and pollen tubes in a high density (4.5 CCP/μm^2) (Emons and Traas, 1986; Blackbourn and Jackson, 1996; Hepler et al., 2001). Nevertheless, the proof of clathrin involvement at the molecular level was missing for decades and other open questions, which deal with the identity of the cargo of endocytic plant CCV, have been addressed only recently.

2.2 Cargo Candidates of Plant Endocytic CCV

The possibility of a complete PM turnover within 30 minutes in rapidly growing plant cells with a polarized organization (Emons and Traas, 1986) strongly implies a function of CCV in the removal of excess plant PM in analogy to the neuronal PM recycling in synapses. Furthermore, in analogy to the down-regulation of the epidermal-growth-factor receptor (EGF-R) and G-protein coupled receptors in mammals (Sorkin and von Zastrow, 2002; Gonzales-Gaitan, 2003), plant CCV might also serve to remove signalling receptors from the plant PM. In this respect, the approx. 500 members of the receptor-like protein kinases (RLKs) in *Arabidopsis* (McCarty and Chory, 2000) are promising candidates as endocytic CCV cargo molecules (Fig. 1; see Russinova and de Vries, 2005). RLKs are type I integral membrane proteins with their cytoplasmic domains harbouring a different number of kinase subdomains. The diverse nature of their extracellular ligand binding domains divides the RLKs into five classes. Thereby, the leucine-rich repeat (LLR) containing group of RLKs assimilates almost half of all RLKs (McCarty and Chory, 2000). While all plant RLKs (with one exception) belong to the serine/threonine kinases, the bulk of their mammalian counterparts, like the EGF-R, belong to the group of tyrosine kinases. From the mammalian RLKs only the minor group of transforming growth factor β-receptors (TGFβ-R) represent serine/threonine kinases (Gonzales-Gaitan, 2003). However, the combination of LLR-repeats with serine/threonine kinase activity is an unique feature of plants (Meyerowitz, 2002). This is true for CLAVATA 1 that plays a crucial role in the maintenance of stem cells during meristem development via a morphological gradient (Bowman and Eshed, 2000). Like CLAVATA 1, some of the RLKs contain the conserved mammalian tyrosine-

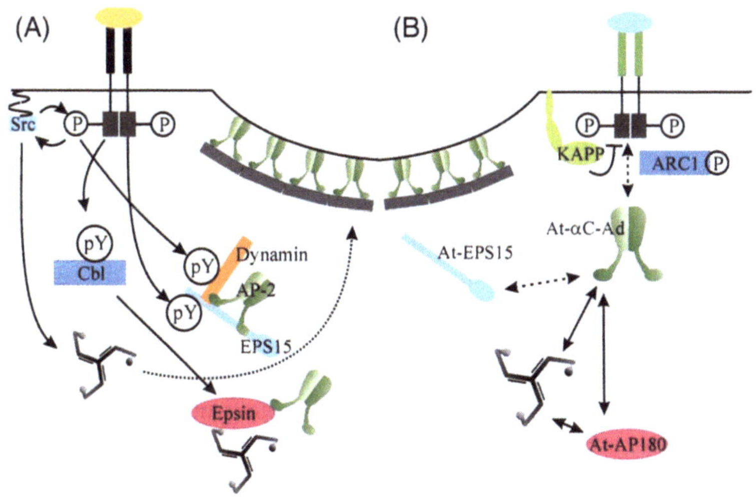

Fig. 1 Model depicting established and putative interactions of mammalian receptor protein kinases (RPKs) and plant receptor-like kinases (RLK) with components of the clathrin endocytosis machinery. **A** Binding of its ligand EGF induces the dimerization of the RPK EGF-R and also its trans-autophosphorylation on tyrosine residues located within its cytoplasmic domain. These in turn represent docking sites for SH2- or phosphotyrosine binding domain containing downstream effector proteins such as the ubiquitin ligase c-Cbl and EPS15, which assemble a signalling network for the regulation of the intracellular response to the ligand. In addition, the EGF-R itself is an active cargo since it modifies directly components of the clathrin endocytosis machinery by covalently phosphorylating their tyrosine residues. **B** The hypothetical model of plant RLK internalization shows similarities to the down-regulation of mammalian RPKs. Plant RLKs also dimerize after ligand binding and some contain the YXXϕ-internalization motif while others might be prone to ubiquitylation. Furthermore, dephosphorylation by the plant-specific phosphatase KAPP is also crucial for their internalization. Several plant homologs of the clathrin endocytosis machinery, required for the downregulation of signalling receptors, have been functionally characterized and their interactions are demonstrated by *solid arrows* (see text) while putative interactions are indicated by *dashed arrows*. An EPS15 homolog has been identified from the *Arabidopsis* database (Holstein 2002)

based internalization motif (YXXϕ) within their cytoplasmic domains, while others, like the S-domain type receptor kinase (SRK) from *Brassica* which is required in the self-incompatibility recognition, might become ubiquitinylated via the E3-ligase activity of the plant specific Armadillo-repeat-containing protein 1 (ARC1) and might thus become endocytosed via the interaction with an A/ENTH protein homolog (Fig. 1) (Holstein, 2002). These features parallel the internalization of the mammalian EGF-R that is dependent on the clathrin endocytosis machinery for internalization after ligand (EGF) binding in some aspects. During EGF-R internalization, CCP formation is promoted by the E3 ubiquitin ligase, the RING-finger protein Cbl, together with the network protein endophilin. Moreover, monoubiquitylation of the

UIM-containing network proteins EPS15 and epsin which are crucial for its internalization are also regulated by the EGF-R (Polo et al., 2002). In addition, monoubiquitylated epsin1, either alone or in a complex together with EPS15, might promote the release of the internalized EGF-R from clathrin adaptors (see Sect. 1.1) (Le Roy and Wrana, 2005). Furthermore, the phosphorylation status of different proteins plays also an important role in EGF-R endocytosis. Both, EPS15 (EGFR-pathway substrate 15) and clathrin are phosphorylated via the membrane-bound kinase Src which in the case of clathrin results in its redistribution to the PM (Wilde et al., 1999). Like the mammalian receptor protein kinases, the internalization of the plant LLR-RLK AtSERK1 (*Arabidopsis thaliana* somatic embryogenesis receptor kinase 1) also depends on its dephosphorylation status by KAPP, a PP2C-type kinase associated protein phosphatase of which the *Arabidopsis* genome contains around 600 genes (Shah et al., 2002). Furthermore, the two plant hormone receptors, the RLKs brassinosteroid insensitive 1 (BRI1) and AtSERK3/BAK1 heterodimerize at the PM in *Arabidopsis* cells. This receptor complex is discussed to transmit the brassinosteroid signal similar to its mammalian counterparts such as the EGF-R or the transforming-growth-factor receptor (TGFβ-R) (Russinova et al., 2004). To date, it is unknown whether endocytosis of plant RLKs is a prerequisite for their signalling as is the case for the mammalian TGFβ-R, which in a trimeric complex with Smad and SARA (Smad anchor for receptor activation) propagates the signal from an early endosomal compartment (Gonzales-Gaitan, 2003). Since plants lack true homologs of Smads and SARA, other transcription factors and FYVE-zinc-finger-domain proteins probably await their discovery.

Besides RLKs, other cargo molecules of plant endocytic CCV come also into consideration. The plant clathrin endocytosis machinery might also play a role in the maintenance of cell polarity, since the auxin efflux carrier PIN1 cycles between an endosomal BFA-sensitive compartment and the basal PM under the influence of the ARF-GTP exchange factor (GEF) GNOM (Geldner et al., 2001, 2003; Friml et al., 2002). Furthermore, the cycling of the syntaxin KNOLLE (SYP111) from the cell plate to an endosomal compartment during cytokinesis could also occur in a clathrin-dependent way, since it contains a YXXϕ-motif within its amino-terminal cytosolic loop (Jürgens 2004). Finally, it is also conceivable that plant endocytic CCV play a role in pressure-driven uptake of the KAT1 inward-rectifying potassium channel that colocalizes with FM4-64 and is constitutively endocytosed from the PM of *Vicia faba* guard cells (Hurst et al., 2004; Meckel et al., 2004).

2.3
Plant Clathrin Coat and Accessory Proteins

Although the first CCV have been isolated from different plant species more than 20 years ago (Mersey et al., 1982; Cole et al., 1987; Coleman et al., 1987;

Demmer et al., 1993) the analysis of their coat proteins was tremendously delayed due to plant-specific problems dealing with vacuolar content leakage during tissue preparation. Especially the isolation of plant clathrin light chains (CLC) proved itself to be difficult and led to controversial results (Lin et al., 1992; Demmer et al., 1993; Robinson et al., 1998). It was only recently that three *Arabidopsis* CLC have been identified via binding to the hub-region of bovine CHC (Scheele and Holstein, 2002). So far, other main coat proteins like the CHC from soybean (Blackbourn and Jackson, 1996) as well as some homologs of AP complex subunits have been identified in plants. The total number of *Arabidopsis* adaptins points to the existence of four plant AP complexes like in mammals (Böhm and Bonifacino, 2001), while other eukaryotes like *Saccharomyces cerevisiae*, *C. elegans* and *Drosophila* lack the fourth AP complex. In this respect, plants resemble mammals even more, since *Arabidopsis* also contains two α-adaptin (Holstein, 2002; Barth and Holstein, 2004), five β-adaptin (Böhm and Bonifacino, 2001; Holstein and Happel, 2002), three γ-adaptin (Schledzewski et al., 1999; Böhm and Bonifacino, 2001), five μ-adaptin (Happel et al., 1997, 2004) and five σ-adaptin genes, while the δ- and ε-adaptins are coded for by single genes (Böhm and Bonifacino, 2001). In addition, like mammalian β1- and β2-adaptins, *Arabidopsis* βB- and βC-adaptins are also highly related (Holstein and Happel, 2000). Since the composition of plant AP complexes is still a mystery, plant adaptins are referred to by letters instead of numbers to avoid confusion until their localization and functions have been unequivocally established.

Adaptins have also been reported from plant species other than *Arabidopsis*, namely a σ1-adaptin from the Chinese medical tree (Maldonado-Mendoza and Nessler, 1996), a σ2-adaptin from maize (Roca et al., 1998), and a β-adaptin from zucchini CCV which was identified using mammalian β1/β2-adaptin antibodies (Holstein et al., 1994). While *Arabidopsis* PM complexes containing auxin transport proteins have been reported as interacting partners of a plant β-adaptin (Murphy et al., 2005), plant homologs of other monomeric mammalian clathrin adaptors, like β-arrestin, Dab2 and ARH have so far not been identified in plant genomes (see Sect. 1.1).

To date, only a limited number of plant adaptins and adaptors have been characterized. In this respect, an *Arabidopsis* γ-adaptin might interact with the dynamin-like protein ADL6 and is probably involved in Golgi-originating vesicle trafficking (Lam et al., 2002). A functional analysis at the molecular level has been performed only for the *Arabidopsis* αC-adaptin (At-αC-Ad), for one of the eight AP180 homologs (Barth and Holstein, 2004; Holstein and Oliviusson, 2005), the CLC (Scheele and Holstein, 2002) and the μA-adaptin (At-μA-Ad) (Happel et al., 2004). The latter has been identified as a receptor-binding partner involved in the Golgi-vacuolar transport pathway rather than in endocytosis (Happel et al., 2004). Thereby its receptor-binding domain (RBD) interacts with the YXXϕ-motifs of the mammalian transmembrane protein TGN38 and also of the pea vacuolar sorting receptor (VSR-PS1) which is the only receptor isolated from plant CCV to date (Happel et al., 2004). Since the five *Arabidopsis* μ-adaptins display highly conserved features within their RBDs (Happel et al., 2004) it is reasonable to assume that a plant μ-adaptin might also interact with the internalization motifs of the RLKs during endocytosis. Like At-μA-Ad, the plant αC-adaptin has also been identified as a subunit of a plant multimeric protein complex that corresponds to the molecular mass of mammalian AP complexes (Barth and Holstein, 2004). In addition, the ear-region of At-αC-Ad functions as a binding site for various network proteins (Barth and Holstein, 2004), while its hinge-region contains an active clathrin binding motif which is probably a plant-specific feature (Holstein, unpublished data). One of the At-αC-Ad binding partners, the monomeric At-AP180 functions in addition as a clathrin assembly protein (Barth and Holstein, 2004) promoting the formation of cages within a narrow size distribution, which is an important feature in respect to plant endocytosis. Surprisingly, the assembly function is dependent on its single DLL-motif which indicates a difference in the fine tuning of clathrin lattice assembly in plants and mammals.

To date, the budding process of plant endocytic CCV remains a mystery at the molecular level. Eight ANTH and three ENTH *Arabidopsis* proteins are promising candidates as AP-2 and clathrin tethering proteins (Holstein and Oliviusson, 2005), of which only one has been functionally characterized so far (Barth and Holstein, 2004). Although, plants contain numerous homologs of the components of the mammalian endocytic clathrin machinery which are required for tethering, budding, scission and uncoating and also serve as linkers to the actin cytoskeleton (Holstein, 2002; Samaij et al., 2004) an important question is still open as to whether plant CME might also be dependent on dynamin as in mammals or whether it is exclusively dependent on the actin cytoskeleton and amphiphysins as in yeast. *Arabidopsis* contains 16 dynamin-related proteins (DRPs) which are clustered in five families (DRP1–5) (Hong et al., 2003). Of these, the only family matching the domain structures of classical mammalian dynamins, is DRP2/ADL6, which is involved in the *trans*-Golgi-network to vacuole transport (Lam et al., 2002). In contrast, the plant specific DRP1 and DRP3 families both lack lipid binding and proline-rich domains (Hong et al., 2003) and ADL1A and ADL1C of DRP1 are involved in cytokinesis rather than in endocytosis (Ueda and Nakano, 2002; Kang et al., 2003). Like the majority of plant DRPs, yeast contains only three dynamin-related proteins which are not involved in endocytosis that rather is dependent on the actin cytoskeleton as well as on amphiphysin homologs (RVS161p/RVS167p) (see Sect. 1.2). However, a true plant amphiphysin homolog has not been discovered in the plant databases, but three Src homology 3 (SH3) domain-containing proteins from *Arabidopsis* can substitute for Rvs167p in yeast deletion mutants, are able to interact with the actin cytoskeleton and in addition can bind to an auxilin-like protein (Lam et al., 2001). So far, the interaction of a plant auxilin homolog with its co-chaperone Hsc70 in the uncoating of plant CCV has not been proven, although the uncoating ATPase from pea has been demonstrated almost a decade ago to be able to work on both plant and mammalian CCV (Kirsch and Beevers, 1993).

3 Conclusions and Future Prospects

The understanding of the plant CME at the molecular level is only in its beginning and the identification and functional characterization of some of its interacting partners is in progress. A conserved uptake mechanism like the clathrin endocytosis machinery already implies the existence of highly conserved plant homologs. Nevertheless, thorough investigations at the molecular level indicate that the fine-tuning of the clathrin machinery during endocytosis in both kingdoms is different, for example the mode of clathrin assembly of the plant At-AP180 compared to its mammalian counterpart. Since the consensus YXXϕ-motif is active in the plant CCV vacuolar sorting receptor, it is reasonable to assume that it might also function as an endocytosis signal in various RLKs such as the meristem organizing receptors ERECTA and CLAVATA which harbour this motif within their cytosolic tails. Further work on plant endocytosis at the molecular level especially with a focus on receptor-clathrin adaptor interactions, will reveal the impact of this important vesicular trafficking pathway in the plant-specific physiological process of cytokinesis and also in cell polarity maintenance and signal transduction during plant development.

References

Bahaji A, Cornejo MJ, Ortiz-Zapater E, Contreras I, Aniento F (2001) Uptake of endocytic markers by rice cells: variations related to growth phase. Eur J Cell Biol 80:178–186

Bahaji A, Aniento F, Cornejo MJ (2003) Uptake of an endocytic marker by rice cells: variations related to osmotic and saline stress. Plant Cell Physiol 44:1100–1111

Bai J, Tucker WC, Chapman ER (2004) PIP2 increases the speed of response of synaptotagmin and steers its membrane-penetration activity toward the plasma membrane. Nat Struct Mol Biol 11:36–44

Baluška F, Šamaj J, Hlavacka A, Kendrick-Jones J, Volkmann D (2004) Actin-dependent fluid-phase endocytosis in inner cortex cells of maize root apices. J Exp Bot 55:463–473

Barth M, Holstein SEH (2004) Identification and functional characterization of *Arabidopsis* AP180, a binding partner of plant αC-adaptin. J Cell Sci 117:2051–2062

Bennett EM, Chen CY, Engqvist-Goldstein AE, Drubin DG, Brodsky FM (2001) Clathrin hub expression dissociates the actin-binding protein Hip1R from coated pits and disrupts their alignment with the actin cytoskeleton. Traffic 2:851–858

Blackbourn HD, Jackson AP (1996) Plant clathrin heavy chain: sequence analysis and restricted localisation in growing pollen tubes. J Cell Sci 109:777–787
Böhm M, Bonifacino JS (2001) Adaptins. The final recount. Mol Biol Cell 12:2907–2920
Bonifacino JS, Traub LM (2003) Signals for sorting of transmembrane proteins to endosomes and lysosomes. Annu Rev Biochem 72:395–447
Borner GHH, Lilley KS, Stevens TJ, Dupree P (2003) Identification of glycosylphosphatidylinositol-anchored proteins in *Arabidopsis*. A proteomic and genomic analysis. Plant Physiol 132:568–577
Bowman JL, Eshed Y (2000) Formation and maintanance of the shoot apical meristem. Trends Plant Sci 5:110–115
Brodsky FM, Chen C-Y, Knuehl C, Towler MC, Wakeham DE (2001) Biological basket weaving: Formation and function of clathrin-coated vesicles. Annu Rev Cell Dev Biol 17:517–568
Chidambaram S, Müllers N, Wiederhold K, Haucke V, Fischer von Mollard G (2004) Specific interaction between SNAREs and ENTH domains of epsin-related proteins in TGN to endosome transport. J Biol Chem 279:4175–4179
Cole L, Coleman JOD, Hawes CR, Horsley D (1987) Antibodies to brain clathrin recognise plant coated vesicles. Plant Cell Rep 6:227–230
Coleman J, Evans D, Hawes C, Horsley D, Cole L (1987) Structure and molecular organization of higher plant coated vesicles. J Cell Sci 88:35–45
Collins BM, McCoy AJ, Kent HM, Evans PR, Owen DJ (2002) Molecular architecture and functional model of the endocytic AP-2 complex. Cell 109:523–535
Conner SD, Schmid SL (2002) Identification of an adaptor-associated kinase AAK1, as a regulator of clathrin-mediated endocytosis. J Cell Biol 156:921–929
Cram WJ (1980) Pinocytosis in plants. New Phytol 84:1–17
Cremona O, Di Paolo G, Wenk MR, Lüthi A, Kim WT, Takei K, Daniell L, Nemoto Y, Shears SB, Flavell RA, McCormick DA, De Camilli P (1999) Essential role of phosphoinositide metabolism in synaptic vesicle recycling. Cell 99:179–188
Crowther RA, Finch JT, Pearse BMF (1976) On the structure of coated vesicles. J Mol Biol 103:785–798
Demmer A, Holstein SEH, Hinz G, Schauermann G, Robinson DG (1993) Improved coated vesicle isolation allows better characterization of clathrin polypeptides. J Exp Bot 44:23–33
Emans N, Zimmermann S, Fischer R (2002) Uptake of a fluorescent marker in plant cells is sensitive to brefeldin A and Wortmannin. Plant Cell 14:71–86
Emons AMC, Traas JA (1986) Coated pits and coated vesicles on the plasma membrane of plant cells. Eur J Cell Biol 154:1209–1223
Engqvist-Goldstein AEY, Drubin DG (2003) Actin assembly and endocytosis: from yeast to mammals. Annu Rev Cell Dev Biol 19:287–332
Friml J, Wisniewska J, Benkova E, Mendgen K, Palme K (2002) Lateral relocation of auxin efflux regulator PIN3 mediates tropism in *Arabidopsis*. Nature 415:806–809
Gaidarov I, Chen Q, Falck JR, Reddy KK, Keen JH (1996) A functional phosphatidylinositol 3,4,5,-triphosphate/phosphoinositide binding domain in the clathrin adaptor AP-2 α-subunit. J Biol Chem 271:20 922–20 929
Geldner N, Friml J, Stierhof YD, Jürgens G, Palme K (2001) Auxin transport inhibitors block PIN1 cycling and vesicle trafficking. Nature 413:425–428
Geldner N, Anders N, Wolters H, Keicher J, Kornberger W, Muller P, Delbarre A, Ueda T, Nakano A, Jürgens G (2003) The *Arabidopsis* GNOM ARF-GEF mediates endosomal recycling, Auxin transport, and auxin-dependent plant growth. Cell 112:219–230

Gonzales-Gaitan M (2003) Signal dispersal and transduction through the endocytic pathway. Nat Rev Mol Cell Biol 4:213–224

Gradmann D, Robinson DG (1989) Does turgor prevent endocytosis in plant cells? Plant Cell Env 12:151–154

Grebe M, Xu J, Möbius W, Ueda T, Nakano A, Geuze HJ, Rook MB, Scheres B (2003) *Arabidopsis* sterol endocytosis involves actin-mediated trafficking via ARA6-positive early endosomes. Curr Biol 13:1378–1387

Gruenberg J (2001) The endocytic pathway: a mosaic of domains. Nat Rev 2:721–730

Happel N, Robinson DG, Holstein SEH (1997) An *Arabidopsis thaliana* cDNA clone (accession No. AF009631) is homologous to the micro-adaptins of clathrin coated vesicle adaptor complexes PGR97-168. Plant Physiol 115:1289

Happel N, Höning S, Neuhaus J-M, Paris N, Robinson DG, Holstein SEH (2004) *Arabidopsis* μA-adaptin interacts in vitro with the tyrosine-motif of the vacuolar sorting receptor VSR-PS1. Plant J 37:678–693

Hepler PK, Vidali L, Cheung AY (2001) Polarized cell growth in higher plants. Annu Rev Cell Dev Biol 17:159–187

Hinrichsen L, Harboth J, Andrees L, Weber K, Ungewickell EJ (2003) Effect of clathrin heavy chain- and α-adaptin specific small interfering RNAs on endocytic accessory proteins and receptor trafficking in HeLa cells. J Biol Chem 278:45 160–45 170

Hinshaw JE (2000) Dynamin and its role in membrane fission. Annu Rev Cell Dev Biol 16:483–519

Holstein SEH, Drucker M, Robinson DG (1994) Identification of a β-type adaptin in plant clathrin-coated vesicles. J Cell Sci 107:945–953

Holstein SEH, Ungewickell H, Ungewickell E (1996) Mechanism of clathrin basket dissociation: separate functions of protein domains of the DnaJ homologue Auxilin. J Cell Biol 135:925–937

Holstein SEH, Happel N (2000) Isolation of clathrin-coated vesicle beta adaptin homologs (Acc. No. AF216385, AF216386, AF216387) from *Arabidopsis thaliana*. Plant Physiol PGR 00-28

Holstein SEH (2002) Clathrin and plant endocytosis. Traffic 3:614–620

Holstein SEH, Oliviusson P (2005) Sequence analysis of *Arabidopsis* E/ANTH-domain-containing proteins: membrane tethers of the clathrin-dependent vesicle budding machinery. Protoplasma (in press)

Hong Z, Bednarek SY, Blumwald E, Hwang I, Jürgens G, Menzel D, Osteryoung KW, Raikhel NV, Shinozaki K, Tsutsumi N, Verma DP (2003) A unified nomenclature for *Arabidopsis* dynamin-related large GTPases based on homology and possible functions. Plant Mol Biol 53:261–265

Huang KM, D'Hondt K, Riezman H, Lemmon SK (1999) Clathrin functions in the absence of heterotetrameric adaptors and AP180-related proteins in yeast. EMBO J 18:3897–3908

Hurst AC, Meckel T, Tayefeh S, Thiel G, Homann U (2004) Trafficking of the plant potassium inward rectifier KAT1 in guard cell protoplasts of *Vicia faba*. Plant J 37:391–397

Jarousse N, Kelly R (2000) Selective inhibition of adaptor complex-mediated vesiculation. Traffic 1:378–384

Jürgens G (2004) Membrane trafficking in plants. Annu Rev Cell Dev Biol 20:481–504

Kalthoff C, Alves J, Urbanke C, Knorr R, Ungewickell EJ (2001) Unusual structure organization of the endocytic proteins AP180 and epsin1. J Biol Chem 277:8209–8216

Kanaseki T, Kadota K (1969) The "vesicle in a basket". A morphological study of the coated vesicle isolated from the nerve endings of the guinea pig brain, with special reference to the mechanism of membrane movements. J Cell Biol 42:202–220

Kang B-H, Busse JS, Bednarek SY (2003) Members of the *Arabidopsis* dynamin-like gene family, ADL1, are essential for plant cytokinesis and polarized cell growth. Plant Cell 15:899–913

Keen JH, Willingham MC, Pastan IH (1979) Clathrin-coated vesicles: isolation, dissociation and factor-dependent reassociation of clathrin baskets. Cell 16:303–312

Kirchhausen T (1999) Adaptors for clathrin-mediated traffic. Annu Rev Cell Dev Biol 15:705–732

Kirchhausen T (2000) Clathrin. Annu Rev Biochem 69:699–727

Kirchhausen T (2000) Three ways to make a vesicle. Nat Rev Mol Cell Biol 1:187–198

Kirsch T, Beevers L (1993) Uncoating of clathrin-coated vesicles by uncoating ATPase from developing peas. Plant Physiol 103:205–212

Krauss M, Kinuta M, Wenk MR, DeCamilli P, Takei K, Haucke V (2003) Arf6 stimulates clathrin/AP-2 recruitment to synaptic membranes by activating phosphatidylinositol phosphate kinase type Iγ. J Cell Biol 162:113–124

Lalanne E, Honys D, Johnson A, Borner GHH, Lilley KS, Dupree P, Grossniklaus U, Twell D (2004) SETH1 and SETH2, two components of the glycosylphosphatidylinositol anchor biosynthetic pathway, are required for pollen germination and tube growth in *Arabidopsis*. Plant Cell 16:229–240

Lam BCH, Sage TL, Bianchi F, Blumwald E (2001) Role of SH3-domain-containing proteins in clathrin-mediated vesicle trafficking in *Arabidopsis*. Plant Cell 13:2499–2512

Lam BCH, Sage TL, Bianchi F, Blumwald E (2002) Regulation of ADL6 activity by its associated molecular network. Plant J 31:565–576

Legendre-Guillemin V, Metzler M, Charbonneau M, Gan L, Chopra V, Philie J, Hayden MR, McPherson PS (2002) HIP1 and HIP12 display differential binding to F-actin, AP-2, and clathrin: Identification of a novel interaction with clathrin light chain. J Biol Chem 277:19 897–19 904

Legendre-Guillemin V, Wasiak S, Hussain NK, Angers A, McPherson PS (2004) ENTH/ANTH proteins and clathrin-mediated membrane budding. J Cell Sci 117:9–18

Le Roy C, Wrana JL (2005) Clathrin- and non-clathrin-mediated endocytic regulation of cell signalling. Nat Rev Mol Cell Biol 6:112–126

Lin H, Harley SM, Butler JM, Beevers L (1992) Multiplicity of clathrin light chain polypeptides from developing pea (Pisum sativum L.). J Cell Sci 103:1127–1137

Liu SH, Towler MC, Chen E, Chen C-Y, Song W, Apodaca G, Brodsky FM (2001) A novel clathrin homolog that co-distributes with cytoskeletal component functions in the trans-Golgi network. EMBO J 20:272–284

Low PS, Legendre L, Heinstein PF, Horn MA (1993) Comparison of elicitor and vitamin receptor-mediated endocytosis in cultured soybean cells. J Exp Bot 44:269–274

Low PS, Chandra S (1994) Endocytosis in plants. Annu Rev Plant Physiol Plant Mol Biol 45:609–631

Maldonado-Mendoza IE, Nessler CL (1996) Cloning and expression of a plant homolog of the small subunit of the Golgi-associated assembly protein AP19 from *Camptotheca acuminata*. Plant Mol Biol 32:1149–1153

Marmor MD, Yarden Y (2004) Role of protein ubiquitylation in regulating endocytosis of receptor tyrosine kinases. Oncogene 23:2057–2070

McCarty DR, Chory J (2000) Conservation and innovation in plant signalling pathways. Cell 103:201–209

McPherson PS (2002) The endocytic machinery at an interface with the actin cytoskeleton: a dynamic, hip intersection. Trends Cell Biol 12:312–315

Meckel T, Hurst AC, Thiel G, Homan U (2004) Endocytosis against high turgor: intact guard cells of Vicia faba constitutively endocytose fluorescently labelled plasma membrane and GFP-tagged K^+-channel KAT1. Plant J 39:182–194

Merrifield CJ, Feldman ME, Wan L, Almers W (2002) Imaging actin and dynamin recruitment during invagination of single clathrin-coated pits. Nat Cell Biol 4:691–698

Mersey BG, Fowke LC, Constable F, Newcomb EH (1982) Preparation of a coated vesicle-enriched fraction from plant cells. Exp Cell Res 141:459–463

Metzler M, Li B, Gan L, Georgiu J, Gutekunst CA, Wang Y, Torre E, Devon RS, Oh R, Legendre-Guillemin V (2003) Disruption of the endocytic protein HIP1 results in neuronal deficits and decreased AMPA receptor trafficking. EMBO J 22:3254–3266

Meyerowitz EM (2002) Plants compared to animals: the broadest comparative study of development. Science 295:1482–1485

Mongrand S, Morel J, Laroche J, Claverol S, Carde JP et al (2004) Lipid rafts in higher plant cells: purification and characterization of Triton X-100-insoluble microdomains from tobacco plasma membrane. J Biol Chem 279:36 277–36 286

Morgan JR, Prasad K, Jin S, Augustine GJ, Lafer EM (2001) Uncoating of clathrin-coated vesicles in presynaptic terminals: role for Hsc70 and Auxilin. Neuron 32:289–300

Motley A, Bright NA, Seaman MNJ, Robinson MS (2003) Clathrin-mediated endocytosis in AP-2-depleted cells. J Cell Biol 162:909–918

Mousavi SA, Malerod L, Berg T, Kjeken R (2004) Clathrin-dependent endocytosis. Biochem J 377:1–16

Munn AL, Stevenson BJ, Geli MI, Riezman H (1995) *end5, end6*, and *end7*: Mutations that cause actin delocalization and block the internalization step of endocytosis in *Saccharomyces cerevisiae.* Mol Biol Cell 6:1721–1742

Murphy AS, Bandyophadhyay A, Holstein SEH, Peer WA (2005) Endocytotic Cycling of PM Proteins. Annu Rev Plant Biol 56:221–251

Newcomb EH (1980) Coated vesicles: their occurence in different plant cell types. In: Ockleford CD, Whyte A (ed) Coated vesicles. Cambridge Univ Press, Cambridge, pp 55–69

Newmyer SL, Christensen A, Sever S (2003) Auxilin-Dynamin interactions link the uncoating ATPase chaperone machinery with vesicle formation. Dev Cell 4:929–940

Nonet ML, Holgado AM, Brewer F, Serpe CJ, Norbeck BA, Holleran J, Wei L, Hartwieg E, Jorgensen EM, Alfonso A (1999) UNC-11, a *Caenorrhabditis elegans* AP180 homologue, regulates the size and protein composition of synaptic vesicles. Mol Biol Cell 10:2343–2360

Otegui MS, Mastronarde DN, Kang BH, Bednarek SY, Staehelin LA (2001) Three-dimensional analysis of syncytial-type cell plates during endosperm cellularization visualized by high resolution electron tomography. Plant Cell 13:2033–2051

Owen DJ, Vallis Y, Noble MEM, Hunter JB, Dafforn TR, Evans PR, McMahon HT (1999) A structural explanation for the binding of multiple ligands by the α-adaptin appendage domain. Cell 97:805–815

Owen DJ, Vallis Y, Pearse BMF, McMahon HT, Evans PR (2000) The structure and function of the β2-adaptin appendage domain. EMBO J 19:4216–4227

Owen DJ (2004) Linking endocytic cargo to clathrin: structural and functional insights into coated vesicle formation. Biochem Soc Trans 32:1–14

Owen DJ, Collins BM, Evans PR (2004) Adaptors for clathrin coats: Structure and function. Annu Rev Cell Dev Biol 20:153–191

Pearse BMF, Robinson MS (1984) Purification and properties of 100-kD proteins from coated vesicles and their reconstitution with clathrin. EMBO J 3:1951–1957

Pearse BMF, Smith CJ, Owen DJ (2000) Clathrin-coat constructions in endocytosis. Curr Opin Struct Biol 10:220–228

Peters BJ, Kent HM, Mills IG, Vallis Y, Butler PJG, Evans PR, McMahon HT (2004) BAR domains as sensors of membrane curvature: the amphiphysin BAR structure. Science 303:495–499

Polo S, Sigismund S, Faretta M, Guidi M, Capua MR, Bossi G, Chen H, De Camilli P, Di Fiore PP (2002) A single motif responsible for ubiquitin recognition and monoubiquitination in endocytic proteins. Nature 416:451–455

Ricotta D, Conner SL, Schmid SL, von Figura K, Höning S (2002) Phosphorylation of the AP-2 μ subunit by AAK1 mediates high affinity binding to membrane protein sorting signals. J Cell Biol 156:791–795

Robinson DG, Depta H (1988) Coated vesicles. Annu Rev Plant Physiol Plant Mol Biol 39:53–99

Robinson DG, Hinz G, Holstein SEH (1998) The molecular characterization of transport vesicles. Plant Mol Biol 38:49–76

Robinson MS (2004) Adaptable adaptors for coated vesicles. Trends Cell Biol 14:167–174

Roca R, Stiefel V, Puigdomenech P (1998) Characterization of the sequence coding for the clathrin coat assembly protein AP17 (σ2) associated with the plasma membrane from *Zea mays* and constitutive expression of its gene. Gene 208:67–72

Rohde G, Wenzel D, Haucke V (2002) A phosphatidylinositol (4,5)-bis-phosphate binding site within μ2-adaptin regulates clathrin-mediated endocytosis. J Cell Biol 158:209–214

Russinova E, de Vries SC (2005) Receptor-mediated endocytosis in plants (in this volume). Springer, Berlin Heidelberg New York

Russinova E, Borst J-W, Kwaaitaal M, Cano-Delgado A, Yin Y, Chory J, de Vries SC (2004) Heterodimerization and endocytosis of *Arabidopsis* Brassinosteroid Receptors BRI1 and AtSERK3 (BAK1). Plant Cell 16:3216–3229

Šamaj J, Baluška F, Voigt B, Volkmann D, Menzel D (2005) Endocytosis and acto-myosin cytoskeleton (in this volume). Springer, Berlin Heidelberg New York

Šamaj J, Baluška F, Voigt B, Schlicht M, Volkmann D, Menzel D (2004) Endocytosis, actin cytoskeleton and signalling. Plant Phys 135:1150–1161

Saxton MJ, Breidenbach RW (1988) Receptor-mediated endocytosis in plants is energetically possible. Plant Physiol 86:993–995

Scheele U, Kalthoff C, Ungewickell E (2001) Multiple interactions of Auxilin1 with clathrin and the AP-2 Adaptor complex. J Biol Chem 276:36 131–35 138

Scheele U, Holstein SEH (2002) Functional evidence for the identification of an *Arabidopsis* clathrin light chain polypeptide. FEBS Lett 514:355–360

Schledzewski K, Brinkmann H, Mendel RR (1999) Phylogenetic analysis of components of the eukaryotic vesicle transport system reveals a common origin of adaptor protein complexes 1, 2, and 3 and the F subcomplex of the coatomer COPI. J Mol Evol 48:770–778

Shah K, Russinova E, Gadella TW Jr, Willemse J, de Vries SC (2002) The *Arabidopsis* kinase-associated protein phosphatase controls internalization of the somatic embryogenesis receptor kinase 1. Genes Dev 16:1707–1720

Shih W, Gallusser A, Kirchhausen T (1995) A clathrin-binding site in the hinge of the β2 chain of mammalian AP-2 complexes. J Biol Chem 270:31 083–31 090

Shih SC, Katzmann DJ, Schnell JD, Sutano M, Emr SD, Hicke L (2002) Epsins and Vps27p/Hrs contain ubiquitin-binding domains that function in receptor endocytosis. Nat Cell Biol 4:389–393

Slepnev VI, De Camilli P (2000) Accessory factors in clathrin-dependent synaptic vesicle endocytosis. Nat Rev 1:161–172

Sorkin A, von Zastrow M (2002) Signal transduction and endocytosis: close encounters of many kinds. Nat Rev Mol Cell Biol 3:600–614

Takei K, Haucke V (2001) Clathrin-mediated endocytosis: membrane factors pull the trigger. Trends Cell Biol 11:385–391

Tebar F, Bohlander SK, Sorkin A (1999) Clathrin assembly lymphoid myeloid leukemia (CALM) protein: localization in endocytic-coated pits, interactions with clathrin, and the impact of overexpression on clathrin-mediated traffic. Mol Biol Cell 10:2687–2702

Traub LM (2003) Sorting it out: AP-2 and alternate clathrin adaptors in endocytic cargo selection. J Cell Biol 163:203–208

Ueda T, Yamaguchi M, Uchimiya H, Nakano A (2001) Ara6, a plant-unique novel type Rab GTPase, functions in the endocytic pathway of *Arabidopsis thaliana*. EMBO J 20:4730–4741

Ueda T, Nakano A (2002) Vesicular traffic: an integral part of plant life. Curr Opin Plant Biol 5:513–517

Ueda T, Uemura T, Sato MH, Nakano A (2004) Functional differentiation of endosomes in *Arabidopsis* cells. Plant Physiol 40:783–789

Umeda A, Meyerholz A, Ungewickell E (2000) Identification of the universal cofactor (auxilin2) in clathrin coat dissociation. Eur J Cell Biol 79:336–342

Ungewickell E, Ungewickell H, Holstein SEH, Lindner R, Prasad K, Barouch W, Martin B, Greene LE, Eisenberg E (1995) Role of auxilin in uncoating clathrin-coated vesicles. Nature 378:632–635

Ungewickell E, Ungewickell H, Holstein SEH (1997) Functional interaction of the auxilin J domain with the nucleotide- and substrate-binding modules of Hsc70. J Biol Chem 272:19 594–19 600

Verstreken P, Koh T-W, Schulze KL, Zhai G, Hiesinger PR, Zhou Y, Mehta SQ, Cao Y, Roos J, Bellen HJ (2003) Synaptojanin is recruited by endophilin to promote synaptic vesicle uncoating. Neuron 40:733–748

Wendland B (2002) Epsins: adaptors in endocytosis? Nat Rev Mol Cell Biol 3:971–977

Wilde A, Beattie EC, Lem L, Riethof DA, Liu SH, Mobley WC, Soriano P, Brodsky FM (1999) EGF receptor signalling stimulates SRC kinase phosphorylation of clathrin, influencing clathrin redistribution and EGF uptake. Cell 96:677–687

Wu X, Zhao X, Baylor L, Kaushal S, Eisenberg E, Greene LE (2001) Clathrin exchange during clathrin-mediated endocytosis. J Cell Biol 155:291–300

Yeung BG, Phan HL, Payne GS (1999) Adaptor complex-independent clathrin function in yeast. Mol Biol Cell 10:3643–3659

Yoshida Y, Kinuta M, Abe T, Liang S, Araki K, Cremona O, Di Paolo G, Moriyama Y, Yasuda T, De Camilli P, Takei K (2004) The stimulatory action of amphiphysin on dynamin function is dependent on lipid bilayer curvature. EMBO J 23:3483–3491

Plant Cell Monogr (1)
J. Šamaj · F. Baluška · D. Menzel: Plant Endocytosis
DOI 10.1007/7089_007/Published online: 2 September 2005

Receptor-Mediated Endocytosis in Plants

Eugenia Russinova · Sacco de Vries (✉)

Laboratory of Biochemistry, Wageningen University, Dreijenlaan 3,
6703 HA Wageningen, The Netherlands
sacco.devries@wur.nl

Abstract Binding of ligands activates cell-surface receptors and triggers a series of signalling events. The activation of the receptors accelerates their internalisation, a process called receptor-mediated endocytosis. Thus, entire receptor–ligand complexes are internalised and processed within the cell. Recent work in a variety of cellular and developmental animal systems further supports the idea that the role of endocytosis extends beyond simply controlling the number of receptors at the cell surface. It has been shown that endocytic transport of the receptor complexes regulates signal transduction and mediates the formation of specialised signalling complexes. Signal transduction events can also modulate specific components of the endocytic machinery. Receptor internalisation in plant cells has recently been demonstrated; however, evidence for receptor-mediated endocytosis in plants is just beginning to emerge. In this review, we highlight the most recent advances in the study of receptor-mediated endocytosis in animals and compare them with what is currently known in plant systems.

1
Receptor-Mediated Endocytosis in Animal Cells

Plasma membrane receptors transduce extracellular information to targets inside the cell. In animal cells, receptors activated upon ligand binding are efficiently internalised and sorted in endosomes, either for recycling back to the plasma membrane or for degradation within lysosomes. Internalisation can occur via different routes, e.g. clathrin-mediated, caveolin-dependent, clathrin- and caveolin-independent endocytosis and phagocytosis. Endocytosis delivers receptors first to the early endosomes, a heterogeneous population of membrane compartments with tubulo-vesicular morphology that is located at the cell periphery (Fig. 1a). Receptors can either be recycled to the plasma membrane from the peripheral and perinuclear endosomes, early and late recycling compartments respectively, or they progress to lysosomes where they are degraded. Fusion and movement of the early endosomes causes internalised receptors to redistribute to larger compartments in the perinuclear area. Those compartments often show the characteristic morphology of multivesicular bodies (MVBs)—large membrane compartments that contain small vesicles in their lumen. Internalised receptors can either be recycled back to the plasma membrane or can be retained and accumulated

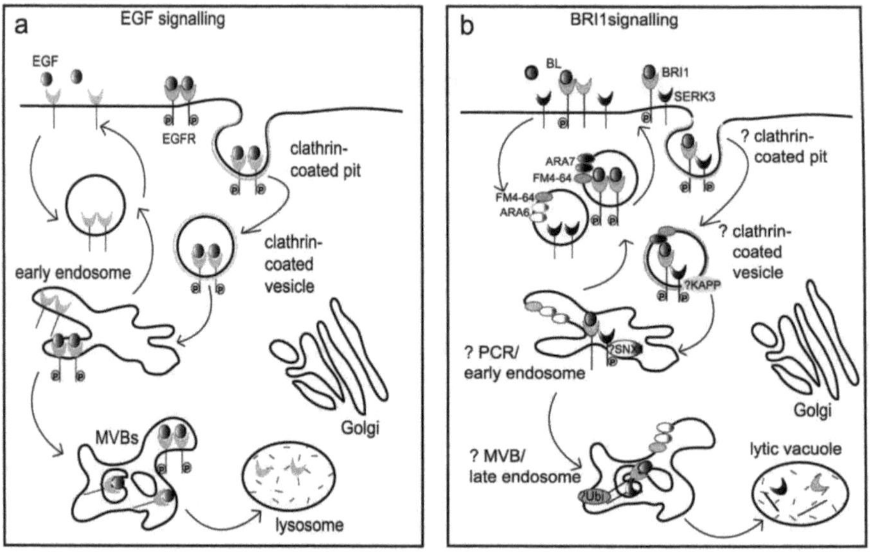

Fig. 1 A comparison of receptor-mediated endocytosis in animal and plant cells. **a** Epidermal growth factor (EGF) receptor is activated by the EGF and endocytosed mainly through clathrin-coated pits. The activated receptor accumulates in early endosomes and multivesicular bodies (MVBs). Ligand-free receptors are almost exclusively recycled to the cell surface. Ligand-bound receptors are sorted to lysosomes for degradation with an increased efficiency compared with that of the ligand-free receptors. EGF receptor remains active in early endosomes and in MVBs, indicated by the presence of phosphate groups. **b** Hypothetical model showing that brassinosteroid receptor complex including BRI1 and AtSERK3 is internalised in FM4-64 positive compartments that are colabelled with Rab5 plant homologues represented by ARA6 or ARA7. Homodimeric combinations of BRI1 and AtSERK3 are internalised and cycle back to the plasma membrane. Heterodimeric combination of BRI1 and AtSERK3 is preferentially internalised for degradation. We propose a general mechanism for degradation of the internalised receptors retained in the early endosomes and MVB compartments that involve KAPP dephosphorylation followed by ubiquitination.

into the MVBs. The multiple invaginations of the MVB membrane serve to trap receptors inside the MVBs, and thus to prevent recycling and to promote their delivery to the lysosomes. In contrast, inactive and ligand-free receptors are almost exclusively recycled back to the cell surface. Ligand-occupied receptors, however, recycle through rapid and slow pathways from early or MVB/late endosomal compartments (reviewed by Sorkin and Zastrow, 2002; Teis and Huber, 2003).

Endocytosis has long been recognised as a means to terminate signalling via degradation of activated receptor complexes after their internalisation from the cell surface. However, it has become clear that the output of the signalling process depends not only on the activation of a particular set of signalling molecules, but also on where or for how long the signal is emitted.

Exciting new findings suggest that the signalling machinery achieves part of its high-order regulation by exploiting the compartmentalisation and functional specialisation of the endocytic pathway. Endocytosis thereby influences signalling far beyond its conventional role in removal of receptors from the cell surface (reviewed by González-Gaitán, 2003; Miaczynska et al., 2004a).

The most extensive studies on receptor-mediated endocytosis in animal cells have been carried out using the epidermal growth factor receptors (EGFRs) as an experimental model. Stimulation of cells with EGF results in rapid clustering of the EGF–receptor complexes in clathrin-coated pits and their translocation into clathrin-coated endocytic vesicles (Carpenter, 2000). The internalisation of EGFR can be effectively blocked by dominant negative mutants of proteins that are essential for clathrin-dependent endocytosis. The first hint that signalling can occur from endosomal compartments was made when activated EGFRs and their downstream signalling components, such as growth-factor-receptor-bound protein 2 (Gbr2), SH-domain-containing transforming protein (Shc) and the Ras guanine–nucleotide exchange factor, son-of-sevenless (SOS), were all found to reside in the early endosomes. Using fluorescence resonance energy transfer (FRET) microscopy it was shown that EGFR interacts in endosomes with Grb2, an initiator of Ras and mitogen-activated protein kinase (MAPK) signalling, demonstrating that signal transduction can continue after endocytosis (Sorkin et al., 2000; reviewed by Sorkin and Zastrow, 2002).

Retention of activated EGFR on the plasma membrane as a result of impaired clathrin-mediated endocytosis via expression of dominant negative mutants of dynamin (Dynamin^{K44A}), a cytoplasmic GTPase that is necessary for the fusion of coated vesicles from the plasma membrane, led to reduced activity of some of the downstream signalling components, such as the MAPK or the phosphatidylinositol 3-kinase (PI3-K) (Miaczynska et al., 2004a). These data suggest that some of the downstream signalling cascades are preferentially activated after internalisation of receptor–ligand complexes into endosomal compartments. The functional significance of EGFR endocytic signalling is that it is believed to allow spatial and temporal separation of signal transduction pathways commencing at the same receptor (Miaczynska et al., 2004a; Holler et al., 2005). This idea has recently been enforced by the findings that in addition to the MAPK signalling endosomes, EGFR also resides in a newly discovered population of early endosomes that carry the Rab5-effector, APPL1 (adaptor protein containing pleckstrin homology (PH) domain, phosphotyrosine-binding (PTB) domain and leucine zipper motif 1). APLL1 undergoes EGF-dependent nuclear shuttling that is required for cell proliferation (Miaczynska et al., 2004b). Another example of endocytic signalling via differential recruitment of activated receptors and numerous adaptor proteins is the transforming growth factor-β (TGF-β) receptor (Di Gullielmo et al., 2003). Recent studies have shown that ligand-induced recruitment to either clathrin-coated compartments or caveolae specifically modulates the signalling output by TGF-β receptors. Activated TGF-β receptor phosphorylates intracellular Smad effectors that translocate to the nucleus and regulate gene transcription. Signal propagation via this pathway depends on the adaptor protein SARA (Smad anchor for receptor activation). On the other hand, signal downregulation via receptor degradation is mediated by the ubiquitin ligase complex, Smad7-Smurf. Upon ligand application, SARA and Smad7-Smurf show mutually exclusive localisation with SARA-linked complexes present in early, autoantigen 1 (EEA1)-positive endosomes originating from clathrin-coated-pits, whereas Smad7-Smurf complex colocalises with caveolin-1, which resides in the caveolae (Di Gullielmo et al., 2003). This partitioning between caveolae and clathrin-associated receptors affects receptor turnover as well as cellular responses.

2
Receptor-Mediated Endocytosis in Plants

The first evidence for the existence of receptor-mediated endocytosis in plants has come from the demonstration of saturable and temperature-dependent uptake of a labelled elicitor fraction into soybean cell suspension (Horn et al., 1989). However, at that time, neither the molecular identity of the receptor nor the exact nature of the elicitor molecule was known. Based on structural characteristics more than 600 proteins in the *Arabidopsis* genome have been assigned to the group of transmembrane receptors. Similar to animal receptors, plant receptor-like kinases (RLKs) have a transmembrane domain and a protein kinase domain located in their cytoplasmic tails. Plant RLKs all have serine/threonine-specific sequence signatures and are divided into 15 subgroups mainly based on the diversity of their extracellular domains (Shiu and Bleecker, 2003). So far, the function and the corresponding ligands for only a few RLKs have been identified. Receptors for the small peptide ligands systemin (Scheer and Rayan, 2002), flagellin (Gomez-Gomez et al., 2001), phytosulfokine (Matsubayashi et al., 2002), clavata 3 (CLV3) (Trotochaud et al., 1999) and the plant steroid hormones called brassinosteroids (Kinoshita et al., 2005) all belong to the leucine-rich repeat (LRR) type of receptor kinases and are involved in defence signalling, cellular differentiation, meristem patterning and plant growth, respectively. The small S-locus cysteine-rich protein (SCR) binds to S-domain type receptor kinase (SRK) and is involved in the *Brassica* self-incompatibility response (Takayama et al., 2001). So far, neither the exact mechanism of receptor activation nor the cellular dynamics of receptor kinase-mediated signalling in any of these pathways is well described (Matsubayashi, 2003; Tichtinsky et al., 2003; Haffani et al., 2004).

Receptor kinase activation in animal cells generally occurs upon binding of a ligand. This stabilises receptor oligomerisation, and thus causes changes in phosphorylation status due to either autophosphorylation or transphosphorylation between the kinase domains within an oligomeric complex. There is considerable evidence that, as seen with animal receptors, many plant receptor kinases undergo autophosphorylation in response to a ligand binding (Takayama et al., 2001; Cabrillac et al., 2001; Wang et al., 2001). There is also evidence that, as in animal cells, oligomerisation of receptors is common and that transphosphorylation within receptor complexes can occur (Shah et al., 2001; Li et al., 2002; Nam and Li, 2002; Russinova et al., 2004). Activated animal receptors are removed from the cell surface by accelerated internalisation towards lysosomes resulting in attenuation of signalling response. For example, inactivation of the EGFR and the platelet-derived growth factor receptor-β (PDGFR) requires the activity of specific protein–tyrosine phosphatases (PTPs). Dephosphorylation of these receptors by their PTPs is only possible after their internalisation by endocytosis and occurs at specific sites called 'dephosphorylation compartments' before targeting to the lysosome or recycling to the plasma membrane (Haj et al., 2002). Thus, during endocytosis a part of the internalised receptors remains dimerised and thereby potentially maintains kinase activity, resulting in protein phosphorylation during their transit to the endosomes. Tyrosine phosphorylation and kinase activity of internalised receptors was shown for EGFRs and insulin receptors and also for other receptor tyrosine kinases (Sorkin and Zastrow, 2002). The extent of receptor phosphorylation after internalisation is determined by the balance between the activities of the receptor kinases and phosphotyrosine phosphatases present either in the endosomes or residing in the internal membranes such as endoplasmic reticulum (ER) (Haj et al., 2002).

Recently, it was demonstrated for the first time in plants that phosphorylation also influences receptor trafficking, suggesting that parallel mechanisms for receptor regulation in plants have evolved (Shah et al., 2002). It was shown that localisation of the LRR receptor-like kinase, *Arabidopsis thaliana* somatic embryogenesis receptor kinase 1 (AtSERK1), is influenced by its ability to interact and to be dephosphorylated by a PP2C-type kinase-associated protein phosphatase (KAPP). It was found that in plant protoplasts, the AtSERK1 localises to the plasma membrane but interacts with KAPP in a phosphorylation-dependent manner only when internalised in FM4-64-positive intracellular vesicles. The FM4-64 styryl dye was used as an endocytic tracer in different cell systems including plants (Ueda et al., 2001; Bolte et al., 2004; Meckel et al., 2004; see the chapter by Šamaj, this volume). When AtSERK1 was mutated at receptor autophosphorylation sites, the mutated receptors were partially sequestered intracellularly in the absence of KAPP. This suggests a role in receptor internalisation for KAPP dephosphorylation of the threonine residues in the AtSERK1 A-loop (Shah et al., 2002). KAPP was found to interact with many plant RLKs including clavata 1 (CLV1), wall-associated receptor kinase 1 (WAK1), SRK and flagellin-sensitive 2 (FLS2), and has been shown to negatively regulate at least a number of these receptors, possibly by dephosphorylation of the active form of the kinases (Braun et al., 1997; Williams et al., 1997; Stone et al., 1998; Gómez-Gómez et al., 2001; Park et al., 2001; Vanoosthuyse et al., 2003). Thus, it appears that dephosphorylation by KAPP and subsequent internalisation could be a universal mechanism of downregulation for plant RLKs.

Additional evidence that plant RLKs are internalised and recycled comes from studies on heterologous cellular systems carried out for the LRR type of RLKs, brassinosteroid-insensitive 1 (BRI1) and AtSERK3, also called BRI1-associated receptor kinase 1 (BAK1) receptors (Russinova et al., 2004). Genetic and molecular data support the notion that BRI1 and AtSERK3 receptors are part of the same brassinosteroid receptor complex in *Arabidopsis* and they initiate brassinosteroid signalling (Li and Chory, 1997; Li et al., 2002; Nam and Li, 2002). BRI1 directly binds the ligand, brassinosteroid hormones (Kinoshita et al., 2005), and heterodimerises with AtSERK3 to transmit the signal downstream (Russinova et al., 2004). How the BRI1-AtSERK3 complex is activated and how it transmits the brassinosteroid (BR) signal is not exactly known, although recent models suggest similarities with either animal tyrosine kinases or transforming growth factor-β (TGF-β) cell-surface receptor activation (Clouse, 2002; Li, 2003).

When transiently expressed in cowpea or in *Arabidopsis* protoplasts, BRI1 and AtSERK3 fused to different green fluorescent protein (GFP) variants localise to the plasma membrane and to small intracellular vesicles. These vesicles were identified as endosomal compartments based on colocalisation experiments with the endocytic tracer FM4-64 (Russinova et al., 2004) and with the *Arabidopsis* Rab5 homologues ARA6 and ARA7 (Russinova E, de Vries S, unpublished results) (Fig. 1b). When the protein synthesis inhibitor cycloheximide (CHX) was applied to protoplasts that expressed a single fluorescently tagged receptor, no changes in the distribution of the fluorescence either in the membrane or in the vesicles were observed, suggesting that the receptors are internalised and recycled back to the plasma membrane. Receptor recycling has been observed for all plant RLKs transiently expressed in the protoplast system, such as AtSERK1, AtSERK2, CLV1 and the BRI1-like proteins, BRL1 and BRL3 (Russinova E, van Breukelen F, de Vries S, unpublished results).

Until now, there has been no evidence that receptor internalisation is a regulatory mechanism for signalling and that ligand binding and receptor complex activation is required in plants. When coexpressed in the presence of CHX in cowpea cells, the endocytosis of heterodimeric BRI1 and AtSERK3 was accelerated compared to the endocytosis of BRI1 homodimers. Fluorescence lifetime imaging microscopy (FLIM) was used to demonstrate that both receptors favour a heterodimeric interaction in the endosomal compartments rather than at the plasma membrane, suggesting a so far unknown mechanism for constitutive redistribution of the receptors within the cell. Internalisation of BRI1 and AtSERK3 in cowpea or in *Arabidopsis* protoplasts was not directly linked to the ligand binding, although elements of the activated BR signalling pathway such as the presence of the unphosphorylated form of the downstream component, *bri*1-EMS-suppressor 1 (BES1) (Yin et al., 2002), have been demonstrated in both cellular systems. Internalisation and recycling of the BRI1 protein was also observed in intact *Arabidopsis* tissues (Russinova et al., 2004). In the root meristem, BRI1 fused to GFP was detected at the plasma membrane and in the vesicle compartments stained with FM4-64. The distribution of the GFP-tagged receptor was not affected by the application of CHX, suggesting that the BRI1-GFP protein is also recycled. BRI1-GFP internalisation *in planta* was not influenced by exogenous application of the ligand (brassinosteroids) or the absence of the AtSERK3 receptor partner when BRI1-GFP fluorescence was examined in brassinosteroid-deficient or *serk3* mutant backgrounds (Russinova et al., 2004). This suggests that at least a part of BRI1 receptors is undergoing a ligand-independent internalisation and recycling similar to what has been described for other plasma membrane proteins.

So far, a few potential components of pathways involved in the recycling of plasma membrane proteins have been identified in plants using brefeldin A (BFA), a fungal inhibitor of vesicular trafficking (Šamaj et al., 2004). These include the ADP ribosylation factor ARF1, the BFA-sensitive ARF-GDP/GTP exchange factor called GNOM (also known as EMB30), as well as members of the Rab family of small GTPases, ARA6 and ARA7 (Baluška et al., 2002; Geldner, 2004; Ueda et al., 2004; Xu and Scheres, 2005). Although a direct link between these proteins and either BRI1 internalisation or BRI1 receptor-kinase function has not yet been provided, it was shown that recycling of BRI1 in *Arabidopsis* roots is BFA sensitive (Russinova et al., 2004). Similar BFA-sensitive recycling in roots was described previously for the auxin efflux facilitator PIN-FORMED1 (PIN1), a plasma membrane (PM) ATPase (Geldner et al., 2001; Baluška et al., 2002), PIN2 (Geldner et al., 2003), PIN3 (Friml et al., 2002), cell wall pectins (Baluška et al., 2002) and sterols (Grebe et al., 2003). In all cases, BFA treatment induced the formation of BFA compartments most likely derived from endosomes and trans-Golgi network (Šamaj et al., 2004). The BFA sensitivity of the endomembrane system is determined by the presence and the abundance of resistant or sensitive ARF-GEFs in respective sites of BFA action. It was shown that PIN1 recycling specifically requires GNOM for internalisation. GNOM itself, however, is not responsible for the internalisation of PIN2 or PM-ATPase, suggesting an involvement of different ARF-GEF complexes or other molecular players in different recycling pathways (Geldner et al., 2003; Jürgens, 2004). Which ARF-GEF is responsible for the BRI1 internalisation is not known. The recycling of several other receptor kinases *in planta*, including AtSERK1 (Kwaaitaal et al., 2005) and the *Arabidopsis* CRINKLY4 (ACR4) RLK (Gifford et al., 2003), was shown to be BFA sensitive. Interestingly, only the extracellular crinkly repeat domain in the ACR4 RLK was absolutely required for internalisation of the protein. Mutated loss of function ACR4 proteins, containing either deletions or point mutations in this domain, were not internalised to the same extent as the wild-type receptor and appeared to be stabilised at the plasma membrane. Internalisation of ACR4 was not dependent on kinase activity of the receptor and no interaction with KAPP was demonstrated. In the case of ACR4, a rapid turnover rather than recycling followed internalisation of the receptors (Gifford et al., 2005). The subsequent fate of the internalised receptors in plants remains unknown. In animals, recycling appears to be a default mechanism and internalised receptors are delivered from the sorting endosomes via recycling endosomes back to the plasma membrane. Receptor proteins such as EGFR are retained in the sorting endosomes through an interaction with a protein called sorting nexin 1 (SNX1), which limits their recycling and possibly targets them for degradation (Kurten et al., 2001; Merino-Trigo et al., 2004). EGFR degradation requires receptor ubiquitination and subsequent lysosomal and proteosomal degradation (Levkowitz et al., 1998; Katzmann et al., 2002). Interestingly, several plant RLKs including SRK and CLV1 have been shown to interact with the *Brassica* homologue of SNX1 in a phosphorylation-dependent manner, suggesting trafficking of those receptors through the endosomal system (Vanoosthuyse et al., 2003). The fact that receptor ubiquitination might be a common mechanism of degradation of plant RLKs is suggested by the isolation of the E3 ubiquitin ligase, arm repeat containing 1 (ARC1), which interacts with the kinase domain of the SRK protein (Gu et al., 1998). ARC1 contains a U domain structurally related to the RING finger domain, raising the possibility that ARC1 and ARC1-related proteins could be involved in the ubiquitination of plant RLKs (Stone et al., 2003). Another plant RLK, the chitinase-related receptor-like kinase 1 (CHRK1), has been shown to interact with a similar U-domain-containing protein in tobacco (Kim et al., 2003).

Aside from a link with KAPP and potentially with ubiquitination machinery, very little is known about the molecular mechanisms regulating internalisation of plant RLKs. The internalisation of plant RLKs requires their connection to the endocytic machinery. Clathrin-mediated endocytosis has been demonstrated in plants and plant homologues to mammalian proteins of the clathrin coat (clathrin, AP-2 complex), and those that perform basic functions in clathrin-coated vesicle budding (AP180, epsin), scission (dynamin) and uncoating (auxillin) events have been identified (Holstein, 2002; see also the chapter by Holstein, this volume). However, direct evidence that plant receptor internalisation requires clathrin-mediated endocytosis is lacking. In animals, the presence of internalisation motifs within the cytoplasmic tails of the receptors that can be recognised by the μ2-adaptin subunit of the adaptor 2 (AP-2) complex or directly by clathrin itself mediated their internalisation (Kurten, 2003). Three distinct sorting signals for selection into clathrin-coated vesicles have been identified in the cytoplasmic tails of certain transmembrane proteins: NPXY, YXXØ (where Ø is a bulky hydrophobic residue) and dileucine (Bonifacino and Traub, 2003). Of these the best characterised is the YXXØ signal, and this sequence binds to the μ subunits of AP complexes in animals and in plants (Robinson and Bonifacino, 2001; Happel et al., 2004). The importance of these motifs for receptor-mediated endocytosis in plants was suggested by reports that show typical acidic or tyrosine-based sorting motifs to be present in cell surface-like receptors encoded by the tomato *Ve* disease-resistant genes (Kawchuk et al., 2001). In addition, some members of the LRR RLKs, such as BRI1, also contain an YXXØ motif within their cytoplasmic tail that binds to the receptor-binding domain of the *Arabidopsis* μA-adaptin (Holstein, 2002; Murphy et al., 2005).

3 Conclusions and Future Prospects

Recent experimental evidence suggests that plant plasma membrane receptors are internalised and recycled via endocytosis. This evidence also suggests that some of the mechanisms underlying endocytic recycling are conserved between plants and animals (see Fig. 1b). However, ligand-dependent receptor-mediated endocytosis as described for animal receptors has not yet been demonstrated for any of the known plant RLKs. It is also not clear how different developmental programmes are regulated by endocytosis of plant RLKs and if endocytosis is functionally as important for signalling as described for animal receptors. There are several limitations that hold back this research. First, the studies on receptor-mediated endocytosis in plants have been hampered by the fact that only a few ligands have been recently identified. The dynamics of the receptor activation and trafficking in vivo is investigated for only a few plant RLKs including AtSERK1, BRI1 and ACR4 (Shah et al., 2002; Russinova et al., 2004; Gifford et al., 2005). So far, mostly genetic approaches have been used to study receptor signalling in plants, and in most cases downstream signalling components are unknown. Second, although rapidly coming of age, knowledge of plant endocytosis and how plant endosomes are organised and functionally differentiated still remains very limited. A significant contribution to the understanding of the endocytic pathway in plants to date has been made by the use of the endocytic Rab proteins, FM4-64 endocytic tracer and the BFA inhibitor. Plant endosomes are not well defined despite the fact that different endosomal compartments have been demonstrated using double labelling experiments and BFA treatments (Geldner et al., 2003; Russinova et al., 2004; Šamaj et al., 2004; Ueda et al., 2004). Furthermore, it still not known whether the internalisation of the plant RLKs is mediated via endocytosis that is dependent on clathrin or lipid rafts. In animals, most of the receptor research is performed on in vitro cultured cell lines that can be useful for generating a ligand-free environment in order to study the induction of the receptor-mediated endocytosis.

In plants, the use of cellular systems, including the transient expression systems in leaf protoplasts, has not yet been optimised to represent ligand-free environments. As in animal cell systems, a major concern is that isolated cell systems may not reflect the physiological situation in the intact organism. This can, in principle, be overcome much more easily in plants by studying the receptor trafficking in the intact organism using advanced fluorescence imaging microspectroscopy techniques.

The results obtained so far support the view that plant RLKs undergo internalisation via endocytosis in living cells. To understand whether this is part of a general plasma membrane protein recycling mechanism or whether it is functionally related to the receptor signalling will be a challenge for the coming years.

References

Baluška F, Hlavacka A, Šamaj J, Palme K, Robinson DG, Matoh T, McCurdy DW, Menzel D, Volkmann D (2002) F-actin-dependent endocytosis of cell wall pectins in meristematic root cells. Insights from brefeldin A-induced compartments. Plant Physiol 130:422–431

Bolte S, Talbot C, Butte Y, Catrice O, Read ND, Satian-Jeunemaitre B (2004) FM dyes as experimental probes for dissecting vesicle trafficking in living plant cells. J Microsc 214:159–173

Bonifacino JS, Traub LM (2003) Signals for sorting of transmembrane proteins to endosomes and lysosomes. Annu Rev Biochem 72:395–447

Braun DM, Stone JM, Walker JC (1997) Interaction of the maize and *Arabidopsis* kinase interaction domain with a subset of receptor-like kinases: implications for transmembrane signalling in plants. Plant J 12:83–95

Cabrillac D, Cock MJ, Dumas C, Gaude T (2001) The S-locus receptor kinase is inhibited by thioredoxin and activated by pollen coat proteins. Nature 410:220–223

Carpenter G (2000) The EGF receptor: a nexus for trafficking and signaling. BioEssays 22:697–707

Clouse SD (2002) Brassinosteroid signal transduction: clarifying the pathway from ligand perception to gene expression. Mol Cell 10:973–982

Di Guglielmo GM, Le Roy C, Goodfellow AF, Wrana JL (2003) Distinct endocytic pathways regulate TGF-β receptor signalling and turnover. Nat Cell Biol 5:410–421

Friml J, Wiśniewska J, Benková E, Mendgen K, Palme K (2002) Lateral relocation of auxin efflux regulator PIN3 mediates tropism in *Arabidopsis*. Nature 415:805–809

Geldner N (2004) The plant endosomal system—its structure and role in signal transduction and plant development. Planta 219:547–560

Geldner N, Anders N, Wolters H, Keicher J, Kornberger W, Muller P, Delbarre A, Ueda T, Nakano A, Jürgens G (2003) The *Arabidopsis* GNOM AFR-GEF mediates endosomal recycling, auxin transport, and auxin-dependent plant growth. Cell 112:219–230

Geldner N, Friml J, Stierhof Y-D, Jurgens G, Palme K (2001) Auxin transport inhibitors block PIN1 cycling and vesicle trafficking. Nature 413:425–428

Gifford ML, Dean S, Ingram GC (2003) The *Arabidopsis* ACR4 gene plays a role in cell layer organisation during the ovule integument and sepal margin development. Development 130:4249–4258

Gifford ML, Robertson F, Soares DC, Ingram GC (2005) ACR4 function, internalisation and turnover are dependent on the extracellular crinkly repeat domain. Plant Cell 17:1154–1166

Gómez-Gómez L, Bauer Z, Boller T (2001) Both the extracellular leucine-rich repeat domain and the kinase activity of FLS2 are required for flagellin binding and signalling in *Arabidopsis*. Plant Cell 13:1155–1163

Gonzáles-Gaitán M (2003) Signal dispersal and transduction through the endocytic pathway. Nat Rev Mol Cell Biol 4:213–224

Grebe M, Xu J, Möbius W, Ueda T, Nakano A, Geuze HJ, Rook MB, Scheres B (2003) *Arabidopsis* sterol endocytosis involves actin-mediated trafficking via ARA6-positive early endosomes. Curr Biol 13:1378–1387

Gu T, Mazzurco M, Sulaman W, Matias DD, Goring DR (1998) Bindings of a novel arm repeat protein to the kinase domain of the S-locus receptor kinase. Proc Natl Acad Sci USA 97:3759–3764

Haffani YZ, Silva NF, Goring DR (2004) Receptor kinase signalling in plants. Can J Bot 82:1–15

Haj FG, Verveer PJ, Squire A, Neel BG, Bastiaens PIH (2002) Imaging sites of receptor dephosphorylation by PTP1B on the surface of the endoplasmic reticulum. Science 295:1708–1711

Happel N, Höning S, Neuhaus J-M, Paris N, Robinson DG, Holstein SEH (2004) *Arabidopsis* μA-adaptin interacts with the tyrosine motif of the vacuolar sorting receptor VSR-PS1. Plant J 37:678–693

Holler D, Volarevic S, Dikic I (2005) Compartmentalization of growth factor receptor signalling. Curr Opin Cell Biol 17:1–5

Holstein SEH (2002) Clathrin and plant endocytosis. Traffic 3:614–620

Holstein SEH (2005) Molecular dissection of the clathrin endocytosis machinery in plants (in this volume). Springer, Berlin Heidelberg New York

Horn MA, Heinstein PF, Low PS (1989) Receptor-mediated endocytosis in plant cells. Plant Cell 1:1003–1009

Jürgens G (2004) Membrane trafficking in plants. Annu Rev Cell Dev Biol 20:481–504

Katzmann DJ, Odorizzi G, Emr SD (2002) Receptor downregulation and multivesicular-body sorting. Nat Rev Mol Cell Biol 3:893–905

Kawchuk LM, Hachey J, Lynch DR, Kulcsar F, van Rooijen G, Waterer DR, Robertson A, Kokko E, Byers R, Howard RJ, Fischer R, Prüfer D (2001) Tomato *Ve* disease resistance genes encode cell surface-like receptors. Proc Natl Acad Sci USA 98:6511–6515

Kim M, Cho H, Kim DO-M, Lee J, Pai H-S (2003) CHRK1, a chitinase-related receptor-like kinase, interacts with NtPUB4, an armadillo repeat protein in tobacco. Biochem Biophys Acta 1651:50–59

Kinoshita T, Cano-Delgado A, Seto H, Hiranuma S, Fujioka S, Yoshida S, Chory J (2005) Binding of brassinosteroids to the extracellular domain of plant receptor kinase BRI1. Nature 433:167–171

Kurten RC (2003) Sorting motifs in receptor trafficking. Adv Drug Del Rev 55:1405–1419

Kurten RC, Eddington AD, Chowdhury P, Smith RD, Davidson AD, Shank B (2001) Self-assembly and binding of a sorting nexin to sorting endosomes. J Cell Sci 114:1743–1756

Kwaaitaal MACJ, de Vries SC, Russinova E (2005) The *Arabidopsis* somatic embryogenesis receptor kinase 1 protein is present in sporophytic and gametophytic cells and undergoes endocytosis. Protoplasma 221:394–405

Levkowitz G, Waterman H, Zamir E, Kam Z, Oved S, Langdom WY, Beguinot L, Geiger B, Yarden Y (1998) C-Cbl/Sli-1 regulates endocytic sorting and ubiquitination of the epidermal growth factor receptor. Gen Dev 12:3663–3674

Li J (2003) Brassinosteroids signal through two receptor-like kinases. Curr Opin Plant Biol 6:494–499

Li J, Chory J (1997) A putative leucine-rich repeat receptor kinase involved in brassinosteroid signal transduction. Cell 90:929–938

Li J, Wen J, Lease KA, Doke JT, Tax FE, Walker JC (2002) BAK1, an *Arabidopsis* LRR receptor-like protein kinase, interacts with BRI1 and modulates brassinosteroid signaling. Cell 110:213–222

Matsubayashi Y (2003) Ligand-receptor pairs in plant peptide signalling. J Cell Sci 116:3863–3870

Matsubayashi Y, Ogawa M, Morita A, Sakagami Y (2002) An LRR receptor kinase involved in perception of a peptide plant hormone, phytosulfokine. Science 296:1470–1472

Meckel T, Hurst AC, Thiel G, Homann U (2004) Endocytosis against high turgor: intact guard cells of *Vicia faba* constitutively endocytose fluorescently labelled plasma membrane and GFP-tagged K^+-channel KAT1. Plant J 39:182–193

Merino-Trigo A, Kerr MC, Houghton F, Linberg A, Mitchell C, Teasdale RD, Gleeson P (2004) Sortin nexin 5 is localized to a subdomain of the early endosomes and is recruited to the plasma membrane following EGF stimulation. J Cell Biol 117:6413–6424

Miaczynska M, Christoforidis S, Giner A, Shevchenko A, Uttenweiler-Joseph S, Habermann B, Wilm M, Patron RG, Zerial M (2004) APPL proteins link Rab5 to nuclear signal transduction via an endosomal compartment. Cell 116:445–456

Miaczynska M, Pelkmans L, Zerial M (2004) Not just a sink: endosomes in control of signal transduction. Curr Opin Cell Biol 16:400–406

Murphy AS, Bandyopadhyay A, Holstein SE, Peer WA (2005) Endocytotic cycling of PM proteins. Annu Rev Plant Biol 56:221–251

Nam KH, Li J (2002) BRI1/BAK1, a receptor kinase pair mediating brassinosteroid signaling. Cell 110:203–212

Park A, Cho SK, Yun U, Jin M, Lee S, Sachetto-Martins G, Park OK (2001) Interaction of the *Arabidopsis* receptor protein kinase Wak1 with glycine-rich protein AtGRP-3. J Biol Chem 276:26688–26693

Robinson MS, Bonifacino JS (2001) Adaptor-related proteins. Curr Opin Cell Biol 13:444–453

Russinova E, Borst J-W, Kwaitaal M, Caño-Delgado A, Yin Y, Chory J, de Vries SC (2004) Heterodimerization and endocytosis of *Arabidopsis* brassinosteroid receptors BRI1 and AtSERK3 (BAK1). Plant Cell 16:3216–3229

Šamaj J (2005) Methods and molecular tools to study endocytosis in plants—an overview (in this volume). Springer, Berlin Heidelberg New York

Šamaj J, Baluška F, Voigt B, Schlicht M, Volkmann D, Menzel D (2004) Endocytosis, actin cytoskeleton and signaling. Plant Physiol 135:1150–1161

Scheer JM, Ryan CA Jr (2002) The systemin receptor SR160 from *Lycopersicon peruvianum* is a member of the LRR receptor kinase family. Proc Natl Acad Sci USA 99:9585–9590

Shah K, Gadella TWJ, van Erp H, Hecht V, de Vries SC (2001) Subcellular localization and dimerization of the *A. thaliana* somatic embryogenesis receptor kinase 1 protein. J Mol Biol 309:641–655

Shah K, Russinova E, Gadella TWJ Jr, Willemse J, de Vries SC (2002) The *Arabidopsis* kinase-associated protein phosphatase controls internalization of the somatic embryogenesis receptor kinase 1. Genes Dev 16:1707–1720

Shiu S-H, Bleecker AB (2003) Expansion of the receptor-like kinase/Pelle gene family and receptor-like proteins in *Arabidopsis*. Plant Physiol 132:530–543

Sorkin A, McClure M, Huang F, Carter R (2000) Interaction of EGF receptor and grb2 in living cells visualized by fluorescence resonance energy transfer (FRET) microscopy. Curr Biol 10:1395–1398

Sorkin A, Von Zastrow M (2002) Signal transduction and endocytosis: close encounters of many kinds. Nat Rev Mol Cell Biol 3:600–614

Stone JM, Trotochaud AE, Walker JC, Clark SE (1998) Control of meristem development by CLAVATA1 receptor kinase and kinase-associated protein phosphatase interactions. Plant Physiol 117:1217–1225

Stone SL, Anderson EM, Mullen R, Goring DR (2003) ARC1 is an E3 ubiquitin ligase and promotes the ubiquitination of proteins during the rejection of self-incompatible *Brassica* pollen. Plant Cell 15:885–898

Takayama S, Shimosato H, Shiba H, Funato M, Che F-S, Watanabe M, Iwano M, Isogal A (2001) Direct ligand-receptor complex interaction controls *Brassica* self-incompatibility. Nature 413:534–538

Teis D, Huber LA (2003) The odd couple: signal transduction and endocytosis. Cell Mol Life Sci 60:2020–2033

Tichtinsky G, Vanoosthuyse V, Cock MJ, Gaude T (2003) Making inroads into plant receptor kinase signalling pathways. Trends Plant Sci 8:231–237

Trotochaud AE, Hao T, Wu G, Yang Z, Clark SE (1999) The CLAVATA1 receptor-like kinase requires CLAVATA3 for its assembly into a signaling complex that includes KAPP and a Rho-related protein. Plant Cell 11:393–406

Ueda T, Yamaguchi M, Uchimiya H, Nakano A (2001) Ara6, a plant-unique novel type Rab GTPase, functions in the endocytic pathway of *Arabidopsis thaliana*. EMBO J 20:4730–4741

Ueda T, Uemura T, Sato MH, Nakano A (2004) Functional differentiation of endosomes in *Arabidopsis* cells. Plant J 40:783–789

Vanoosthuyse V, Tichtinsky G, Dumas C, Gaude T, Cock MJ (2003) Interaction of calmodulin, a sorting nexin and kinase-associated protein phosphatase with the *Brassica oleracea* S locus receptor kinase. Plant Physiol 133:919–929

Wang Z-Y, Seto H, Fujioka S, Yoshida S, Chory J (2001) BRI1 is a critical component of a plasma-membrane receptor for plant steroids. Nature 410:380–383

Williams RW, Wilson JM, Meyerowitz EM (1997) A possible role for kinase-associated protein phosphatase in the *Arabidopsis* CLAVATA1 signaling pathway. Proc Natl Acad Sci USA 94:10467–10472

Xu J, Scheres B (2005) Dissection of *Arabidopsis* ADP-Ribosylation Factor 1 function in epidermal cell polarity. Plant Cell 17:525–536

Yin Y, Wang Z-Y, Mora-Garcia S, Li J, Yoshida S, Asami T, Chory J (2002) BES1 accumulates in the nucleus in response to brassinosteroids to regulate gene expression and promote stem elongation. Cell 109:181–191

Plant Cell Monogr (1)
J. Šamaj · F. Baluška · D. Menzel: Plant Endocytosis
DOI 10.1007/7089_008/Published online: 27 September 2005

Sterol Endocytosis and Trafficking in Plant Cells

Miroslav Ovečka[1,2] (✉) · Irene K. Lichtscheidl[2]

[1]Institute of Botany, Slovak Academy of Sciences, Dubravska cesta 14, 84523 Bratislava, Slovakia
miroslav.ovecka@savba.sk

[2]Institution of Cell Imaging and Ultrastructure Research, University of Vienna, Althanstrasse 14, 1090 Vienna, Austria
miroslav.ovecka@savba.sk, irene@pflaphy.pph.univie.ac.at

Abstract Structural sterols are integral components of biological membranes. They regulate membrane permeability and fluidity, and they influence the activity of membrane proteins. In *Arabidopsis*, their composition is critical for normal plant development. The endocytosis and recycling of plasma membrane sterols display similar pathways as some polarly distributed proteins, and thus sterol-dependent trafficking can be an integral part of the polarity establishment in plants. Here, we summarise recent data about sterol endocytosis and sterol trafficking within endocytic pathways in different aspects of cell development in plants.

1 Introduction

Biological membranes are dynamic supramolecular and multicomponent structures that form boundaries of the cell and of intracellular compartments. They act as barriers, and in addition they play an active role in functional processes like signal transduction and intracellular trafficking.

Lipids belong to structural components of membranes and at the same time serve as a solvent for membrane proteins. The composition of lipids and their arrangement in the lipid bilayer therefore decide the physical properties of the membrane as well as the functionality of membrane proteins. Additionally, membrane-associated functions like the recruitment and assembly of membrane-bound multicomponent complexes of cytosolic proteins depend on the structure and distribution of membrane lipids (reviewed in van Meer and Sprong, 2004).

The behaviour of the membrane as a supramolecular continuum is maintained during membrane fusion, fission and all kinds of trafficking events mediated by vesicles including endocytosis. Intracellular membrane flow through vesicular trafficking redistributes not only a plethora of different cargoes, but also regulates the exchange and renovation of the membranes of different compartments within the cell. By endocytosis, an energy-dependent process conserved in all eukaryotic cells, substances of different nature become internalised through the cell surface. Thus, both intracellular transport via exo- and endocytosis and the maintenance of the identity and functional diversity of distinct organellar membranes in the cell are highly regulated.

Structural sterols are essential components of the plant plasma membrane supporting not only the structure of the membrane, but also influencing considerably the physical and physiological properties. Genetic studies of the sterol biosynthetic pathway in plants revealed, in addition, that they regulate and modulate different aspects of plant development (Clouse, 2002). Proper understanding of membrane dynamics requires a basic clarification of how different membranes are specialised and how plasma membrane-resident processes are regulated. Thus, an understanding of endocytosis and recycling of sterols as well as sterol-mediated internalisation of different cargoes is an important issue in the study of the membrane physiology of plant cells. In this chapter, we therefore address the role of sterols in dynamic membrane trafficking processes, and we summarise data about sterols and their structural and physiological functions in plant cell membranes.

2 General Concept of Endocytosis: Cooperation of Lipids and Lipid-Associated Proteins

Membrane lipids play a pivotal role in most aspects of plant life such as growth, cell differentiation and response to the environment. They are involved in signal transduction, membrane trafficking and organisation of the cytoskeleton. The composition and distribution of phospholipids and of sterols in the plasma membrane are also significant for endocytosis.

2.1 Phospholipids Promote Curvature of Membranes

Curvature of the membrane is a first step and an important physical prerequisite for vesicle formation. The structure, size and chemical modification of phospholipids can induce tension on the membrane. The resulting inward or outward curvature depends on the accumulation and combination of lipid species in certain membrane domains (Kooijman et al., 2003). It is promoted by the asymmetric distribution of lipids in the bilayer (Farge et al., 1999), as was shown for instance for aminophospholipids like phosphatidylethanolamine and phosphatidylserine within the cytoplasmic leaflet of Golgi, endosomal and plasma membranes. These lipids contribute to the vesicle budding competence of diverse membranes (Pomorski et al., 2003). For establishing asymmetry, special aminophospholipid translocators like flippases translocate phospholipids through the membrane. In yeast, transmembrane lipid translocation and subsequent proper vesicle budding requires the activity of Drs2p, a potential phospholipid translocase from the family of P-type ATPases. Yeast expresses five members of this protein family, and Drs2p is implicated in clathrin-dependent budding of vesicles from the *trans*-Golgi network (Chen et al., 1999, reviewed by Graham, 2004). Two other P-type ATPases closely related to Drs2p, namely Dnf1p and Dnf2p, were localised to the plasma membrane in yeast. Interestingly, their inactivation in *dnf1dnf2* mutant cells affected endocytosis; at lower temperatures (15 °C) these mutant cells displayed a significant delay in the internalisation of the endocytosis-specific dye FM4-64 (Pomorski et al., 2003). A related gene, *ALA1*, is expressed in the *Arabidopsis* genome, and encodes for a protein with aminophospholipid translocase activity (Gomès et al., 2000).

2.2 Sterols Form Microdomains in Membranes

Structural sterols are lipids of the isoprenoid family that are essential for all eukaryotic membranes, although their composition is different in mammalian, yeast and plant cells. In artificial membranes, cholesterol accumulates in sterol-rich domains and significantly determines the asymmetric properties of membranes. This phase separation together with the type of sterols controls the direction of the curvature. As a result, the membrane swells and vesicles pinch off (Bacia et al., 2005). However, the precise mechanisms of how sterols can control the budding and direction of the curvature remain to be resolved.

In biological membranes, similar phase separation of membrane lipids was observed. Sterols, mainly cholesterol and sphingolipids, accumulate in microdomains in the extracellular leaflet of yeast and mammalian plasma membranes that were described as detergent-resistant lipid rafts (Simons and Toomre, 2000). Plant sphingolipids are represented by ceramides and glycosylceramides from the group of phosphosphingolipids (Sperling et al., 2004). Ceramides are also important components of plant cells that are thought to be involved in the formation of membrane functional microdomains (Norberg and Liljenberg, 1991). Lipid rafts are suggested to be a basic structural principle for the recruitment and enrichment of proteins, for spatial regulation of protein–protein interactions within the membrane, for the compartmentalisation of cell signalling and for protein- and lipid-based sorting in vesicular trafficking pathways (reviewed by Simons and Ikonen, 1997). Endocytosis and sorting of cargo molecules can be monitored and/or controlled by the sterol composition of the budding vesicle.

2.3 Proteins Associate with Lipid Rafts

Lipid-anchored proteins are associated with the membrane via domains embedded in the lipid layer. One example of such an anchor is glycosylphosphatidylinositol (GPI). GPI-anchored proteins are enriched at the outer surface of the plasma membrane. Other membrane-associated proteins modified by acylation or prenylation are enriched at the inner cytoplasmic surface (van Meer and Sprong, 2004).

Studies in mammalian cells using membrane extraction by cold Triton X-100 have suggested that GPI-anchored proteins are non-uniformly distributed over the surface of cell membranes. They are enriched in cholesterol-stabilised domains and thus their sorting mechanism seems to be sensitive to cholesterol and sphingolipids (Pralle et al., 2000; Mayor and Maxfield, 1995; Hao et al., 2001). In plants, an abundance of GPI-anchored proteins was confirmed by genomic and proteomic analysis (Borner et al., 2003; Lalanne et al., 2004).

3
Recycling and Turnover of Cell Surface Components by Endocytosis

Lipid rafts of the plasma membrane are constantly turned over through endocytosis and membrane flow. Assembly of cholesterol-rich lipid rafts probably occurs as early as during membrane biogenesis in the endoplasmic reticulum from where they are further transported via Golgi to the plasma membrane, as was suggested in mammalian cells (Heino et al., 2000). These sterol domains also remain intact during redistribution of the plasma membrane by endocytosis. In animal cells, they are recycled through defined early endocytic compartments (Simonds and Toomre, 2000).

3.1
Evidence of Functional Sterol Recycling

The general concept of membrane recycling is that molecules internalised at the plasma membrane reach early endosomes, from where they are either sorted towards late endosomes and lysosomes for degradation or recycled back to the plasma membrane passing through recycling endosomes. The *trans*-Golgi network (TGN) could be part of this recycling, as was shown for yeast and mammalian cells, where early and late endosomes exchange cargo with the TGN (Lemmon and Traub, 2000). Similarly, GPI-anchored proteins enriched in lipid rafts of the plasma membrane can enter the TGN and Golgi during internalisation and recycling in mammals (Nichols et al., 2001).

Evidence for cholesterol trafficking from the plasma membrane comes from studies of fluorescent lipid analogues, exogenously inserted into the plasma membrane of mammalian cells. BODIPY-LacCer, a lipid fluorescent analogue of lactosylceramide from the group of glycosphingolipids, is endocytosed and targeted to late endosomes and lysosomes in fibroblasts derived from patients with multiple types of sphingolipid storage diseases (Puri et al.,

1999). These fibroblasts are defective in cholesterol homeostasis, leading to an accumulation of cholesterol and sphingolipids in late endosomes instead of Golgi apparatus, which is the target in normal situations. After cholesterol depletion from these cells, BODIPY-LacCer trafficking was shown to be directed back to the Golgi apparatus. Additional evidence comes from loading of cholesterol into normal fibroblasts, which induced mistargeting, so that BODIPY-LacCer was delivered again to late endosomes and lysosomes (Puri et al., 1999). These results clearly indicate that endocytosis is regulated by local arrangement of sterols within the plasma membrane.

4 Sterols in Mammalian Cells

4.1 Sterol Content

Cholesterol, a typical member of the sterol family, is an essential component of mammalian membranes. It is involved in different membrane functions such as membrane transport and receptor activity.

4.2 Localisation and Distribution of Sterols

Cholesterol is transported bi-directionally between the endoplasmic reticulum (ER) and Golgi through vesicles. In the *trans*-Golgi, cholesterol-enriched membranes are present in luminal membrane leaflets. It is thought to be the main source of lipid rafts in the plasma membrane. Thus, cholesterol in mammalian cells is enriched in the *trans*-Golgi and TGN, endosomes and the plasma membrane (van Meer and Sprong, 2004).

4.3 Endocytosis and Function of Sterols

Endocytic uptake of cholesterol and further trafficking can be monitored by fluorescent cholesterol analogues like dehydroergosterol, or cholesterol-binding antibiotic filipin having fluorescent properties. Microscopic imaging of fluorescence in living cells shows that a major cellular pool of cholesterol is present in the endocytic recycling compartments as well as in the TGN (Mukherjee et al., 1998). This was also confirmed in cultured human lymphoblastoid cells by electron microscopy using the biotinylated derivative of the cholesterol-binding toxin perfringolysin O (Möbius et al., 2002). In addition, cholesterol-enriched membrane microdomains were isolated from late endosomes (Fivaz et al., 2002) and multivesicular bodies (Möbius et al.,

2003), where only internal vesicles but not the limiting membrane contained cholesterol (Möbius et al., 2002). Endocytosed GPI-anchored proteins remain about three times longer in the endocytic recycling compartments compared to other transmembrane proteins. After cholesterol depletion, the GPI-anchored proteins can recycle at the same rate as other proteins (Mayor et al., 1998).

Another example of plasma membrane proteins that are localised in cholesterol-enriched lipid rafts but released from this compartment upon cholesterol depletion are epidermal growth factor receptors (EGFR). Moreover, cholesterol depletion also inhibits EGF internalisation and downregulation of the EGFR (Pike and Casey, 2002). These results clearly show that functional recycling of lipids and lipid-anchored proteins depends on the sterol content in the plasma membrane.

5 Sterols in Yeast

5.1 Sterol Content and Distribution

The sterol composition of the yeast endomembrane system is well characterised. Ergosterol, a typical yeast sterol, is highly accumulated in the plasma membrane whereas other membranes contain intermediates of the sterol biosynthetic pathway but only little ergosterol (Zinser et al., 1993). The role of sterols in yeast is comparable with their role in other organisms. They are involved in the modulation of membrane fluidity, permeability and activity of membrane-associated proteins (Lees and Bard, 2004).

5.2 Endocytosis, Trafficking and Function of Sterols

Fluorescently visualised sterols in the *Schizosaccharomyces pombe* were localised in the cleavage furrow during cytokinesis and in growing polar tips. After sterol inhibition by filipin, the actomyosin ring was abnormally positioned during cytokinesis (Wachtler et al., 2003).

During the formation of a mating projection, reorganisation of the plasma membrane and accumulation of lipid rafts result in the retention of certain proteins at the tip of the mating projection (Bagnat and Simons, 2002). This is a morphogenic prerequisite for polar development, as evidenced in sphingolipid and ergosterol biosynthetic mutants. They fail to target proteins to the tip, and thus they are defective in mating (Bagnat and Simons, 2002). In addition, sterols seem to be involved in the regulation of protein and lipid trafficking through early endosomes (Heese-Peck et al., 2002).

6
Sterols in Plants

6.1
Sterol Content

Plants contain different sterols in their free form as free sterols, and conjugated as steryl esters and steryl glucosides (Benveniste, 2005). Important for the function of membranes are free sterols with a free 3-β-hydroxyl group, such as sitosterol, stigmasterol and 24-methylcholesterol (Hartmann, 2004). Cholesterol is a minor component in plants, but a tissue-specific increase can be observed in special developmental stages of plant ontogeny, such as that in meristematic tissues during floral development (Hobbs et al., 1996) or in potato tubers (Arnqvist et al., 2003). In general, the production rate of sterols varies in different cell types. In proliferating and actively growing tissues like plant shoots and root meristems, sterol production is high as compared to that in mature tissues (Hartmann, 2004).

6.2
Localisation and Distribution of Sterols

Sterol-rich domains contain phytosterols, which are considered as structural and functional key elements of membranes. The concept of lipid rafts tends to be accepted in plants (Mongrand et al., 2004; Borner et al., 2005). Sterols in plant membrane domains may also be responsible for retention of signalling molecules and selection of exo-endocytic cargo molecules.

Lipid rafts have been discovered in membranes of plant cells with the same methodological approach as in animal cells, using their resistance to non-ionic detergents. Important parameters are the number, distribution and arrangement of lipid-associated proteins. Glycosylphosphatidylinositol (GPI)-anchored proteins are preferentially enriched in lipid rafts in yeast and mammalian cells. Thus, characterisation of plant-specific GPI-anchored proteins represents an analogous approach, and major candidates from this class were identified in the genome of *Arabidopsis* (Borner et al., 2002, 2003). Putative GPI-anchored (Borner et al., 2002, 2003) and glycosylinositol phosphorylceramide (GIPC)-anchored (Thomson and Okuyama, 2000; Oxley and Bacic, 1999) proteins are present in plant cells.

Asymmetric distribution of sterol-rich microdomains was documented in tobacco cells using their enhanced detergent resistance (Peskan et al., 2000; Takos et al., 1997). GPI anchors of the beta subunit of membrane-bound heterotrimeric G protein are associated with these microdomains in the plasma membrane of tobacco leaves (Pescan et al., 2000). The question remains of whether sterol-rich microdomains are based on the same structural principle as in mammalian and yeast cells, and whether they anchor and arrange a similar species of proteins. In vitro studies of the membrane biogenesis using idealised membranes should clarify this point. For phytosterols, it was confirmed by Xu and co-workers (2001).

6.3 Function of Sterols

The main function of plant sterols is their structural role in the membrane. Sterols interact with phospholipid fatty acyl chains and proteins in the membrane and restrict the motion of fatty acyl chains in the bilayer by side-chain interactions. Thereby they affect the membrane fluidity and the permeability to water, as was documented in model membranes (Schuler et al., 1990, 1991). Complexation of sterols by cyclic polyene antibiotics like filipin makes membranes permeable to ions and small molecules (Kinsky, 1970), which is connected with morphological damage of membranes (Kinsky et al., 1967, Fig. 1). In addition, certain types of sterols can modulate the width of the membrane bilayer (Marsan et al., 1996, 1998). These properties of sterols are important for the efficient membrane flexibility required in vesicle budding and fusion. Moreover, sitosterol determines the integrity of the plasma membrane (Schuler et al., 1991) as well as the synthesis of cellulose (Peng, 2002). When administered to mammalian cells, it induces functional modulation of receptors coupled to G proteins at the plasma membrane (Gimpl et al., 1997). Stigmasterol modulates the activity of plasma membrane H^+-ATPase, and thus regulates efflux of protons (Grandmougin-Ferjani et al., 1997). PHABULOSA (PHB), a homeodomain protein containing a sterol/lipid-binding domain, is involved in the radial patterning of the shoot in *Arabidopsis*. The activity of PHB most probably depends on its binding to sterols at the membrane (McConnell et al., 2001).

7 Study of Sterol Function in Plants: Lessons from *Arabidopsis* Mutants

The study of sterols requires a functional characterisation of their biosynthetic pathways. In plants, most of the biosynthetic steps (Benveniste, 1986) and the genes responsible for the production of catalysing enzymes (Bach and Benveniste, 1997) have been described. They produce brassinosteroids (BR), steroid-like molecules with signalling function, and the structural sterols discussed above. Comprehensive data about plant sterol biosynthesis were summarised in specialised reviews (Benveniste, 2002; Benveniste, 2005; Hartmann, 2004; Schaller, 2004).

Mutations of the genes involved result in a change of the mixture of sterols, which in turn influences development and morphogenesis considerably. Concerning the biosynthesis of brassinosteroids, some mutants of brassinosteroid

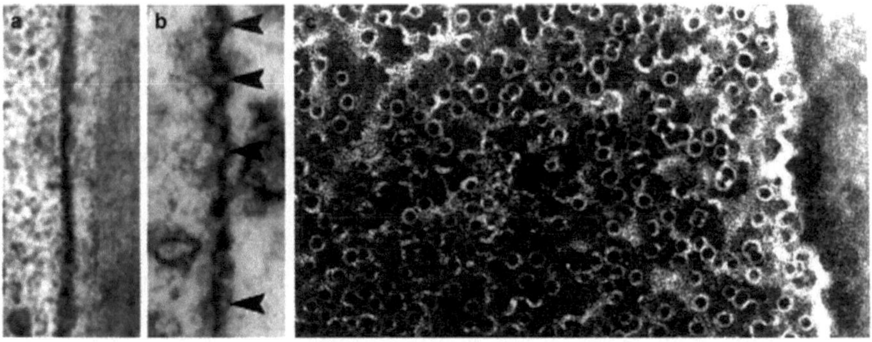

Fig. 1 Ultrastructure of the plasma membrane in cells treated by filipin, the antibiotic binding to 3-β-hydroxysterols. **a-b**, epidermal root cells of *Arabidopsis thaliana*. Complexation of phytosterols by filipin forms 20–30 nm protuberances at the plasma membrane (**b**)which are not present in control cells (**a**). Reprinted with permission from Grebe et al. (2003). Reprinted from Grebe et al. (2003) Curr Biol 13:1378–1387, with permission from Elsevier. **c**, Lesions of 20–25 nm in the plasma membrane of human erythrocyte cells induced by filipin. Plasma membrane is insensitive to filipin after cholesterol depletion (Verkleij et al. 1973), which indicates that lesions in the membrane are induced by formation of sterol–filipin complexes. Reprinted with permission from Kinsky et al. (1967). Reprinted from Kinsky et al. (1967) Biochim Biophys Acta 135:844–861, with permission from Elsevier. Note the similarity of the morphological disruption in the structure of membrane bilayer by filipin in different cell systems

precursors produce dwarf phenotypes due to defects in post-embryonic cell elongation. These defects can be rescued partly by external addition of brassinolide, the most biologically active BR in many plants. Other mutants have several embryonic and post-embryonic patterning defects, and exogenous brassinosteroids are unable to rescue them (reviewed by Clouse, 2002).

As concerns the synthesis of structural sterols, mutants with altered sterol levels and embryonic defects are *smt1, smt2, fackel* and *hydra1*. Sterol methyltransferase (SMT) is a multiallelic gene with several known mutations. The SMT enzyme balances the ratio between cholesterol and other typical phytosterols (sitosterol, stigmasterol and 24-methylcholesterol) that is necessary for normal cell growth and membrane integrity (Schaeffer et al., 2001; Diener et al., 2000). The *smt1* mutant accumulates cholesterol and mutation of the *SMT1* locus therefore shows pleiotropic defects, such as absence of asymmetric cell division in the globular embryos, and developmental defects in apical shoot meristem and in cotyledons (Diener et al., 2000). Suppression of *SMT2* reduced growth and fertility (Scheaffer et al., 2001; Sitbon and Jonsson, 2001). Another mutation of *SMT2*, the *cvp1* mutant, is enriched in campesterol and depleted in sitosterol. This unbalanced composition of structural sterols attributes a role for sterols, but not for brassinosteroids, in vascular tissue patterning (Carland et al., 2002).

Fackel is a dwarf mutant with defects in the integral membrane protein sterol C-14 reductase (Jang et al., 2000; Schrick et al., 2000). These defects cannot be rescued by brassinosteroids. Since the mutation affects embryo and plant body organisation the *FACKEL* gene seems to be required for cell division and expansion. *Hydra2* is an allelic mutant to *fackel* showing affected patterning of root cell files and root hair formation. Auxin- and ethylene-dependent regulation pathways are involved in these patterning defects (Souter et al., 2002). A mutant with strong depletion of camposterol and sitosterol is *hydra1* encoding for sterol C-8,7 isomerase. It has aberrant morphogenesis with non-regulated cell size and shape (Souter et al., 2002).

The mechanisms of how sterols alter development and morphogenesis are not known, but seedlings treated with brassinosteroids or sterols displayed important changes in the expression of various genes involved in plant cell division and growth. Interestingly, fenpropimorph, an inhibitor of the sterol C-14 reductase biosynthetic step, can phenocopy the *fackel* mutant, and causes changes in gene expression (Schaller et al., 1994; He et al., 2003).

One of the proposed effects of altered sterol biosynthesis on plant morphogenesis could be a direct link between sterols in the plasma membrane and cellulose synthesis (Schrick et al., 2004), because sitosterol-β-glucoside is a precursor for cellulose biosynthesis (Peng, 2002). This could explain why *fackel, hydra1* and *smt1/cph* mutants possess reduced cellulose content, incomplete cell walls and cell wall thickenings. The same defects can be mimicked by the sterol inhibitors 15-azosterol and fenpropimorph (Schrick et al., 2004). Apparently, the alterations of the structural sterols lead to destabilisation of the plasma membrane and therefore affect cellulose synthesis.

8
Sterols in Endocytosis

Although the concept of endocytosis had been established for a long time in animal and yeast cells, membrane internalisation by endocytosis was accepted only later as a basic process in plant cells. It is held responsible for physiological changes of the cell surface and exchange of membrane components, and it also plays a key role in the turnover of plasma membrane sterols and further sterol trafficking.

8.1
Sterol Function in Endocytosis

Sterols have numerous diverse functions in endocytosis. In animal cells, cholesterol depletion affects the formation of clathrin-coated pits (Rodal et al., 1999; Subtil et al., 1999). In yeast, several sterol mutants with impaired endocytosis were characterised. Very well described are mutants encoding ergosterol biosynthetic enzymes. Cells of the *erg2* mutant defective in sterol C-8 isomerase fail in the internalisation step of endocytosis, because ergosterol in the plasma membrane is replaced by a different mixture of sterol intermediates and thus the unique function of ergosterol required for endocytosis is missing. As a result, the *erg2* mutant and some double mutants in the *erg2* background are defective in both fluid-phase endocytosis and receptor-mediated endocytosis. Interestingly, intracellular trafficking and degradation in the vacuole were not affected (Munn et al., 1999). Another mutation of the *ERG4* and *ERG5* loci (encoding for sterol C-24 reductase and sterol C-22 desaturase, respectively) in double mutant combination does not prevent normal internalisation, but further endocytic trafficking of markers to the vacuole was delayed (Heese-Peck et al., 2002). These experiments shed more light on possible mechanisms of endocytosis and the involvement of sterols in this process. The whole scenario still needs to be elucidated, but so far it appears that several functional steps of endocytosis depend on sterols.

8.2
Internalisation of Plant Sterols

Mutations of the sterol biosynthetic pathway in *Arabidopsis* show direct effects of sterols on plant morphogenesis, but on the cellular level we have still

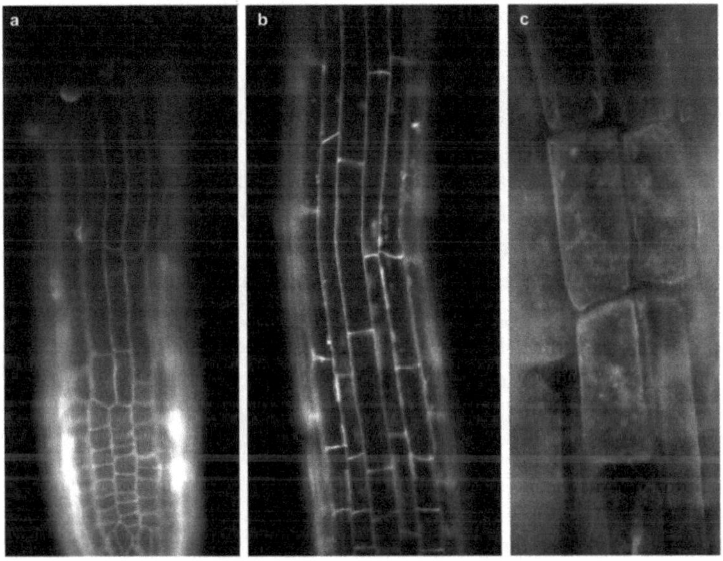

Fig. 2 Sterol-binding antibiotic filipin is used as a vital probe for fluorescent detection of plant 3-β-hydroxysterols. In *Arabidopsis* roots, sterols labelled with filipin for 30 min are visualised in all epidermal cells of the root meristem (**a**) and the elongation zone (**b**). After 2 h, filipin–sterol complexes are detected in intracellular compartments (**c**), indicating endocytic internalisation of sterols from the plasma membrane

only few data about endocytosis of structural sterols. In order to understand the roles of endocytosis and trafficking of sterols, they were visualised by fluorescence labelling using the polyene sterol-binding antibiotic filipin. This vital probe for fluorescent detection of plant 3-β-hydroxysterols revealed the sterol location, endocytic uptake and trafficking in young epidermal cells of *Arabidopsis* roots (Grebe et al., 2003, Fig. 2). Filipin–sterol complexes were labelled at the plasma membrane, but interestingly, these complexes also internalised into intracellular structures indicating endocytosis (Fig. 2). This was proven by co-localisation of sterols with the non-specific fluorescent marker of the plasma membrane, FM4-64 (Grebe et al., 2003). This styryl dye is fluorescent only in the hydrophobic environment of the lipid membrane, and became an important tool in the study of endocytosis (Betz et al., 1992; Šamaj et al., 2005; see also the chapter by Šamaj, this volume). Internalisation and the staining pattern of the intracellular structures are time-dependent and allow the endocytic pathway and different endocytic organelles to be identified (Betz et al., 1996; Meckel et al., 2004).

9 Intracellular Sterol Trafficking

9.1 Biosynthetic Sterol Secretion

Membrane trafficking occurs in two pathways: secretion and targeting, and endocytic internalisation and recycling. Sterols in plant cells, like in animals and yeast, are distributed by vesicular transport to target membranes. Inhibition of such conventional secretory transport using the well-known fungal metabolite brefeldin A (BFA) altered the distribution of free sterols (Moreau et al., 1998a), as well as their biosynthesis (Merigout et al., 2002). A similar mistargeting of the intracellular transport of lipids was induced by monensin, a Golgi-disturbing agent preventing secretion (Dinter and Berger, 1998). As the lipid content in the Golgi fraction was enhanced by the treatment (Moreau et al., 1988a), it was concluded that Golgi-mediated vesicular transport is responsible for the sorting of lipids and sterols to the plasma membrane in plant cells (Moreau et al., 1988a, 1988b; Moreau et al., 1998b).

9.2 Non-vesicular Sterol Transport

In addition to "classical" vesicular exo-endocytic pathways, non-vesicular lipid exchange through short-range close contacts of membranes can also be involved in sterol trafficking. In soybean cells, formation of such domains between the plasma membrane and ER was found (Grabski et al., 1993). Close association of the ER with the plasma membrane in plant cells was described in detail (Lichtscheidl and Url, 1990; Hepler et al., 1990). The role of such non-vesicular transport and recycling between ER and the plasma membrane was described for sterols in yeast (Li and Prinz, 2004).

9.3 Endocytic Vesicular Recycling of Sterols

Although high attention has been paid to phosphoinositides in vesicular trafficking of plant cells (Moreau and Cassagne, 1994; Moreau et al., 1998b, Bessoule and Moreau, 2004), the situation for sterols in plants is much less clear. It is not known whether sterol components of the plasma membrane arranged in lipid rafts are packed together for endocytosis and recycling in plant cells. However, sterols labelled in the plasma membrane of root meristematic cells of *Arabidopsis* were internalised and accumulated in early endocytic compartments. These compartments partly co-localised with the *trans*-Golgi markers, but the majority of the structures were associated with the marker of plant endosomes, Ara6 (Grebe et al., 2003; Ueda et al., 2001). Ara6 is localised in endosomes, and together with Ara7 and Rha1 they represent plant-specific molecular markers for endosomes (see the chapter by Nielsen, this volume). They all belong to the family of Rab5 proteins from the group of small GTPases (Ueda and Nakano, 2002; Vernoud et al., 2003).

In animal cells, Rab proteins control trafficking, exocytosis, endocytosis and endosome fusion, and act as organisers of local membrane domains (Sonnichsen et al., 2000). Localised production of PI(3)P enables the recruitment of Rab5 effectors which bind this lipid domain through the FYVE domains (Nielsen et al., 2000). Tandem FYVE construct tagged with DsRed or green fluorescent protein (GFP) was successfully used as a marker for plant endosomes (Šamaj et al., 2004a, Voigt et al., 2005).

9.4 Turnover of the Plasma Membrane Proteins

Recent data show that the recycling of plasma membrane proteins by endocytosis is an integral part of functional plasma membrane turnover. It was documented for putative auxin efflux carriers of the PIN family, such as PIN1 (Geldner et al., 2001; Baluška et al., 2002) and PIN2 (Grebe et al., 2003), the putative auxin influx carrier AUX1 (Grebe et al., 2002), a plasma membrane H^+ ATPase (Geldner et al., 2001; Baluška et al., 2002) and the K^+-inward rectifying channel (Homann and Thiel, 2002; Hurst et al., 2004). In addition, the recycling of structural components of the plant cell wall, such as arabinogalactan proteins and calcium- and boron-crosslinked pectins, was found (Šamaj et al., 2000; Baluška et al., 2002; Yu et al., 2002).

Proteins of the PIN family are involved in the definition of bilateral polarity due to polar localisation in cells of the root meristem from *Arabidopsis* (Friml et al., 2002, 2003; see the chapter by Chen and Masson, this volume). Their recycling may be based on the functional distribution of structural plant sterols. Endocytosis of sterol-enriched membrane domains can play a sorting role in the recycling of membrane proteins. In *Arabidopsis* roots, sterols and PIN2 recycle within the common endocytic pathway (Grebe et al., 2003).

Sterols accumulate in early endosomes, as was detected by partial co-localisation with the early endosome marker Ara6 in BFA experiments (Grebe et al., 2003). This finding was further supported by identification of GNOM, a BFA-sensitive, membrane-associated guanine–nucleotide exchange factor (GEF) of ARF in *Arabidopsis*. This protein mediates proper PIN1 localisation and endosomal accumulation upon BFA treatment, and represents a natural target of BFA action in living cells (Geldner et al., 2003).

9.5 Sterols and Polarity

Mutation of the *SMT1* gene responsible for correct bulk sterol biosynthesis in *Arabidopsis* reduces polar auxin transport, and induces defects in cell polarity and in PIN localisation (Willemsen et al., 2003). Application of synthetic auxin to *smt1* mutant can rescue defects in the polarity of epidermal cells (Willemsen et al., 2003). It shows the relation of sterol homeostasis to the establishment and maintenance of cell polarity. Similarities in the trafficking of sterols and the presumptive auxin transport protein PIN2 (Grebe et al., 2003) suggest a possible involvement of sterol-based trafficking and sorting mechanisms in the basic processes of plant cell polarity. It must still be proven, however, in other processes involved in establishing polarity, such as in polarly growing root hairs. Growing tips of root hairs possess functional networks of actin cytoskeleton and vesicular trafficking, which are interconnected with different signalling pathways (Šamaj et al., 2002; Šamaj et al., 2004a, 2004b). Our experiments revealed rapid tip-localised endocytosis in growing root hairs of *Arabidopsis* (Ovečka et al., 2005). Internalisation of plasma membrane structural sterols in the tip, which is indispensable for the progression of tip growth, indicates that some sterol-based trafficking pathways are involved in the maintenance of plant cell polarity (Ovečka et al., manuscript in preparation).

10 Conclusions and Future Prospects

In summary, genetic approaches so far have improved considerably our scope of sterol functions. In plant mutants, sterols are not depleted, but the final ratio is altered. Physiological processes in cells have different sensitivities to altered sterol composition, and small changes in the ratio of sterols have physiological and developmental consequences. Thus, the explanation of the phenotype needs comprehensive study. Alteration of the sterol content in the membranes is clearly the reason for some defects, but the exact spatial distribution in the membrane of sterol mutants is not known. Molecular mechanisms of developmental changes by modification of sterol composition are therefore largely elusive. It remains to be determined whether altered sterol composition also affects the processes of endocytosis, sorting and recycling in sterol biosynthetic mutants.

Molecular and cell biological approaches must be focused on the understanding of lipid and sterol trafficking, the functional assembly of cell membranes, the molecular mechanisms of cell signalling, the adaptation to the environment and the expression of developmental cues in plants. More data are needed to get a complete picture of membrane biogenesis in plant cells. There are only a few characterised receptors in plants, whereas mammalian cells have been extensively studied as regards cell signalling and receptor-mediated endocytosis. However, a systemic approach to reveal a complex scenario of membrane biogenesis in plants has been started and is gradually accelerating. Large-scale characterisation and sequence analysis of Arf genes (Gebbie et al., 2005) and SNARE genes (Uemura et al., 2004) in *Arabidopsis*, and their study in vesicular trafficking and recycling, are the first conceptual results.

Another perspective of future research is in the determination of the macromolecular dynamics responsible for the assembly of multi-protein complexes involved in signalling, and in modulation of the cytoskeleton and the cell wall, and thus in determining all developmental processes. There is no doubt that the characterisation of structural sterols and their roles in cell membranes will facilitate our understanding of endocytosis in living plant cells.

Acknowledgements This work was supported by the European Commission, Research Project of Human Potential Programme TIPNET (HPRN-CT-2002-00265), and by grant No. 2/5085/25 from the Grant Agency VEGA.

References

Arnqvist L, Dutta PC, Jonsson L, Sitbon F (2003) Reduction of cholesterol and glycoalkaloid levels in transgenic potato plants by overexpression of a type 1 sterol methyltransferase cDNA. Plant Physiol 131:1792–1799

Bach TJ, Benveniste P (1997) Cloning of cDNAs or genes encoding enzymes of sterol biosynthesis from plants and other eukaryotes: heterologous expression and commentation analysis of mutations for functional characterization. Prog Lipid Res 36:197–226

Bacia K, Schwille P, Kurzchalia T (2005) Sterol structure determines the separation of phases and the curvature of the liquid-ordered phase in model membranes. Proc Natl Acad Sci USA 102:3272–3277

Bagnat M, Simons K (2002) Cell surface polarization during yeast mating. Proc Natl Acad Sci USA 99:14183–14188

Baluška F, Hlavačka A, Šamaj J, Palme K, Robinson DG, Matoh T, McCurdy DW, Menzel D, Volkmann D (2002) F-actin-dependent endocytosis of cell wall pectins in meristematic root cells: insights from brefeldin A-induced compartments. Plant Physiol 130:422–431

Benveniste P (1986) Sterol biosynthesis. Ann Rev Plant Physiol 37:275–307

Benveniste P (2002) Sterol metabolism. In: Somerville CR, Meyerowitz EM (eds) The *Arabidopsis* book. American Society of Plant Biologists, Rockville, MD, DOI 10.1199/tab.0004, p 31

Benveniste P (2005) Prenyllipids and their derivatives: sterols, prenylquinones, carotenoids and terpenoids. In: Murphy DJ (ed) Plant lipids: biology, utilization and manipulation. Blackwell/CRC, Oxford, 353–387

Bessoule J-J, Moreau P (2004) Phospholipid synthesis and dynamics in plant cells. In: Daum G (ed) Lipid metabolism and membranebiogenesis. Top Curr Genet, vol 6. Springer, Berlin Heidelberg New York, p 89–124

Betz WJ, Mao F, Bewick GS (1992) Activity-dependent fluorescent staining and destaining of living vertebrate motor nerve terminals. J Neurosci 12:363–375

Betz WJ, Mao F, Smith CB (1996) Imaging exocytosis and endocytosis. Curr Opin Neurobiol 6:365–371

Borner GHH, Sherrier DJ, Stevens TJ, Arkin IT, Dupree P (2002) Prediction of glycosylphosphatidylinositol-anchored proteins in *Arabidopsis*. A genomic analysis. Plant Physiol 129:486–499

Borner GHH, Lilley K, Stevens TJ, Dupree P (2003) Identification of glycosylphosphatidylinositol-anchored proteins in *Arabidopsis*. A proteomic and genomic analysis. Plant Physiol 132:568–577

Borner GHH, Sherrier DJ, Weimar T, Michaelson LV, Hawkins ND, MacAskill A, Napier JA, Beale MH, Lilley KS, Dupree P (2005) Analysis of detergent-resistant membranes in *Arabidopsis*. Evidence for plasma membrane lipid rafts. Plant Physiol 137:104–116

Carland FM, Fujioka S, Takatsuto S, Yoshida S, Nelson T (2002) The identification of CVP1 reveals a role for sterols in vascular patterning. Plant Cell 14:2045–2058

Chen CY, Ingram MF, Rosal PH, Graham TR (1999) Role for Drs2p, a P-type ATPase and potential aminophospholipid translocase, in yeast late Golgi function. J Cell Biol 147:1223–1236

Chen R, Masson PH (2005) Auxin transport and recycling of PIN proteins in plants (in this volume). Springer, Berlin Heidelberg New York

Clouse SD (2002) *Arabidopsis* mutants reveal multiple roles for sterols in plant development. Plant Cell 14:1995–2000

Diener AC, Li H, Zhou WX, Whoriskey WJ, Nes WD (2000) Sterol methyltransferase 1 controls the level of cholesterol in plants. Plant Cell 12:853–870

Dinter A, Berger EG (1998) Golgi-disturbing agents. Histochem Cell Biol 109:571–590

Farge E, Ojcius DM, Subtil A, Dautry-Varsat A (1999) Enhancement of endocytosis due to aminophospholipid transport across the plasma membrane of living cells. Am J Physiol 276:C725–C733

Fivaz M, Vilbois F, Thurnheer S, Pasquali C, Abrami L, Bickel PE, Parton RG, van der Goot FG (2002) Differential sorting and fate of endocytosed GPI-anchored proteins. EMBO J 21:3989–4000

Friml J, Benkova E, Blilou I, Wisniewska J, Hamann T, Ljung K, Woody S, Sandberg G, Scheres B, Jürgens G, Palme K (2002) AtPIN4 mediates sink-driven auxin gradients and root patterning in *Arabidopsis*. Cell 108:661–673

Friml J, Vieten A, Sauer M, Weijers D, Schwarz H, Hamann T, Offringa R, Jürgens G (2003) Efflux-dependent auxin gradients establish the apical–basal axis of *Arabidopsis*. Nature 426:147–153
Gebbie LK, Burn JE, Hocart CH, Williamson RE (2005) Genes encoding ADP ribosylation factors in *Arabidopsis thaliana* L. Heyn.; genome analysis and antisense suppression. J Exp Bot 56:1079–1091
Geldner N, Friml J, Stierhof Y-D, Jürgens G, Palme K (2001) Auxin transport inhibitors block PIN1 cycling and vesicle trafficking. Nature 413:425–428
Geldner N, Anders N, Wolters H, Keicher J, Kornberger W, Muller P, Delbarre A, Ueda T, Nakano A, Jürgens G (2003) The *Arabidopsis* GNOM ARF-GEF mediates endosomal recycling, auxin transport, and auxin-dependent plant growth. Cell 112:219–230
Gimlp G, Burger K, Fahrenholz F (1997) Cholesterol as modulator of receptor function. Biochemistry 36:10959–10974
Gomès E, Jakobsen MK, Axelsen KB, Geisler M, Palmgren MG (2000) Chilling tolerance in *Arabidopsis* involves ALA1, a member of a new family of putative aminophospholipid translocases. Plant Cell 12:2441–2453
Grabski S, de Feijter A, Schindler M (1993) Endoplasmic reticulum forms a dynamic continuum for lipid diffusion between contiguous soybean root cells. Plant Cell 5:25–38
Graham TR (2004) Flippases and vesicle-mediated protein transport. Trends Cell Biol 14:670–677
Grandmougin-Ferjani A, Schuler-Muller I, Hartmann MA (1997) Sterol modulation of the plasma membrane H^+-ATPase activity from corn roots reconstituted into soybean lipids. Plant Physiol 113:163–174
Grebe M, Friml J, Swarup R, Ljung K, Sandberg G, Terlou M, Palme K, Bennett MJ, Scheres B (2002) Cell polarity signaling in *Arabidopsis* involves a BFA-sensitive auxin influx pathway. Curr Biol 12:329–334
Grebe M, Xu J, Möbius W, Ueda T, Nakano A, Geuze HJ, Rook MB, Scheres B (2003) *Arabidopsis* sterol endocytosis involves actin-mediated trafficking via ARA6-positive early endosomes. Curr Biol 13:1378–1387
Hao M, Mukherjee S, Maxfield F (2001) Cholesterol modulation induces large-scale domain segregation in living cell membranes. Proc Natl Acad Sci USA 98:13072–13077
Hartmann M-A (2004) Sterol metabolism and functions in higher plants. In: Daum G (ed) Lipid metabolism and membrane biogenesis. Top Curr Genet, vol 6. Springer, Berlin Heidelberg New York, p 183–211
He J-X, Fujioka S, Li T-C, Kang SG, Seto H, Takatsuto S, Yoshida S, Jang J-C (2003) Sterols regulate development and gene expression in *Arabidopsis*. Plant Physiol 131:1258–1269
Heese-Peck A, Pichler H, Zanolari B, Watanabe R, Daum G, Riezman H (2002) Multiple functions of sterols in yeast endocytosis. Mol Biol Cell 13:2664–2680
Heino S, Somerharju P, Ehnholm C, Olkkonen E, Ikonen E (2000) Dissecting the role of the Golgi complex and lipid rafts in biosynthetic transport of cholesterol to the cell surface. Proc Natl Acad Sci USA 97:8375–8380
Hepler PK, Palevitz BA, Lancelle SA, McCauley MM, Lichtscheidl IK (1990) Cortical endoplasmic reticulum in plant cells. J Cell Sci 96:355–373
Hobbs DH, Hume JH, Rolph CE, Cooke DT (1996) Changes in lipid composition during floral development of *Brassica campestris*. Phytochemistry 42:335–339
Homann U, Thiel G (2002) The number of K^+ channels in the plasma membrane of guard cell protoplasts changes in parallel with the surface area. Proc Natl Acad Sci USA 99:10215–10220
Hurst AC, Meckel T, Tayefeh S, Thiel G, Homann U (2004) Trafficking of the plant potassium inward rectifier KAT1 in guard cell protoplasts of *Vicia faba*. Plant J 37:391–397

Jang JC, Fujioka S, Tasaka M, Seto H, Takatsuto S, Ishii A, Aida M, Yoshida S, Sheen J (2000) A critical role of sterols in embryonic patterning and meristem programming revealed by the *fackel* mutants of *Arabidopsis thaliana*. Genes Dev 14:1485–1497

Kinsky SC, Luse SA, Zopf D, van Deenen LLM, Haxby J (1967) Interaction of filipin and derivatives with erythrocyte membranes and lipid dispersions: electron microscopic observations. Biochim Biophys Acta 135:844–861

Kinsky SC (1970) Antibiotic interaction with model membranes. Ann Rev Pharmacol 10:119–142

Kooijman EE, Chupin V, de Kruijff B, Burger KN (2003) Modulation of membrane curvature by phosphatidic acid and lysophosphatidic acid. Traffic 4:162–174

Lalanne E, Honys D, Johnson A, Borner GH, Lilley KS, Dupree P, Grossniklaus U, Twell D (2004) SETH1 and SETH2, two components of the glycosylphosphatidylinositol anchor biosynthetic pathway, are required for pollen germination and tube growth in *Arabidopsis*. Plant Cell 16:229–240

Lees ND, Bard M (2004) Sterol biochemistry and regulation in the yeast *Saccharomyces serevisiae*. In: Daum G (ed) Lipid metabolism and membrane biogenesis. Top Curr Genet, vol 6. Springer, Berlin Heidelberg New York, p 213–240

Lemmon SK, Traub LM (2000) Sorting in the endosomal system in yeast and animal cells. Curr Opin Cell Biol 12:457–466

Li Y, Prinz WA (2004) ATP-binding cassette (ABC) transporters mediate nonvesicular, raft-modulated sterol movement from the plasma membrane to the endoplasmic reticulum. J Biol Chem 279:45226–45234

Lichtscheidl IK, Url WG (1990) Organization and dynamics of cortical endoplasmic reticulum in inner epidermal cells of onion bulb scales. Protoplasma 157:203–215

Marsan MP, Muller I, Milon A (1996) Ability of clionasterol and poriferasterol (24-epimers of sitosterol and stigmasterol) to regulate membrane lipid dynamics. Chem Phys Lipids 84:117–121

Marsan MP, Bellet-Amalric E, Muller I, Zaccai G, Milon A (1998) Plant sterols: a neutron diffraction study of sitosterol and stigmasterol in soybean phosphatidylcholine membranes. Biophys Chem 75:45–55

Mayor S, Maxfield F (1995) Insolubility and redistribution of GPI-anchored proteins at the cell surface after detergent treatment. Mol Biol Cell 6:929–944

Mayor S, Sabharanjak S, Maxfield FR (1998) Cholesterol-dependent retention of GPI-anchored proteins in endosomes. EMBO J 17:4626–4638

McConnell JR, Emery J, Eshed Y, Bao N, Bowman J, Barton K (2001) Role of PHABULOSA and PHAVOLUTA in determining radial patterning in shoots. Nature 411:709–713

Meckel T, Hurst AC, Thiel G, Homann U (2004) Endocytosis against high turgor: intact guard cells of *Vicia faba* constitutively endocytose fluorescently labelled plasma membrane and GFP-tagged K^+-channel KAT1. Plant J 39:182–193

Mérigout P, Képés F, Perret AM, Satiat-Jeunemaitre B, Moreau P (2002) Effects of brefeldin A and nordihydroguaiaretic acid on endomembrane dynamics and lipid synthesis in plant cells. FEBS Lett 518:88–92

Mongrand S, Morel J, Laroche J, Claverol S, Carde J-P, Hartmann M-A, Bonneu M, Simon-Plas F, Lessire R, Bessoule J-J (2004) Lipid rafts in higher plant cells. Purification and characterization of Triton X-100-insoluble microdomains from tobacco plasma membrane. J Biol Chem 279:36277–36286

Möbius W, Ohno-Iwashita Y, van Donselaar EG, Oorschot VMJ, Shimada Y, Fujimoto T, Heijnen HFG, Geuze HJ, Slot JW (2002) Immunoelectron microscopic localization of cholesterol using biotinylated and non-cytolytic perfringolysin O. J Histochem Cytochem 50:43–55

Möbius W, van Donselaar E, Ohno-Iwashita Y, Shimada Y, Heijnen HFG, Slot JW, Geuze HJ (2003) Recycling compartments and the internal vesicles of multivesicular bodies harbor most of the cholesterol found in the endocytic pathway. Traffic 4:221–231

Moreau P, Bertho P, Juguelin H, Lessire R (1988a) Intracellular transport of very long chain fatty acids in etiolated leek seedlings. Plant Physiol Biochem 26:173–178

Moreau P, Juguelin H, Lessire R, Cassagne C (1988b) Plasma membrane biogenesis in higher plants: in vivo transfer of lipids to the plasma membrane. Phytochem 27:1631–1638

Moreau P, Cassagne C (1994) Phospholipid trafficking and membrane biogenesis. Biochim Biophys Acta 1197:257–290

Moreau P, Hartmann MA, Perret AM, Sturbois-Balcerzak B, Cassagne C (1998a) Transport of sterols to the plasma membrane of leek seedlings. Plant Physiol 117:931–937

Moreau P, Bessoule JJ, Mongrand S, Testet E, Vincent P, Cassagne C (1998b) Lipid trafficking in plant cells. Prog Lipid Res 37:371–391

Mukherjee S, Zha X, Tabas I, Maxfield FR (1998) Dehydroergosterol as a fluorescent cholesterol analog. Biophys J 75:1915–1925

Munn AL, Heese-Peck A, Stevenson BJ, Pichler J, Riezman H (1999) Specific sterols required for the internalization step of endocytosis in yeast. Mol Biol Cell 10:3943–3957

Nielsen E, Christoforidis S, Uttenweiler-Joseph S, Miaczynska M, Dewitte F, Wilm M, Hoflack B, Zerial M (2000) Rabenosyn-5, a novel Rab5 effector, is complexed with hVPS45 and recruited to endosomes through a FYVE finger domain. J Cell Biol 151:601–612

Nielsen E (2005) Rab GTPases in plant endocytosis (in this volume). Springer, Berlin Heidelberg New York

Nichols BJ, Kenworthy AK, Polishchuk RS, Lodge R, Roberts TH, Hirschberg K, Phair RD, Lippincott-Schwartz J (2001) Rapid cycling of lipid raft markers between the cell surface and Golgi complex. J Cell Biol 153:529–542

Norberg P, Liljenberg C (1999) Lipids of plasma membranes prepared from oat root cells. Plant Physiol 96:1136–1141

Ovečka M, Lang I, Baluška B, Ismail A, Illeš P, Lichtscheidl IK (2005) Endocytosis and vesicle trafficking during tip growth of root hairs. Protoplasma (in press)

Oxley D, Bacic A (1999) Structure of glycosylphosphatidylinositol anchor of an arabinogalactan protein from *Pyrus communis* suspension-cultured cells. Proc Natl Acad Sci USA 96:14246–14251

Peng L, Kawagoe I, Hogan P, Delmer D (2002) Sitosterol-β-glucosidase as a primer for cellulose synthesis in plants. Science 295:147–150

Peskan T, Westermann M, Oelmuller R (2000) Identification of low-density Triton X-100-insoluble plasma membrane microdomains in higher plants. Eur J Biochem 267:6989–6995

Pike LJ, Casey L (2002) Cholesterol levels modulate EGF receptor-mediated signaling by altering receptor function and trafficking. Biochemistry 41:10315–10322

Pomorski T, Lombardi R, Riezman H, Devaux PF, Van Meer G, Holthuis JC (2003) Drs2p-related P-type ATPases Dnf1p and Dnf2p are required for phospholipid translocation across the yeast plasma membrane and serve a role in endocytosis. Mol Biol Cell 14:1240–1254

Pralle A, Keller P, Florin E, Simons K, Horber JK (2000) Sphingolipid–cholesterol rafts diffuse as small entities in the plasma membrane of mammalian cells. J Cell Biol 148:997–1008

Puri V, Watanabe R, Dominguez M, Sun X, Wheatley CL, Marks DL, Pagano RE (1999) Cholesterol modulates membrane traffic along the endocytic pathway in sphingolipid-storage diseases. Nat Cell Biol 1:386–388

Rodal SK, Skretting G, Garred O, Vilhardt F, van Deurs B, Sandvig K (1999) Extraction of cholesterol with methyl-β-cyclodextrin perturbs formation of clathrin-coated endocytic vesicles. Mol Cell Biol 10:961–974

Šamaj J, Šamajová O, Peters M, Baluška F, Lichtscheidl I, Knox JP, Volkmann D (2000) Immunolocalization of LM2 arabinogalactan protein epitope associated with endomembranes of plant cells. Protoplasma 212:186–196

Šamaj J, Ovečka M, Hlavačka A, Lecourieux F, Meskiene I, Lichtscheidl I, Lenart P, Salaj J, Volkmann D, Bögre L, Baluška F, Hirt H (2002) Involvement of the mitogen-activated protein kinase SIMK in regulation of root hair tip growth. EMBO J 21:3296–3306

Šamaj J, Baluška F, Voigt B, Schlicht M, Volkmann D, Menzel D (2004a) Endocytosis, actin cytoskeleton, and signaling. Plant Physiol 135:1150–1161

Šamaj J, Baluška F, Menzel D (2004b) New signalling molecules regulating root hair tip growth. Trends Plant Sci 9:217–220

Šamaj J, Read ND, Volkmann D, Menzel D, Baluška F (2005) The endocytic network in plants. Trends Cell Biol 15:425–437

Šamaj J (2005) Methods and molecular tools to study endocytosis in plants—an overview (in this volume). Springer, Berlin Heidelberg New York

Schaeffer A, Bronner R, Benveniste P, Schaller H (2001) The ratio of campesterol to sitosterol that modulates growth in *Arabidopsis* is controlled by sterol methyltransferase 2;1. Plant J 25:605–615

Schaller H, Gondet L, Maillot-Vernier P, Benveniste P (1994) Sterol overproduction is the biochemical basis of resistance to a triazole in calli from a tobacco mutant. Planta 194:295–305

Schaller H (2004) New aspects of sterol biosynthesis in growth and development of higher plants. Plant Physiol Biochem 42:465–476

Schrick K, Mayer U, Horrichs A, Kuhnt C, Bellini C, Dangl J, Schmidt J, Jürgens G (2000) Fackel is a sterol C-14 reductase required for organized cell division and expansion in *Arabidopsis* embryogenesis. Genes Dev 14:1471–1484

Schrick K, Fujioka S, Takatsuto S, Stierhof Y-D, Stransky H, Yoshida S, Jürgens G (2004) A link between sterol biosynthesis, the cell wall, and cellulose in *Arabidopsis*. Plant J 38:227–243

Schuler I, Duportail G, Glasser N, Benveniste P, Hartmann MA (1990) Soybean phosphatidylcholine vesicles containing plant sterols: a fluorescence anisotropy study. Biochim Biophys Acta 1028:82–88

Schuler I, Milon A, Nakatami Y, Ourisson G, Albrecht AM, Benveniste P, Hartmann MA (1991) Differential effects of plant sterols on water permeability and on acyl chain ordering of soybean phosphatidylcholine bilayers. Proc Natl Acad Sci USA 88:6926–6930

Simons K, Ikonen E (1997) Functional rafts in cell membranes. Nature 387:569–572

Simons K, Toomre D (2000) Lipid rafts and signal transduction. Nat Rev Mol Cell Biol 1:31–39

Sitbon F, Jonsson L (2001) Sterol composition and growth of transgenic tobacco plants expressing type 1 and type 2 sterol methyltransferases. Planta 212:568–572

Sonnichsen B, De Renzis S, Nielsen E, Rietdorf J, Zerial M (2000) Distinct membrane domains on endosomes in the recycling pathway visualized by multicolor imaging of Rab4, Rab5, and Rab11. J Cell Biol 149:901–914

Souter M, Topping J, Pullen M, Friml J, Palme K, Hackett R, Grierson D, Lindsey K (2002) *Hydra* mutants of *Arabidopsis* are defective in sterol profiles and auxin and ethylene signaling. Plant Cell 14:1017–1031

Sperling P, Warnecke D, Heinz E (2004) Plant sphingolipids. In: Daum G (ed) Lipid metabolism and membrane biogenesis. Top Curr Genet, vol 6. Springer, Berlin Heidelberg New York, p 337–381

Subtil A, Gaidarov I, Kobylarz K, Lamson MA, Keen JH, McGraw TE (1999) Acute cholesterol depletion inhibits clathrin-coated pit budding. Proc Natl Acad Sci USA 96:6775–6780

Takos AM, Dry IB, Soole KL (1997) Detection of glycosylphosphatidylinositol-anchored proteins on the surface of the *Nicotiana tabacum* protoplasts. FEBS Lett 405:1–4

Thompson GA, Okuyama H (2000) Lipid-linked proteins of plants. Prog Lipid Res 39:19–39

Ueda T, Yamaguchi M, Uchimiya H, Nakano A (2001) Ara6, a plant-unique novel type Rab GTPase, functions in the endocytic pathway of *Arabidopsis thaliana*. EMBO J 17:4730–4741

Ueda T, Nakano A (2002) Vesicular traffic: an integral part of plant life. Curr Opin Plant Biol 5:513–517

Uemura T, Ueda T, Ohniwa RL, Nakano A, Takeyasu K, Sato MH (2004) Systematic analysis of SNARE molecules in *Arabidopsis*: dissection of the post-Golgi network in plant cells. Cell Struct Funct 29:49–65

Van Meer G, Sprong H (2004) Membrane lipids and vesicular traffic. Curr Opin Cell Biol 16:373–378

Verkleij AJ, de Kruijff B, Gerritsen WF, Demel RA, van Deenen LLM, Ververgaert PHJ (1973) Freeze-etch electron microscopy of erythrocytes, *Acholeplasma laidlawii* cells and liposomal membranes after the action of filipin and amphotericin B. Biochim Biophys Acta 291:577–581

Vernoud V, Horton AC, Yang ZB, Nielsen E (2003) Analysis of the small GTPase gene superfamily of *Arabidopsis*. Plant Physiol 131:1191–1208

Voigt B, Timmers A, Šamaj J, Hlavačka A, Ueda T, Preuss M, Nielsen E, Mathur J, Emans N, Stenmark H, Nakano A, Baluška F, Menzel D (2005) Actin-based motility of endosomes is linked to the polar tip growth of root hairs. Eur J Cell Biol 84:609–621

Wachtler V, Rajagopalan S, Balasubramanian MK (2003) Sterol-rich plasma membrane domains in the fission yeast *Schizosaccharomyces pombe*. J Cell Sci 116:867–874

Willemsen V, Friml J, Grebe M, van den Toorn A, Palme K, Scheres B (2003) Cell polarity and PIN protein positioning in *Arabidopsis* require sterol methyltransferase 1 function. Plant Cell 15:612–625

Xu X, Bittman R, Duportail G, Heissler D, Vilcheze C, London E (2001) Effect of the structure of natural sterols and sphingolipids on the formation of ordered sphingolipid/sterol domains (rafts). Comparison of cholesterol to plant, fungal, and disease-associated sterols and comparison of sphingomyelin, cerebrosides, and ceramide. J Biol Chem 276:33540–33546

Yu Q, Hlavačka A, Matoh T, Volkmann D, Menzel D, Goldbach HE, Baluška F (2002) Short term boron deprivation inhibits endocytosis of cell wall pectins in meristematic cells of maize and wheat root apices. Plant Physiol 130:415–421

Zinser E, Paltauf F, Daum G (1993) Sterol composition of yeast organelle membranes and subcellular distribution of enzymes involved in sterol metabolism. J Bacteriol 175:2853–2858

Plant Cell Monogr (1)
J. Šamaj · F. Baluška · D. Menzel: Plant Endocytosis
DOI 10.1007/7089_009/Published online: 22 September 2005

Auxin Transport and Recycling of PIN Proteins in Plants

Rujin Chen[1] (✉) · Patrick H. Masson[2]

[1]Plant Biology Division, Samuel Roberts Noble Foundation, 2510 Sam Noble Parkway, Ardmore, OK 73401, USA
rchen@noble.org

[2]Laboratory of Genetics, University of Wisconsin-Madison, 425-G Henry Mall, Madison, WI 53706, USA

Abstract Polar transport of the phytohormone auxin is mediated by plasma-membrane and endosome localized carrier proteins. PIN proteins are the best studied auxin efflux components implicated in the establishment of the auxin gradient required for growth and patterning in plants. Emerging models postulate a role for vesicular trafficking and protein phosphorylation and dephosphorylation in the regulation of PIN protein subcellular localization and auxin transport activity, providing a conceptual framework for our understanding of auxin transport and its role in plant development.

1 Introduction

Long before the discovery of its chemical structure in the mid-1930s, indole-3-acetic acid (IAA), the prominent endogenous form of the plant growth regulator auxin, was implicated in tropic responses to light and gravity (reviewed by Lomax et al., 1995). It is widely accepted that foci of IAA biosynthesis are located in young leaves and developing leaf primordia, and in the meristematic region of the primary root tip as well as in the tips of emerged lateral roots (Ljung et al., 2005; for a recent review see Woodward and Bartel, 2005). IAA is polarly transported to other tissues where it plays a regulatory role in cell division, differentiation, and elongation. Increasing evidence suggests that this directional auxin movement/distribution is closely linked to the diverse effects of IAA on plant development (Leyser, 2001; Swarup and Bennett, 2003; Zazimalova and Napier, 2003; Berleth et al., 2004; Weijers and Jurgen, 2004, 2005; Kramer, 2004; Kepinski and Leyser, 2005). Both cellular uptake and efflux mediated by protein carriers are involved. This review provides an overview of a group of recently identified transmembrane proteins of the PIN family representing auxin efflux components. In particular, emphasis is placed on emerging models of how biological activities of PIN proteins are regulated. Notably, another class of proteins belonging to the multidrug resistance ABC-type transporters has been postulated to play a role in polar auxin transport (Muday and Murphy, 2002; Murphy et al., 2004). Involvement of this type of protein in auxin transport is reviewed separately in the chapter by Blakeslee et al. in this volume.

Polar transport of auxin is a cell-to-cell process that has been best described in a modified chemiosmotic model for auxin transport (reviewed by Lomax et al., 1995). Due to differences in the acidity between apoplastic and intracellular space, IAA (mostly as the protonated form in the acidic apoplastic space) enters cells via passive diffusion and/or by the activity of a $2H^+$-IAA^- symporter (auxin influx carrier) driven by the proton motive force across the plasma membrane. Once inside cells where the pH is neutral, IAA is in the ionized form which is impermeable through the plasma membrane. The ionic form of IAA has to exit cells via auxin efflux carriers whose activity is driven by the membrane potential. This model postulates that the polarity of auxin transport is mediated by asymmetrical localization of auxin influx and efflux carriers on opposite ends of cells. Recent immunolocalization studies of the putative auxin influx and efflux component/facilitator proteins strongly support this model. Several candidates have been identified as the auxin influx carrier. Among them are AUX1 (for AUXIN RESISTANCE 1, Bennett et al., 1996) and members of the AUX1 amino acid/auxin:proton symport permease (AAAP) family (Swarup et al., 2000; Swarup and Bennett, 2003). On the other hand, PIN proteins have been identified as the auxin efflux component/facilitator proteins.

2 Pin-Formed (PIN) Protein Family

Molecular genetics studies have led to the discovery of two founding members (*pin-formed 1* and *agravitropic 1*; Okada et al., 1991; Bell and Maher, 1990, Fig. 1) of the putative auxin efflux component/facilitator PIN family (for reviews see Palme and Galweiler, 1999; Friml and Palme, 2002; Friml, 2003; Paponov et al., 2005). Facilitated by the completed genome sequence information of *Arabidopsis thaliana* and the availability of sequence-indexed T-DNA insertional mutants (http://www.arabidopsis.org), six more closely related PIN proteins and seven PIN-like proteins have been identified (Table 1). Biological functions of five PIN proteins have been recently characterized while functions of others remain to be elucidated.

2.1 PIN1 (Pin-Formed 1)

When *Arabidopsis* plants are grown in the presence of polar auxin transport (PAT) inhibitors, such as 1-naphthylphthalamic acid (NPA), inflorescence stems develop into pin-shaped structures lacking flowers and other lateral organs (Okada et al., 1991; Fig. 1). The lack of lateral organ development is likely

Table 1 *PIN* and *PIN*-like genes in *Arabidopsis thaliana*

Gene	AGI model	T-DNA insertion lines	Function	Reference
PIN genes				
PIN1	At1g73590	Salk_047613, Salk_097144	Embryogenesis, phototropism	Galweiler et al. 1998
PIN2	At5g57090	Salk_025934, Salk_139657	Gravitropism, basipetal auxin transport	Chen et al. 1998, Luschnig et al. 1998, Muller et al. 1998, Utsuno et al. 1998
PIN3	At1g70940	Salk_005544, Salk_036969, Salk_038609, Salk_113246, Salk_022278	Gravitropism, phototropism	Friml et al. 2002a
PIN4	At2g01420	SGT1708-3-3, SGT4961-5-3	Organogenesis	Friml et al. 2002b
PIN5	At5g16530	Salk_021738, Salk_042994, Salk_051354	TBD	Paponov et al. 2005
PIN6	At1g77110	Salk_046393, Salk_095142, Salk_021147	TBD	Benkova et al. 2003
PIN7	At1g23080	Salk_098657, Salk_044687, Salk_059225, Salk_048791, Salk_113246, Salk_12675, Salk_022278, Salk_075529	Embryogenesis	Friml et al. 2003
PIN8	At5g15100	Salk_107965, Salk_044651	TBD	Paponov et al. 2005

TBD = to be determined

Table 1 (continued)

Gene	AGI model	T-DNA insertion lines	Function	Reference
PIN-like genes				
PIN9	At2g17500	Salk_072996, Salk_078134	TBD	Shin HS and Chen R, unpublished
PIN10	At5g01990	Salk_074076, Salk_074172, Salk_130335, Salk_130431, Sail_1246.co2.v1	TBD	Shin HS and Chen R, unpublished
PIN11	At5g65980	Salk_068682, Salk 069610, Salk_069485	TBD	Shin HS and Chen R, unpublished
PIN12	At1g20925	N.A.	TBD	Shin HS and Chen R, unpublished
PIN13	At1g71090	Salk_125391, Salk_024808	TBD	Shin HS and Chen R, unpublished
PIN14	At1g76520	N.A.	TBD	Shin HS and Chen R, unpublished
PIN15	At1g76530	Salk_065131	TBD	Shin HS and Chen R, unpublished

TBD = to be determined

Fig. 1 Inflorescence stems of wild-type plant (**a**) and *pin1* mutant (**b**) and seedlings of wild-type (**c**) and *agr1* mutant (**d**). *pin1* plants exhibit defective lateral organ development, proliferation of vasculature, and fusion of cotyledons. Roots of *agr1* mutant form irregular waving on a tilted agar surface

due to reduced auxin accumulation at the tip of inflorescence stems resulting from interrupted polar auxin transport by NPA. Inflorescence stems of the *pin-formed 1* (*pin1*) mutant resemble those of wild-type plants treated with polar auxin transport inhibitors (Okada et al., 1991). Exogenous application of auxins on the tips of the *pin1* inflorescence stems restores the development of lateral organs, supporting the notion that the defect is due to limited auxin supply to the tips (Reinhardt et al., 2000; Vernoux et al., 2000; Benjamins et al., 2001; Aida et al., 2002).

Molecular cloning and characterization of the *PIN1* gene from transposon-tagged *pin1* mutant lines suggest that the encoded PIN1 protein is likely a candidate of the auxin efflux carrier sharing amino acid sequence homologies limited to the transmembrane-spanning domains with some bacterial transporters of the transporter superfacilitator family (Galweiler et al., 1998). Even though biochemical evidence of auxin transport activity is still lacking for PIN1 protein, immunolocalization studies clearly demonstrated that PIN1 protein is asymmetrically localized to the basal end of vascular cells in stems and to the apical end in roots (facing the root apex), suggesting a role for PIN1 in basipetal and acropetal auxin transport in shoots and roots, respectively. In agreement with the immunolocalization studies, *pin1* mutants exhibit reduced basipetal auxin transport activity in stems and altered auxin maximum in the root tip (Galweiler et al., 1998; Sabatini et al., 1999; Reinhardt et al., 2000; Vernoux et al., 2000; Aida et al., 2002; Furutani et al., 2004).

2.2 AGRAVITROPIC 1 (EIR1/PIN2/WAV6)

Another class of mutants has been isolated from separate screenings for defective root gravitropism (*agr1-1* to *agr1-8*, Bell and Maher, 1990; Chen et al., 1998; Utsuno et al., 1998) and thigmotropism (*wav6-52*, Okada and Shimura, 1990), as well as increased resistance to the inhibitory effect of ethylene on root elongation (*eir1*, Roman et al., 1995; Luschnig et al., 1998). This class represents an allelic series of mutants defective in the same *AGR1/EIR1/PIN2/WAV6* locus (Chen et al., 1998; Luschnig et al., 1998; Muller et al., 1998; Utsuno et al., 1998). Mutant phenotypes include an impaired gravity response in both roots and etiolated hypocotyls (Chen et al., 1998). The gravity defects are remarkably similar to the phenotypes of wild-type plants treated with PAT inhibitors, suggesting that the defects result from an impaired auxin distribution in the affected tissues. Root basipetal auxin transport is significantly reduced in the mutant compared to the wild type (Chen et al., 1998; Rashotte et al., 2000; Shin et al., 2005). Mutant roots are also resistant to the inhibitory effect of ethylene and PAT inhibitors (Chen et al., 1998).

Consistent with the observed mutant phenotypes, *AGR1/PIN2* transcripts are detected in the root distal and central elongation zones, but absent in the root cap region (Chen et al., 1998). Immunolocalization studies using AGR1/PIN2-specific antibodies suggest that AGR1/PIN2 protein is localized at the basal end (facing the root base or root–shoot junction) of the root epidermal and lateral root cap cells, coinciding with the direction of basipetal auxin transport in root peripheral tissues (Muller et al., 1998; Bonsirichai et al., 2004; Shin et al., 2005). On the other hand, localization of AGR1/PIN2 in root cortical cells is controversial, since both apical (facing the root tip) and basal localizations are reported (Muller et al., 1998; Friml et al., 2004). A close examination of the localization pattern of a PIN2-eGFP reporter (Blilou et al., 2005) reveals two opposite localization patterns of the fusion protein in the cortical cell file, being apical in the cell division zone and basal in the elongation zone (Fig. 2, R. Chen et al., unpublished results). The boundary of the transition in polarity appears to be sharp, coinciding with the cells leaving the division zone (Fig. 2, R. Chen et al., unpublished results). This bilateral local-

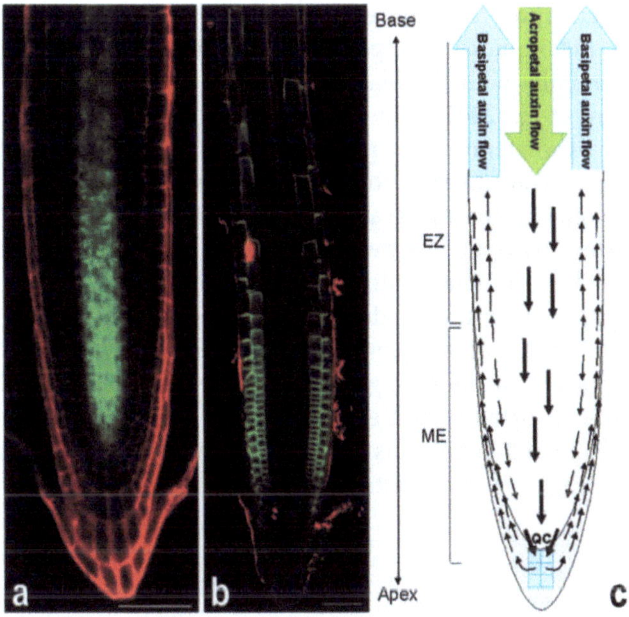

Fig. 2 Localization of **a** PIN1-eGFP and **b** PIN2-eGFP fusion proteins, and **c** schematic drawing of basipetal and acropetal auxin flows (*arrows*) and auxin maximum (*shaded squares*) in *Arabidopsis* roots (ME, meristematic region; EZ, elongation zone). The auxin maximum underneath the quiescent center (QC) may result from a net balance between auxin arrival and departure from the tip. The polarity of PIN2 coincides with basipetal auxin flow in the epidermal and upper part of the cortical cells, as well as with the acropetal auxin flow in the lower part of cortical cells. Scale bars, 50 μm

ization of PIN2 in the subepidermal cell file raises an interesting possibility that auxin transport is directed to two opposing directions in the root cortical cells (Fig. 2).

2.3 PIN3

PIN3 is the first PIN protein reported for having its intracellular localization both regulated by vesicular trafficking and modified by changes in environmental cues. The *PIN3* gene is expressed in the gravity-perceiving cells (statocytes) of roots, hypocotyls, and inflorescence stems, in addition to the root pericycle cells. In hypocotyls and inflorescence stems, the PIN3 protein is located in the plasma membrane at the inner periclinal side of endodermal cells, although it is occasionally found evenly distributed or associated with the basal membrane within these cells (Friml et al., 2002). When roots are oriented vertically, PIN3 is distributed uniformly within the statocyte columella cells, associating with both the plasma membrane and vesicles of

70-nm diameter. Upon root-tip reorientation within the gravity field (gravistimulation), PIN3 quickly changes its distribution within these cells, associating preferentially with the plasma membrane on the new lower side of the cells (Friml et al., 2002). This process is proposed to contribute to the establishment of a lateral auxin gradient across the root cap upon gravistimulation. Upon transmission to the elongation zone, this auxin gradient is largely responsible for the gravitropic curvature response (Friml et al., 2002). Consistent with an involvement of PIN3 in gravity signal transduction, *pin3* knockout mutant seedlings were reported to display altered root and hypocotyl gravitropism, in addition to having shorter primary organs in the presence of light, and an open apical hook (Friml et al., 2002). However, it should be noted that the root gravitropic defect of various *pin3* mutants, including those reported (Friml et al., 2002), is weak or nonexistent under different experimental conditions (B. Harrison and P. Masson, unpublished results), which suggests that substantial functional redundancy may exist.

2.4 PIN4

PIN4 protein is detected as early as the globular stage embryos at the cell surface of hypophysis and the basal end of the uppermost suspensor cell (Friml et al., 2002). As embryogenesis progresses, the PIN4 signal is confined to the basal end of provascular cells. In the developing root, PIN4 accumulates at the apical end (facing the root tip) of the quiescent center, surrounding initials, and their daughter cells (Friml et al., 2002). Consistent with its localization and putative function as an auxin efflux facilitator, loss-of-function *pin4* mutants display specific changes in auxin distribution and pattern aberrations, including changes in rate and plane of cell divisions during early embryogenesis and root development, cell fate determination (revealed by five different cell-type markers), and polar cell expansion (excessive columella differentiation) (Friml et al., 2002).

2.5 PIN6 and PIN7

Even though the function of PIN6 protein remains to be elucidated, promoter activity analysis reveals that the *PIN6* gene is expressed in the earliest stages during the development of lateral root primordia. Subsequently, PIN6:GUS activity is restricted to the primordial margins, where *PIN4* and *PIN7* genes are also expressed (Benkova et al., 2003). Similar to several other *pin* single mutants, *pin6* plants exhibit no obvious growth defects under the various experimental conditions tested (H. Shin and R. Chen, unpublished results).

PIN7 protein is localized in the basal cell of the two-cell stage embryos, both in endomembranes and the apical boundary facing the smaller apical cell. The asymmetry of PIN7 protein in the two-cell stage embryo reflects the earliest polarization of the basal cell, supporting the notion that PIN7 mediates an auxin flux into the smaller apical cell in the early embryo. At the 32-cell stage, the asymmetric localization of PIN7 suddenly reverses, switching to the basal end of the suspensor cells. The reversal of PIN7 localization is accompanied by an apical-to-basal reversal of the auxin gradient revealed by both immunolocalization of auxin using specific anti-IAA antibodies and the expression of auxin-responsive reporters *DR5::GUS* or *DR5::GFP* (Friml et al., 2003).

Consistent with its polar and reversible localization during early embryogenesis, loss-of-function mutations in *PIN7* gene result in misspecification of hypophysis cells and failure in the establishment of the auxin gradient in early embryos (Friml et al., 2003). However, the defects observed in *pin7* mutants are transient and corrected during later stages of embryogenesis, likely due to functional complementation by other PIN proteins such as PIN4 (Friml et al., 2003). Indeed, double, triple, and quadruple mutants of *PIN* genes display more severe defects than single mutants, in agreement with the phenotype of wild-type embryos treated with NPA (Liu et al., 1991). In some combinations, seedling lethal phenotypes are observed (Friml et al., 2003; Blilou et al., 2005).

PIN genes and their corresponding proteins exhibit spatial- and temporal-specific expression/localization patterns. However, these patterns are not static and appear to vary in response to both internal and external factors (Peer et al., 2004). For example, in single, double, and triple *pin* mutants ectopic expression of existing PIN proteins has been observed, which explains why many single mutants exhibit only mild defects if at all (Blilou et al., 2005; Paponov et al., 2005). On the other hand, dynamic changes in *PIN* gene expression and protein localization may reflect the multiple points of regulation needed for establishment of auxin gradients, including the effect of auxin accumulation on the establishment and maintenance of polar auxin transport routes, as suggested by the canalization hypothesis (Sachs, 1991, 2000).

3 Regulatory Mechanisms of Polar Auxin Transport

3.1 Regulation of PIN Protein Plasma-Membrane Recycling by Vesicular Trafficking

Brefeldin A (BFA), a fungal toxin that specifically inhibits guanine–nucleotide exchange factor for ADP-ribosylation factors (ARF GEF), selectively blocks secretory pathways from endosomes to the plasma membrane (PM), while it still allows endocytosis to occur (for recent reviews see Neubenführ et al., 2002; Šamaj et al., 2004). When the plant samples are treated with BFA, the plasma-membrane localized PIN1 protein is shifted to intracellular BFA compartments (Steinmann et al., 1999; Geldner et al., 2001, 2003; Baluška et al., 2002). These BFA compartments are also decorated with the endocytic marker FM4-64 dye, suggesting that PIN1 protein is endocytosed (Geldner et al., 2004). Pretreatments with the actin depolymerization drug cytochalasin D (CytD) prevented PIN1 internalization in the presence of BFA, suggesting that an intact actin cytoskeleton is required for this process. The effect of BFA on PIN1 trafficking is reversible, as shown in BFA washout experiments, which suggests that PIN1 and some other plasma-membrane proteins such as H^+-ATPase and syntaxin KNOLLE rapidly recycle between the plasma membrane and endosomes via endo- and exocytosis (Steinman et al., 1999). Again, in the presence of CytD, the reversible effect of BFA on PIN1 recycling is interrupted, revealing that the endosomal secretion/exocytosis of PIN1 is actin-dependent as well. Investigation of the cellular components involved in PIN1 internalization and cycling reveals that vesicular trafficking mediated by BFA-sensitive ARF GEF GNOM is important, since PIN1 polar localization is interrupted in *gnom* early embryos, with concomitant defects in cell polarity and patterning (Geldner et al., 2001, 2003, 2004).

The plasma-membrane localized PIN2 protein is rapidly internalized in root epidermal and cortical cells in the cell division and elongation zones when treated with the vesicular trafficking inhibitor BFA, suggesting that a similar mechanism to that regulating PIN1 vesicular trafficking also operates in this case (Grebe et al., 2003; Boonsirichai et al., 2003; Blilou et al., 2005; Xu et al., 2005; Shin et al., 2005). Investigation of the role of ADP-ribosylation factors (ARFs) in the vesicular trafficking of PIN2 provides some new insights into the regulatory process. It has been shown that the small GTPase ARF1 is localized to PM and BFA-induced endosomal compartments (Baluška et al., 2002). Furthermore, a recent study by Xu and Scheres (2005) indicated that overexpression of the wild-type ARF1 does not obviously affect the endocytosis of PIN2 and other plasma-membrane proteins. However, overexpression of dominant negative ARF1-Q71L (GTP-locked) and ARF1-T31N (GDP-locked) caused slow internalization of a PIN2-eGFP fusion protein when compared with changes in Golgi and endocytic markers, suggesting that ARF1 also influences vesicular cycling of the auxin efflux carrier PIN2. The discrepancy in kinetics between BFA effects on PIN2 and that of ARF dominant negative mutations suggests that ARF1 mediates PIN2 internalization in an indirect way. Like GNOM, ARF1 belongs to a gene family. It is likely that other ARFs, along with GNOM and/or other BFA-sensitive ARF GEFs, may be directly involved in PIN2 recycling (Xu and Scheres, 2005).

PIN3 protein localization in root columella cells appears to undergo rapid cycling between the plasma membrane and endosomal vesicles. This BFA-sensitive and actin-dependent cycling has been proposed to contribute to the relocalization of PIN3 to the apical PM upon gravistimulation (Friml et al., 2002). In general, vesicular trafficking regulated by both ARFs and ARF GEFs appears to be directly and/or indirectly involved in the recycling of various PIN proteins between the PM and endosomal compartments (Geldner et al., 2003, 2004; Grebe et al., 2003; Xu and Scheres, 2005). However, this mechanism appears to be more general since it also regulates other plasma-membrane proteins including AUX1, PM-ATPase, KNOLLE, and DnaJ-domain-containing protein ARG1 (Steinman et al., 1999; Baluška et al., 2002; Grebe et al., 2002, 2003; Geldner et al., 2003, 2004; Boonsirichai et al., 2003). On the contrary, serine/threonine protein kinases and phosphatases have been implicated in specific regulation of the polarity and function of PIN proteins.

3.2 Regulation of PIN Protein Localization and Activity by Protein Kinases and Phosphatases

The molecular mechanisms underlying the polarity of the auxin transport machinery remain to be elucidated. Recently, both protein kinases and phosphatases have been implicated in the regulation of auxin transport. Loss-of-function mutations in the *RCN1* gene (RCN=root curling in the presence of NPA) result in plants defective in root curling, apical hook formation, and root and shoot elongation (Garbers et al., 1996; Deruere et al., 1999). Furthermore, mutant plants also exhibit an elevated level of root basipetal auxin transport (Rashotte et al., 2001), and an impaired sensitivity to the polar auxin transport inhibitor NPA of auxin accumulation in hypocotyls (Garbers et al., 1996; Rashotte et al., 2001). The *Arabidopsis RCN1* gene encodes one of three A regulatory subunits of protein phosphatase 2A, a heterotrimeric serine/threonine protein phosphatase (DeLong et al., 2002; Luan, 2003; Zhou et al., 2004).

Protein phosphatase 2A (PP2A) has been implicated in regulating PIN2 auxin efflux activity in the root elongation zone, since interruption of PP2A activity by high doses of PP1 and PP2A inhibitors such as cantharidin and okadaic acid inhibits PIN2-mediated root basipetal auxin transport and an auxin response in the root elongation zone, resulting in inhibition of the root gravitropic response (Shin et al., 2005). PIN2 protein appears to be a direct target of the high doses of cantharidin treatments, since the reduced root basipetal auxin transport in *agr1* mutant plants is not further inhibited by cantharidin (Shin et al., 2005). Regulation of auxin transport by protein phosphatases appears to be complex. Treatment with low doses of cantharidin or okadaic acid, similar to *rcn1* mutations, result in an elevated level of basipetal auxin transport. These dosage-dependent opposing effects of cantharidin and okadaic acid could be explained by the existence of over 250 potential isoforms of PP2A heterotrimeric complexes in *Arabidopsis*, whose activities may be differentially targeted by chemical inhibitors in a dosage-dependent manner or by mutations in regulatory subunits of PP2A (Zhou et al., 2004; Shin et al., 2005)

Loss-of-function mutations in the *PINOID* (*PID*) gene, which encodes a serine/threonine protein kinase, result in pin-shaped inflorescence stems similar to that of the *pin-formed* (*pin1*) mutant (Christensen et al., 2000; Benjamins et al., 2001; Furutani et al., 2004). While *pid* mutants do not exhibit obvious root defects, constitutive overexpression of *PID* (*35S::PID*), but not of the mutated kinase-negative MPID (*35S::MPID*), causes defects in hypocotyl and root gravitropic responses, and eventually leads to loss of the primary root meristem function in transgenic plants (Christensen et al., 2000; Benjamins et al., 2001). These various defects are accompanied by a reduced auxin accumulation in the primary root tip and apical-to-basal shift in the polarity of several PIN proteins, including PIN1, PIN2, and PIN4 in subepidermal cells (Friml et al., 2004). Furthermore, when *PID* is overexpressed in young embryos (driven by *RPS5A* promoter), polar localization of PIN1 is interrupted. Consistent with the role of PIN1, PIN4, and PIN7 in auxin transport and embryo patterning (Friml et al., 2003), *RPS5A* ≫ *PID* globular and heart-stage embryos are incapable of establishing an auxin maximum in the hypophysis leading to patterning defects (Friml et al., 2004).

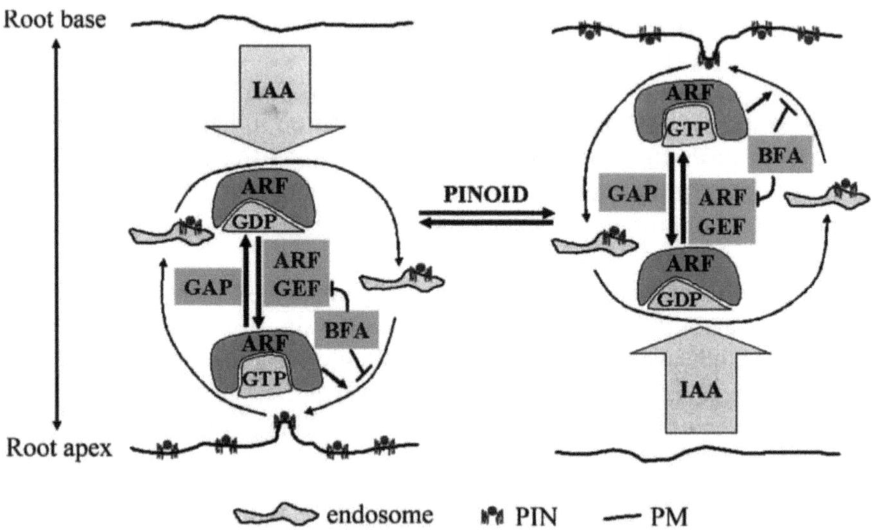

Fig. 3 Regulation of the polarity of PIN proteins by vesicle trafficking and serine/threonine protein kinase PINOID. The small GTPase ADP-ribosylation factor (ARF) undergoes transformation changes to the inactive GDP-binding or active GTP-binding form, regulated by GTPase-activating protein (GAP) and guanine–nucleotide exchange factor (ARF GEF), respectively. BFA inhibits BFA-sensitive ARF GEF, resulting in accumulation of PIN proteins in endosomes. When PINOID is constitutively expressed, the apical localization (facing the root apex) of certain PIN proteins in roots shifts to a basal position (facing the root–shoot junction), thus reversing the direction of auxin flow (*arrow*). Auxin may be involved in a feedback regulation of its own transport by positively regulating the transcription of *PINOID*

In the pin-shaped inflorescence apices of loss-of-function *pid* mutants, a functional PIN1-eGFP fusion protein is localized to the basal end of epidermal cells, in contrast to the apical localization in wild-type plants (Friml et al., 2004). This apical-to-basal shift of PIN1 localization is specific to the loss-of-function mutations in the *PID* locus, since it was not observed in morphologically comparable pin-shaped inflorescence apices of wild-type plants induced by NPA or in *35S::PID* plants. Thus, PID plays a more specific role in PIN localization than the vesicle trafficking regulator GNOM. Furthermore, PID may act as a binary switch of PIN protein localization, with suboptimal levels (as observed in *pid* loss-of-function mutants) leading to basal and apical PIN1 localization in stems and roots, respectively, whereas superoptimal PID levels (as observed in constitutive overexpression plants) result in an opposite PIN1 distribution in the same organs (Friml et al., 2004; Fig. 3).

4 PIN Proteins and Gravisensing in Plants

4.1 PIN Proteins and DnaJ-Domain Proteins

PIN3 is the first PIN protein localized in the gravity-sensing statocytes and implicated in regulating tropic responses. While gravity signal transduction within the root cap statocytes remains poorly understood, it appears to involve a perception of gravity through the sedimentation or weight exerted by starch-filled amyloplasts on sensitive membranes (Chen et al., 2002; Blancaflor and Masson, 2003). This action may promote the opening of mechanosensitive channels and the activation of a gravity signal transduction pathway leading to rapid cytoplasmic alkalinization of statocytes and changes in PIN3 distribution within these cells. Several genes have been isolated through genetic screens, whose predicted protein products likely contribute directly to gravity signal transduction. Amongst these are the ARG1 and ARL2 proteins that appear to be excellent candidates as signal transducers. Indeed, knockout mutations in these genes result in specific defects in root and hypocotyl gravitropism with no pleiotropic phenotypes. Furthermore, the expression of *ARG1* in root cap statocytes or hypocotyl endodermal cells of the *arg1* mutant results in complete rescue of the gravitropic defect in the corresponding organ (Boonsirichai et al., 2003).

ARG1 and ARL2 encode paralogous J-domain proteins with a predicted coiled coil domain at their C-terminus, and a hydrophobic spacer domain in the middle (Sedbrook et al., 1999; Boonsirichai et al., 2003; Guan et al., 2003). Interestingly, ARG1 was found to be a peripheral membrane protein that associates with components of the vesicular trafficking pathway, which is consistent with its involvement in PIN3 vesicular trafficking between the endosomal compartments and plasma membrane of root cap statocytes. Analysis of the expression of the auxin-responsive *DR5::GUS* reporter suggests that ARG1 and ARL2 are needed for proper auxin redistribution across the cap. In mutant roots, *DR5::GUS* expression extends to two external columns of columella cells relative to wild type. Furthermore, upon gravistimulation, mutant root tips do not undergo lateral activation of *DR5::GUS* expression in the bottom cells of the lateral cap, in agreement with a defect in lateral polarity development in response to gravistimulation. This process is accompanied by an inability of mutant root tip statocytes to undergo PIN3 relocalization in response to gravistimulation (B. Harrison and P. Masson, unpublished results). Together, these data are consistent with an involvement of ARG1 and ARL2 in controlling PIN3 cycling between the plasma membrane and endosomes of both vertical and gravistimulated root tips, although a possible role for these proteins in controlling the activity of auxin transporters cannot be excluded at this time (Boonsirichai et al., 2003). Work is under way to determine how ARG1 and ARL2 modulate PIN3 localization within the root cap statocytes.

4.2
PIN Proteins and Actin Cytoskeleton

As discussed earlier, the intact actin cytoskeleton is required for both endocytosis and exocytosis of PIN proteins (Steinman et al., 1999). The actin cytoskeleton has also been implicated in regulating gravitropism, since interruption of the intact actin microfilaments by the actin depolymerization drug Latrunculin B (LatB, 100 nM) enhances the gravitropic responses in several different plant species (Hou et al., 2004). Detailed analyses of the effect of LatB indicate that it interrupts the fine actin filament network in different regions of the treated roots, alters the dynamics of amyloplast sedimentation in the columella cells, and prolongs the intracellular alkalinization response. These abnormalities are accompanied by an early and persistent lateral auxin gradient in the root cap, as revealed by the auxin response *DR5::GUS* reporter expression (Hou et al., 2004).

Together, the results of these pharmacological studies are difficult to reconcile with a simple role of actin microfilaments in vesicular trafficking of PIN proteins within the gravity-perceiving root cap cells, and their relocalization in response to gravistimulation. However, it should be noticed that different concentrations of LatB (100 nM) or cytochalasin D (20 μM) were used in the gravitropism and PIN recycling studies (Hou et al., 2004; Steinman et al., 1999), and it is reasonable to speculate that these drugs may have different effects/targets when used at different concentrations. It is also possible that the actin microfilaments are inhibitory to the signal transduction pathway at the later phase of graviresponse before roots overshoot the vertical (Hou et al., 2004). In the presence of LatB, this inhibitory effect could be alleviated by efficient endocytosis of PIN3 proteins from the lower plasma membrane of gravistimulated statocytes, leading to an enhanced graviresponse (for the function of actin in endocytosis, see the chapter by Šamaj et al., this volume).

5 Conclusions and Future Prospects

Increasing evidence supports an emerging model of the PIN efflux system-dependent auxin gradient in the control of growth and patterning in roots and other organs of *Arabidopsis thaliana* (Benkova et al., 2003; Blilou et al., 2005). A central player in the determination of PIN protein polarity, and hence auxin gradient establishment/maintenance, is the serine/threonine protein kinase PINOID. Auxin may be involved in a feedback regulation of its own transport through positive regulation of *PINOID* transcription, or Ca^{2+} signaling (Benjamins et al., 2001, 2003; Friml et al., 2004; for a recent review see Leyser, 2005). Another regulator of auxin transport is the heterotrimeric serine/threonine protein phosphatase 2A, which likely regulates the auxin efflux rate rather than polarity, since it modulates the auxin efflux activity of PIN2 (Shin et al., 2005). However, the direct targets of these protein kinases and phosphatases remain to be identified.

Vesicular trafficking regulated by small GTPases of the ARF subfamily, GTPase-activating protein (GAP) and ARF GEF GNOM, plays an important role in the dynamic cycling of PIN proteins between the plasma membrane and endosomal compartments. It provides a potential mechanism for rapid retargeting of PIN proteins to different regions of the plasma membrane in response to developmental and environmental cues (Muday and Murphy, 2003; Jurgens, 2004; Molendijk et al., 2004; Murphy et al., 2004). Potential membrane targeting mechanisms may include protein–protein interactions between PIN and NPA-binding MDR-type ABC transporters and protein–lipid interactions targeting PIN proteins to lipid raft membrane microdomains, a mechanism used for controlling membrane receptor signaling in mammalian cells (Grebe et al., 2003; Willemsen et al., 2003; Murphy et al., 2004; Le Roy and Wrana, 2005).

PIN genes have been identified and the corresponding proteins localized in both monocot and dicot plants (Baluška et al., 2002; Ni et al., 2002; Šamaj et al., 2004; Schnabel and Frugoli, 2004; Paponov et al., 2005). They represent a unique class of proteins implicated in auxin transport regulation in plants. Future challenges lie in the elucidation of the biological functions of these PIN and PIN-like proteins in *Arabidopsis* and other plant species, and in the identification of regulators and cellular networks, i.e., Ca^{2+}-binding proteins (Benjamins et al., 2003), actin cytoskeleton (Muday, 2000), J-domain proteins, NPA-binding MDR-type ABC transporters (Noh et al., 2003), protein kinases and phosphatases (DeLong et al., 2002), and the Calossin-like BIG protein (Gil et al., 2001), which control the polarity and activities of PIN and PIN-like proteins.

Acknowledgements RC would like to thank Jeffery Long (Salk), Ben Scheres, and Jian Xu (Utrecht University) for providing the PIN1-eGFP and PIN2-eGFP lines, respectively. We thank Dr. Jozef Šamaj for comments on the manuscript. Research was supported by a startup fund from the Noble Foundation to RC, and by NSF, NASA, and HATCH funds to PHM.

References

Aida M, Vernoux T, Furutani M, Traas J, Tasaka M (2002) Roles of PIN-FORMED1 and MONOPTEROS in pattern formation of the apical region of the *Arabidopsis* embryo. Development 129:3965–3974

Baluška F, Hlavacka A, Šamaj J, Palme K, Robinson DG, Matoh T, McCurdy DW, Menzel D, Volkmann D (2002) F-actin-dependent endocytosis of cell wall pectins in meristematic root cells. Insights from brefeldin A-induced compartments. Plant Physiol 130:422–431

Bell CJ, Maher EP (1990) Mutants of *Arabidopsis thaliana* with abnormal gravitropic responses. Mol Gen Genet 220:289–293

Benjamins R, Quint A, Weijers D, Hooykaas P, Offringa R (2001) The PINOID protein kinase regulates organ development in *Arabidopsis* by enhancing polar auxin transport. Development 128:4057–4067

Benjamins R, Ampudia CS, Hooykaas PJ, Offringa R (2003) PINOID-mediated signaling involves calcium-binding proteins. Plant Physiol 132:1623–1630

Benkova E, Michniewicz M, Sauer M, Teichmann T, Seifertova D, Jurgens G, Friml J (2003) Local, efflux-dependent auxin gradients as a common module for plant organ formation. Cell 115:591–602

Bennett MJ, Marchant A, Green HG, May ST, Ward SP, Millner PA, Walker AR, Schulz B, Feldmann KA (1996) *Arabidopsis* AUX1 gene: a permease-like regulator of root gravitropism. Science 273:948–950

Berleth T, Krogan NT, Scarpella E (2004) Auxin signals—turning genes on and turning cells around. Curr Opin Plant Biol 7:553–563

Blakeslee JJ, Peer WA, Murphy AS (2005) MDR/PGP auxin transport proteins and endocytic cycling (in this volume). Springer, Berlin Heidelberg New York

Blancaflor EB, Masson PH (2003) Plant gravitropism. Unraveling the ups and downs of a complex process. Plant Physiol 133:1677–1690

Blilou I, Xu J, Wildwater M, Willemsen V, Paponov I, Friml J, Heidstra R, Aida M, Palme K, Scheres B (2005) The PIN auxin efflux facilitator network controls growth and patterning in *Arabidopsis* roots. Nature 433:39–44

Boonsirichai K, Sedbrook JC, Chen R, Gilroy S, Masson PH (2003) ALTERED RESPONSE TO GRAVITY is a peripheral membrane protein that modulates gravity-induced cytoplasmic alkalinization and lateral auxin transport in plant statocytes. Plant Cell 15:2612–2625

Chen R, Hilson P, Sedbrook J, Rosen E, Caspar T, Masson PH (1998) The *Arabidopsis thaliana* AGRAVITROPIC 1 gene encodes a component of the polar-auxin-transport efflux carrier. Proc Natl Acad Sci USA 95:15 112–15 117

Chen R, Guan C, Boonsirichai K, Masson PH (2002) Complex physiological and molecular processes underlying root gravitropism. Plant Mol Biol 49:305–317

Christensen SK, Dagenais N, Chory J, Weigel D (2000) Regulation of auxin response by the protein kinase PINOID. Cell 100:469–478

DeLong A, Mockaitis K, Christensen S (2002) Protein phosphorylation in the delivery of and response to auxin signals. Plant Mol Biol 49:285–303

Deruere J, Jackson K, Garbers C, Soll D, Delong A (1999) The RCN1-encoded A subunit of protein phosphatase 2A increases phosphatase activity in vivo. Plant J 20:389–399

Friml J, Wisniewska J, Benkova E, Mendgen K, Palme K (2002) Lateral relocation of auxin efflux regulator PIN3 mediates tropism in *Arabidopsis*. Nature 415:806–809

Friml J, Benkova E, Blilou I, Wisniewska J, Hamann T, Ljung K, Woody S, Sandberg G, Scheres B, Jurgens G, Palme K (2002) AtPIN4 mediates sink-driven auxin gradients and root patterning in *Arabidopsis*. Cell 108:661–673

Friml J, Palme K (2002) Polar auxin transport—old questions and new concepts? Plant Mol Biol 49:273–284

Friml J, Vieten A, Sauer M, Weijers D, Schwarz H, Hamann T, Offringa R, Jurgens G (2003) Efflux-dependent auxin gradients establish the apical-basal axis of *Arabidopsis*. Nature 426:147–153

Friml J (2003) Auxin transport—shaping the plant. Curr Opin Plant Biol 6:7–12

Friml J, Yang X, Michniewicz M, Weijers D, Quint A, Tietz O, Benjamins R, Ouwerkerk PB, Ljung K, Sandberg G, Hooykaas PJ, Palme K, Offringa R (2004). A PINOID-dependent binary switch in apical-basal PIN polar targeting directs auxin efflux. Science 306:862–865

Furutani M, Vernoux T, Traas J, Kato T, Tasaka M, Aida M (2004). PIN-FORMED1 and PINOID regulate boundary formation and cotyledon development in *Arabidopsis* embryogenesis. Development 131:5021–5030

Galweiler L, Guan C, Muller A, Wisman E, Mendgen K, Yephremov A, Palme K (1998) Regulation of polar auxin transport by AtPIN1 in *Arabidopsis* vascular tissue. Science 282:2226–2230

Garbers C, DeLong A, Deruere J, Bernasconi P, Soll D (1996) A mutation in protein phosphatase 2A regulatory subunit A affects auxin transport in *Arabidopsis*. EMBO J 15:2115–2124

Geldner N, Friml J, Stierhof YD, Jurgens G, Palme K (2001) Auxin transport inhibitors block PIN1 cycling and vesicle trafficking. Nature 413:425–428

Geldner N, Anders N, Wolters H, Keicher J, Kornberger W, Muller P, Delbarre A, Ueda T, Nakano A, Jurgens G (2003) The *Arabidopsis* GNOM ARF-GEF mediates endosomal recycling, auxin transport, and auxin-dependent plant growth. Cell 112:219–230

Geldner N, Richter S, Vieten A, Marquardt S, Torres-Ruiz RA, Mayer U, Jurgens G (2004) Partial loss-of-function alleles reveal a role for GNOM in auxin transport-related, post-embryonic development of *Arabidopsis*. Development 131:389–400

Gil P, Dewey E, Friml J, Zhao Y, Snowden KC, Putterill J, Palme K, Estelle M, Chory J (2001) BIG: a calossin-like protein required for polar auxin transport in *Arabidopsis*. Genes Dev 15:1985–1997

Guan C, Rosen E, Boonsirichai K, Poff K, Masson P (2003). The *ARG1-LIKE2* (*ARL2*) gene of *Arabidopsis thaliana* functions in a gravity signal transduction pathway that is genetically distinct from the PGM pathway. Plant Physiol 133:100–112

Hou G, Kramer V, Wang Y, Chen R, Perbal G, Gilroy S, Blancaflor E (2004) The promotion of gravitropism in *Arabidopsis* roots upon actin disruption is coupled with the extended alkalinization of the columella cytoplasm and a persistent lateral auxin gradient. Plant J 39:113–125

Jurgens G (2004) Membrane trafficking in plants. Annu Rev Cell Dev Biol 20:481–504

Kepinski S, Leyser O (2005) Plant development: auxin in loops. Curr Biol 15:R208–210

Kramer EM (2004) PIN and AUX/LAX proteins: their role in auxin accumulation. Trends Plant Sci 9:578–582

Le Roy C, Wrana JL (2005) Clathrin- and non-clathrin-mediated endocytic regulation of cell signalling. Nat Rev Mol Cell Biol 6:112–126

Leyser O (2001) Auxin signalling: the beginning, the middle and the end. Curr Opin Plant Biol 4:382–386

Liu C, Xu Z, Chua NH (1993) Auxin polar transport is essential for the establishment of bilateral symmetry during early plant embryogenesis. Plant Cell 5:621–630

Ljung K, Hull AK, Celenza J, Yamada M, Estelle M, Normanly J, Sandberg G (2005) Sites and regulation of auxin biosynthesis in *Arabidopsis* roots. Plant Cell 17:1090–1104

Lomax TL, Muday GK, Rubery PH (1995) Auxin transport. In: Davies PJ (ed) Plant hormones: physiology, biochemistry, and molecular biology. Kluwer, Dordrecht, The Netherlands, pp 509–530

Luan S (2003) Protein phosphatases in plants. Annu Rev Plant Biol 54:63–92

Luschnig C, Gaxiola RA, Grisafi P, Fink GR (1998) EIR1, a root-specific protein involved in auxin transport, is required for gravitropism in *Arabidopsis thaliana*. Genes Dev 12:2175–2187

Molendijk AJ, Ruperti B, Palme K (2004) Small GTPases in vesicle trafficking. Curr Opin Plant Biol 7:694–700

Muday GK (2000) Maintenance of asymmetric cellular localization of an auxin transport protein through interaction with the actin cytoskeleton. J Plant Growth Regul 19:385–396

Muday GK, Murphy AS (2002) An emerging model of auxin transport regulation. Plant Cell 14:293–299

Murphy AS, Bandyopadhyay A, Holstein SE, Peer WA (2004) Endocytotic cycling of PM proteins. Annu Rev Plant Biol 56:221–251

Nebenführ A, Ritzenthaler C, Robinson DG (2002) Brefeldin A: deciphering an enigmatic inhibitor of secretion. Plant Physiol 130:1102–1108

Ni WM, Chen XY, Xu ZH, Xue HW (2002) Isolation and functional analysis of a *Brassica juncea* gene encoding a component of auxin efflux carrier. Cell Res 12:235–245

Noh B, Bandyopadhyay A, Peer WA, Spalding EP, Murphy AS (2003) Enhanced gravi- and phototropism in plant mdr mutants mislocalizing the auxin efflux protein PIN1. Nature 423:999–1002

Okada K, Shimura Y (1990) Reversible root tip rotation in *Arabidopsis* seedlings induced by obstacle-touching stimulus. Science 250:274–276

Okada K, Ueda J, Komaki MK, Bell CJ, Shimura Y (1991) Requirement of the auxin polar transport system in early stages of *Arabidopsis* floral bud formation. Plant Cell 3:677–684

Palme K, Galweiler L (1999) PIN-pointing the molecular basis of auxin transport. Curr Opin Plant Biol 2:375–381

Paponov IA, Teale WD, Trebar M, Blilou I, Palme K (2005) The PIN auxin efflux facilitators: evolutionary and functional perspectives. Trends Plant Sci 10:170–177

Peer WA, Bandyopadhyay A, Blakeslee JJ, Makam SN, Chen RJ, Masson PH, Murphy AS (2004) Variation in expression and protein localization of the PIN family of auxin efflux facilitator proteins in flavonoid mutants with altered auxin transport in *Arabidopsis thaliana*. Plant Cell 16:1898–1911

Rashotte AM, Brady SR, Reed RC, Ante SJ, Muday GK (2000) Basipetal auxin transport is required for gravitropism in roots of *Arabidopsis*. Plant Physiol 122:481–490

Rashotte AM, DeLong A, Muday GK (2001) Genetic and chemical reductions in protein phosphatase activity alter auxin transport, gravity response, and lateral root growth. Plant Cell 13:1683–1697

Reinhardt D, Mandel T, Kuhlemeier C (2000) Auxin regulates the initiation and radial position of plant lateral organs. Plant Cell 12:507–518

Roman G, Lubarsky B, Kieber JJ, Rothenberg M, Ecker JR (1995) Genetic analysis of ethylene signal transduction in *Arabidopsis thaliana*: five novel mutant loci integrated into a stress response pathway. Genetics 139:1393–1409

Sabatini S, Beis D, Wolkenfelt H, Murfett J, Guilfoyle T, Malamy J, Benfey P, Leyser O, Bechtold N, Weisbeek P, Scheres B (1999) An auxin-dependent distal organizer of pattern and polarity in the *Arabidopsis* root. Cell 99:463–472

Sachs T (1991) Cell polarity and tissue patterning in plants. Development Suppl 1:83–93

Sachs T (2000) Integrating cellular and organismic aspects of vascular differentiation. Plant Cell Physiol 41:649–656

Šamaj J, Baluška F, Voigt B, Schlicht M, Volkmann D, Menzel D (2004) Endocytosis, actin cytoskeleton, and signaling. Plant Physiol 135:1150–1161

Šamaj J, Baluška F, Voigt B, Volkmann D, Menzel D (2005) Endocytosis and actomyosin cytoskeleton (in this volume). Springer, Berlin Heidelberg New York

Schnabel EL, Frugoli J (2004) The PIN and LAX families of auxin transport genes in *Medicago truncatula*. Mol Genet Genomics 272:420–432

Sedbrook J, Chen R, Masson P (1999) ARG1 (Altered Response to Gravity) encodes a DnaJ-like protein that potentially interacts with the cytoskeleton. Proc Natl Acad Sci USA 96:1140–1145

Shin H, Shin HS, Guo Z, Blancaflor EB, Masson PH, Chen R (2005) Complex regulation of *Arabidopsis* AGR1/PIN2-mediated root gravitropic response and basipetal auxin transport by cantharidin-sensitive protein phosphatases. Plant J 42:188–200

Steinmann T, Geldner N, Grebe M, Mangold S, Jackson CL, Paris S, Galweiler L, Palme K, Jurgens G (1999) Coordinated polar localization of auxin efflux carrier PIN1 by GNOM ARF GEF. Science 286:316–318

Swarup R, Marchant A, Bennett MJ (2000) Auxin transport: providing a sense of direction during plant development. Biochem Soc Trans 28:481–485

Swarup RA, Bennett MJ (2003) Auxin transport: the fountain of life in plants? Dev Cell 5:824–826

Utsuno K, Shikanai T, Yamada Y, Hashimoto T (1998) Agr, an Agravitropic locus of *Arabidopsis thaliana*, encodes a novel membrane-protein family member. Plant Cell Physiol 39:1111–1118

Vernoux T, Kronenberger J, Grandjean O, Laufs P, Traas J (2000) PIN-FORMED 1 regulates cell fate at the periphery of the shoot apical meristem. Development 127:5157–5165

Weijers D, Jurgens G (2004) Funneling auxin action: specificity in signal transduction. Curr Opin Plant Biol 7:687–693

Weijers D, Jurgens G (2005) Auxin and embryo axis formation: the ends in sight? Curr Opin Plant Biol 8:32–37

Willemsen V, Friml J, Grebe M, van den Toorn A, Palme K, Scheres B (2003) Cell polarity and PIN protein positioning in *Arabidopsis* require STEROL METHYLTRANSFERASE1 function. Plant Cell 15:612–625

Woodward AW, Bartel B (2005) Auxin: regulation, action, and interaction. Ann Bot (Lond) 95:707–735

Zazimalova E, Napier RM (2003) Points of regulation for auxin action. Plant Cell Rep 21:625–634

Zhou HW, Nussbaumer C, Chao Y, DeLong A (2004) Disparate roles for the regulatory A subunit isoforms in *Arabidopsis* protein phosphatase 2A. Plant Cell 16:709–722

Plant Cell Monogr (1)
J. Šamaj · F. Baluška · D. Menzel: Plant Endocytosis
DOI 10.1007/7089_010/Published online: 22 September 2005

MDR/PGP Auxin Transport Proteins and Endocytic Cycling

Joshua J. Blakeslee · Wendy Ann Peer · Angus S. Murphy (✉)

Department of Horticulture, Purdue University, West Lafayette, IN 47907, USA
murphy@purdue.edu

Abstract Auxin is an essential regulator of plant growth and development. Polarized transport of auxin is responsible for apical dominance, tropic growth, and organ development. Previous studies have demonstrated that the polarized movement of auxin is dependent upon the action of polarly localized, endocytotically cycled PIN auxin efflux facilitator proteins. More recently, plant orthologs of mammalian multidrug-resistance (MDR)/P-glycoprotein (PGP) type ABC transporters have been shown to function in auxin transport. In this review, the PGP nomenclature/numbering system established by Martinoia et al., (*Planta* 214:345–355,2002) is used, as there is increasing evidence that in plants MDR/PGPs function as PGPs and not as multiple specificity MDR proteins. Defects in *PGP1* and *PGP19* (*MDR1*)genes result in decreased auxin transport and reduced growth phenotypes in *Arabidopsis* (*pgp1*, *pgp19*), maize (*br2*), and sorghum (*dw3*). Further, dwarf phenotypes are more severe in *Arabidopsis* double mutants, indicating that PGPs have overlapping functions. More recently, MDR/PGPs have been shown to function as ATP-activated hydrophobic anion transporters capable of auxin transport. Further, MDR/PGPs have been shown to stabilize PIN1 in detergent-resistant membrane microdomains, and synergistic MDR/PGP-PIN interactions have been shown to increase the rate and specificity of MDR/PGP-mediated auxin transport. Several lines of evidence indicate that, like their mammalian counterparts, *Arabidopsis* MDR/PGPs are regulated via endocytic cycling. Here we review the evidence for endocytic cycling of MDR/PGPs *in planta* and provide a model by which this cycling could occur.

1
Introduction to Auxin Transport

The plant hormone auxin is an essential regulator of plant growth and development. The primary auxin, indole-3-acetic acid (IAA), is synthesized primarily in young tissues at the shoot tip and transported to the root tip, where it is redirected basipetally through root cortical and epidermal tissues. Polarized auxin transport provides directional and positional information for developmental processes such as vascular differentiation, apical dominance, organ development and tropic growth (Benkova et al., 2003; Blancaflor and Masson, 2003; Blilou et al., 2005). The importance of auxin transport in normal plant growth is demonstrated by the severe developmental defects exhibited by plants treated with auxin transport inhibitors or carrying mutations that reduce auxin transport (reviewed in Friml, 2003) Polar transport of IAA is regulated at the cellular level, and is best described by a chemiosmotic model in which plasma membrane (PM) ATPases generate an H^+ gradient between the neutral cytoplasm and the acidic extracellular space (Lomax et al., 1995; Swarup and Bennett, 2003). Lipophilic absorption of apoplastic IAAH into the PM is augmented by a tissue-specific, gradient-driven H^+ symport activity characterized by the AUX1 family of proteins (Lomax et al., 1995; Swarup and Bennett, 2003). In the neutral cytoplasm, IAA is found almost exclusively in the lipid insoluble anionic form and can only exit the cell via efflux carriers. The polar bias of IAA efflux is attributed to highly regulated, polarly localized efflux complexes characterized by the PIN family of proteins (Friml, 2003; Friml and Palme, 2002; Chen and Masson, this volume).

2
Auxin Efflux Inhibitors

Auxin efflux inhibitors (AEIs) are important tools for auxin transport studies. AEIs such as triiodobenzoic acid (TIBA) compete with IAA at the site of cellular efflux, while AEIs such as 1-*N*-naphthylpthalamic acid (NPA) and cyclopropyl propane dione are noncompetitive inhibitors that interact with regulatory sites of the auxin efflux complex (Muday et al., 2003). NPA has been used extensively to characterize auxin efflux from plant tissues and membrane vesicles (Muday et al., 2003). NPA binding is seen in all tissues and has long been used as a PM marker (Katekar and Geissler, 1979; Lomax et al., 1995). In etiolated zucchini hypocotyls, both integral and peripheral NPA-binding sites have been identified on the PM (Bernasconi et al., 1996; Dixon et al., 1996; Jacobs and Rubery, 1988). Other studies identified high- and low-affinity NPA-binding sites in these same membranes and noted that NPA binding correlates poorly with auxin transport inhibition, especially in the presence of observed *in planta* NPA amidase activity (Geissler et al., 1985; Michalke et al., 1992; Murphy and Taiz, 1999a,b). Further, some morphological changes observed in *Arabidopsis thaliana* seedlings exposed to NPA are only seen at levels of NPA treatment that are higher than those needed to inhibit auxin transport (Geissler et al., 1985, Michalke et al., 1992; Murphy and Taiz, 1999a, b). In *Arabidopsis thaliana* seedlings, NPA has been shown to accumulate and undergo light-dependent hydrolysis in the root/shoot apices and the root-shoot transition zone (Murphy and Taiz, 1999a, b). NPA hydrolysis was shown to be sensitive to aminopeptidase and ABC transporter inhibitors and requires two components: a high-affinity NPA-binding component and an NPA amidase activity (Murphy and Taiz, 1999a, b). High-affinity NPA binding was attributed to an integral membrane protein, while NPA amidase activity was found primarily in peripheral membrane fractions (Murphy and Taiz, 1999a, b). Studies demonstrating colocalization of flavonols in *A. thaliana* and displacement of NPA from microsomal vesicles by quercetin and kaempferol suggest that NPA mimics the activity of natural flavonoid inhibitors of auxin transport (Bernasconi et al., 1996; Jacobs and Rubery, 1988; Murphy et al., 2000). Flavonols have subsequently been shown to negatively regulate auxin transport in *A. thaliana* (Brown et al., 2001; Peer et al., 2001).

3 The PIN Family of Auxin Efflux Facilitators

PIN proteins, named for the pin-formed inflorescence phenotype of the *pin1* mutant (Okada et al., 1991), align with the vector of auxin transport, and are necessary for normal polarized plant development and auxin movement (Benkova et al., 2003; Blilou et al., 2005; Chen and Masson, this volume). Further, *pin* mutants exhibit altered auxin transport and phenotypes that can be phenocopied by AEI treatment (Friml and Palme, 2002). The *A. thaliana* PIN family consists of eight members that have distinct, yet overlapping, functions in auxin transport (Benkova et al., 2003; Blilou et al., 2005; Friml, 2003; Friml and Palme, 2002). PIN1 plays an essential role in the basipetal transport of auxin in shoot tissues, and acropetal transport in root tissues (Galweiler et al., 1998; Palme and Galweiler, 1999). PIN2/AGR1/EIR2 is expressed in roots, is required for gravitropic bending, and functions in both ethylene responses and basipetal auxin transport from the root tip (Chen et al., 1998; Muller et al., 1998). PIN2, which is most abundant in the root apex and elongation zone, exhibits basal localization in epidermal cells, but primarily apical localization in cortical cells (Chen et al., 1998; Muller et al., 1998; Friml et al., 2004). PIN3 displays a lateral localization in the shoot endodermis, but a more apolar localization in root columella and pericycle cells (Friml et al., 2002b). Upon gravitropic stimulation, PIN3 is laterally relocalized in root tissues, and is hypothesized to play a role in the lateral redistribution of auxin involved in gravitropism in roots and in both phototropic and gravitropic growth in shoots (Friml et al., 2002). PIN4 is hypothesized to function as an auxin sink located basally to the quiescent center of the root apical meristem (Friml et al., 2002). PIN7 has been implicated in the formation of apical–basal auxin gradients essential for the establishment and maintenance of embryonic polarity (Friml et al., 2003). Normal plant development is dependent upon multiple PIN family members functioning together in order to create localized auxin gradients responsible for organ/meristem formation and patterning (Benkova et al., 2003; Blilou et al., 2005).

4 Endocytic Cycling of PIN Proteins

Auxin efflux complexes characterized by PIN1 and PIN3 appear to be functionally regulated by endocytic cycling and undergo rapid, actin-dependent shuttling between the PM and endomembrane compartments (Friml, 2003; Geldner et al., 2001, 2003, Chen and Masson, this volume). When treated with the membrane trafficking inhibitor brefeldin A (BFA), PIN1 and PIN3 reversibly aggregate in endocytic juxtanuclear compartments (Friml et al., 2002; Geldner et al., 2001), in a fashion similar to that observed with both the PM H^+-ATPase (Geldner et al., 2001; Baluška et al., 2002) and the auxin efflux carrier AUX1 in protophloem cells (Grebe et al., 2002). Additionally, in epidermal cells, BFA treatment resulted in a more diffuse, cytoplasmic localization of the AUX1 protein (Grebe et al., 2002). The BFA-sensitive endosomal cycling of PIN1 was further investigated and was confirmed in maize where a PIN1 ortholog was localized to endosomal BFA compartments together with internalized cell wall pectins (Baluška et al., 2002; Šamaj et al., 2004). BFA inhibition of PIN1 cycling is directly mediated by the GNOM/EMB30 ADP-ribosylation factor (ARF) GDP/GTP exchange factor (ARF-GEF) (Geldner et al., 2003), and GNOM is required for proper PIN1 localization and function. Additionally, BFA washout with TIBA or high concentrations of NPA results in continued internalization of PIN1 (Geldner et al., 2001). However, the concentrations of NPA required to inhibit internalized PIN1 restoration to the PM are substantially higher (2 orders of magnitude) than those needed to inhibit auxin transport and do not result in disruptions in the actin cytoskeleton or structure of the endoplasmic reticulum, arguing against NPA action being solely the result of altered vesicular trafficking (Geldner et al., 2001; Muday et al., 2003; Petrasek et al., 2003).

5 Isolation of NPA-Binding Proteins from *A. thaliana*

NPA affinity chromatography in conjunction with aminopeptidase enzyme activity and flavonoid binding assays was recently used to identify NPA-binding protein complexes in detergent-resistant membrane microdomains from *A. thaliana* seedlings (Murphy et al., 2002). Consistent with NPA binding and hydrolysis studies conducted on whole seedlings and microsomes, NPA affinity chromatography identified both high- and low-affinity NPA binding fractions (Murphy et al., 2002). High-affinity NPA binding fractions were found to contain plant orthologs of mammalian multidrug-resistance (MDR)/P-glycoproteins (PGPs) and smaller quantities of the glycosylphosphatidylinositol (GPI)-anchored immunophilin, TWD1 (Murphy et al., 2002). MDR/PGPs have been implicated in auxin transport in *A. thaliana*, maize, and sorghum (Geisler et al., 2003; Multani et al., 2003; Noh et al., 2001). Low-affinity NPA binding fractions were characterized by APM1, a member of the gluzincin dual function aminopeptidase/protein trafficking family of which the mammalian insulin responsive aminopeptidase (IRAP) and aminopeptidase N (APN) are the best-characterized members (Murphy et al., 2002).

IRAP is essential for the vesicular cycling of the asymmetrically localized mammalian glucose transporter GLUT4 (Baumann and Saltiel, 2001) and human APN is a modulator of cellular sterol uptake (Kramer et al., 2005). APM1 is a 103-kDa transmembrane aminopeptidase with functionally distinct membrane trafficking and catalytic domains and, is found both on the PM and in low-density endomembrane vesicles that do not colocalize with standard Golgi PM trafficking markers (Murphy et al., 2002; Bandyopadhyay and Murphy, unpublished results). Recent evidence that the trafficking domain of APN is the specific target of the sterol endocytosis inhibitor ezetimibe (Kramer et al., 2005) suggests that APN/aminopeptidase M (APM) mediation of sterol endocytosis may regulate the formation of these light vesicles (Kramer et al., 2005). APM1 was copurified with multiple proteins associated with membrane protein secretion and vesicular cycling, including patellin (Peterman et al., 2004), β-adaptin, the dynamin-like protein ADL1a/DRP1a (Kang et al., 2003), HSP70p, and calreticulin (Murphy et al., 2002).

6 The Role of APM1 in Vesicular Cycling

The IRAP-mediated vesicular trafficking mechanism of GLUT4 in mammals bears a striking similarity to the dynamic cycling of asymmetrically localized plant auxin transport proteins such as PIN1 and AUX1 (Geldner et al., 2001; Muday and Murphy, 2002; Swarup et al., 2001), suggesting that APM1 may regulate the subcellular localization of PM transport proteins, including auxin efflux carriers. The trafficking of MDR1, MDR2, and SPGP in mammalian hepatic canalicular cells is thought to be mediated by a mechanism similar to IRAP-GLUT4 trafficking (Kipp and Arias, 2002). Direct protein–protein interactions observed between APM1, TWD1, PGP1, and PGP19/MDR1 (Geisler et al., 2003; Lee and Murphy, unpublished results) suggest that APM1 regulates vesicular trafficking of MDR/PGP proteins in a similar manner in *A. thaliana*. Elucidation of the mechanisms underlying the endocytic cycling of mammalian MDR/PGPs may, therefore, provide a useful model for the regulation of plant ABC transporters of the MDR/PGP subclass.

7 The Role of MDR/PGPs in Auxin Transport

There is increasing evidence that plant orthologs of mammalian MDR/PGPs are components of auxin efflux and, possibly, influx complexes (Geisler et al., 2003, 2005; Noh et al., 2001, 2003; Terasaka et al., 2005). MDR/PGPs are members of the ABC transporter superfamily and generally consist of two homologous halves, each containing a nucleotide-binding fold/ATP-hydrolysis site and six transmembrane helices, joined by a flexible linker region (Ambudkar et al., 1999). In *A. thaliana*, the MDR/PGP subfamily consists of 21 members with overlapping patterns of expression (Jasinski et al., 2003; Martinoia et al., 2002) (Fig. 1). Defects in *PGP1* and *PGP19* (*MDR1*) genes result in decreased auxin transport and reduced-growth phenotypes in *Arabidopsis* (*pgp1*, *pgp19*), maize (*br2*), and sorghum (*dw3*) (Geisler et al., 2003, 2005; Multani et al., 2003; Noh et al., 2001). Dwarf phenotypes are more severe in *Arabidopsis* double mutants, indicating that PGPs have overlapping functions (Geisler et al., 2003).

7.1 MDR/PGPs Stabilize PIN1 Localization in the PM in Detergent-Resistant Membrane Microdomains

A mechanistic explanation for MDR/PGP function in auxin transport was suggested when PIN1 was found to be mislocalized in xylem parenchyma cells of hypertropic *Arabidopsis atpgp19* hypocotyls. PIN1 localization was not different from that in the wild type in *pgp1* roots (Geisler et al., 2003) and only slightly disturbed in *atpgp19* roots that exhibit only marginally altered graviresponsiveness; PIN2 localization was not different from that in the wild type in these same roots (Bandyopadhyay and Murphy, unpublished; Noh et al., 2003). These results suggest that PGP19/MDR1 regulates auxin transport by functioning in either the distribution and trafficking of PIN1 or the stabilization of PIN1 on the PM. It has been suggested that it is more likely that PGP19 mediates PIN1 stability at the PM rather than trafficking of PIN1 to the membrane (Blakeslee et al., 2005). In Noh et al., (2003), PIN1 immunolocalizations utilized a high concentration of the detergent Triton X-100 as a permeabilizing antibody carrier. Reduction of Triton X-100 concentrations or substitution of Triton X-100 with other nonionic detergents in the antibody treatments resulted in apparently normal basal localization of PIN1 in the xylem parenchymal cells of *pgp19* hypocotyls; conversely, when Triton X-100 concentrations used in root immunofluorescence visualizations were increased, the result was disruption of PIN1 localization in *pgp19*, but not in the wild type (Bandyopadhyay and Murphy, unpublished).

The identification of PGPs in sterol-rich, detergent-resistant, lipid-raft-like microdomains (DRMs; Mongrand et al., 2004; Shogomori and Brown, 2003; Oveèka and Lichtscheidl, this volume) in *A. thaliana* (Murphy et al., 2002; Borner et al., 2005) is consistent with the observed effects of different detergent treatments on PIN1 immunolocalization in *pgp19* hypocotyls. In animals, DRM complexes involved in the formation and stabilization of multiprotein complexes in the PM are biochemically defined by their insolubility in relatively high concentrations of Triton X-100 (Shogomori and Brown, 2003). Human MDR1/PGP1 has been localized to DRMs and caveolae (Lavie et al.,

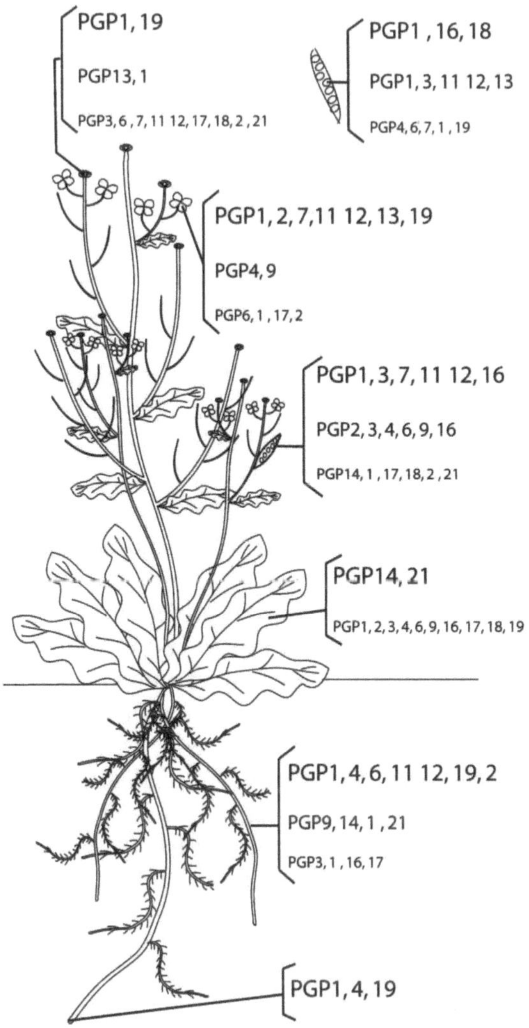

Fig. 1 Members of the multidrug-resistance (*MDR*)/P-glycoprotein (*PGP*) family have overlapping expression patterns in *Arabidopsis*. *In silico* investigation of the expression pattern of the 21 members of the MDR/PGP subfamily of ABC transporters in *Arabidopsis* reveals that these proteins have distinct, yet overlapping, tissue-specific expression patterns. In each organ, one to five MDR/PGPs are expressed at high levels (*large font*). Additionally, several other MDR/PGPs are expressed at either moderate (*medium-sized font*) or low (*small font*) levels in each organ

1998; Lavie and Liscovitch, 2000; Luker et al., 2000; Troost et al., 2004), where MDR/PGPs have been shown to maintain cholesterol distribution and membrane stability (Garrigues et al., 2002; Wang et al., 2000). MDR/PGPs are thought to stabilize DRMs as a result of their inherent membrane stability and, in some cases, by their phospoholipid "flippase" function (Borst et al., 2000; vanHelvoort et al., 1996).

Several lines of evidence suggest that PGP19 plays a similar role in stabilizing PIN1 in DRM-associated complexes in *Arabidopsis*. DRM fractions containing PGP1 and PGP19 were found to be enriched in C18 fatty acids, glucosyl ceramide, and sitosterol, correlating well with the sterol-rich composition of mammalian and plant DRMs (Blakestee and Murphy, unpublished; Borner et al., 2005; Mongrand et al., 2004). These fractions also contained the GPI-anchored proteins, FAGP2 and the immunophilin-like protein TWD1, which has been shown to directly interact with PGP1 and PGP19 (Murphy et al., 2002; Geisler et al., 2003). TWD1 is thought to be essential for PGP1 and PGP19 function, as *twd1* mutants exhibit auxin-deficient phenotypes and deficiencies in auxin transport more severe than those observed in *pgp1 pgp19* double mutants (Geisler et al., 2003). In animal cells GPI-anchored proteins play a role in the formation of DRMs (Brown and London, 1998), although similar function has not yet been demonstrated in plant tissues. A recent proteomic characterization of *Arabidopsis* DRMs did not detect GPI-anchored proteins (Borner et al., 2005), but the DRMs utilized were derived from cultured cell lines which often display altered protein expression or localization. Reciprocal coimmunoprecipitation of PGP19 and PIN1 from membrane fractions further substantiated PGP19–PIN1 interactions in a DRM complex (Lee and Murphy, unpublished), as did evidence that PIN1 associated with DRMs derived from wild-type seedlings but not with DRMs derived from *pgp19* mutant seedlings (Blakeslee and Murphy, unpublished). Direct interactions between PGP19 and PIN1 are also possible, as interactions between the C-terminus of PGP19 and the soluble loop of PIN1 were identified using yeast two-hybrid assays (Lee and Murphy, unpublished).

7.2 Endocytic Cycling of Mammalian MDR/PGPs

MDR/PGPs in both mammals and *A. thaliana* appear to be regulated by endocytic cycling. Human MDR1 has been localized in DRMs that have been implicated in endocytic membrane trafficking and protein sorting (reviewed in Brown and London, 1998). Human MDR/PGPs in hepatic canalicular cells have been shown to cycle between a relatively large endomembrane pool and the PM in response to taurocholate or cyclic AMP stimulus (Kipp and Arias, 2002) and, in multidrug-resistant cancer cells, MDR1 is actively cycled between the PM and an endosomal membrane compartment via a clathrin-dependent mechanism (Kim et al., 1997). Cycling of MDR/PGPs was also found to be dependent on the action of phosphoinositol-3-kinases (PI3K), which are essential for vesicle trafficking in yeast, animal, and plant cells (Fruman et al., 1998). Additionally, MDR1 labeled with enhanced green fluorescent protein has been shown to undergo vesicular trafficking, disruption of which led to retention of PGP in Golgi bodies (Fu et al., 2004). Trafficking of human MDR1 was found to be dependent on a dileucine protein–protein interaction motif at the C-terminus of the protein, but was restored by treatment with cyclosporin A, suggesting an interaction with an immunophilin-like protein (Loo et al., 2005). Consistent with this hypothesis, the immunophilin-like protein FKBP12 has been shown to play an essential role in MDR/PGP function unrelated to FKBP12 enzymatic peptide-prolyl isomerase (PPIase) activity (Hemenway and Heitman, 1996; Mealey et al., 1999). It is likely that FKBP12 regulates MDR/PGP function via a protein–protein interaction, which can be mimicked by treating C-terminally truncated MDR1 with cyclosporin A.

7.3 Evidence for Endocytic Cycling in MDR/PGPs in *A. thaliana*

In *Arabidopsis*, PGP1 and PGP19 interactions with the FKBP42 immunophilin-like protein TWD1 suggest that similar mechanisms may regulate PGP trafficking in plants. It has been proposed that TWD1 functions in both stabilization of multiprotein auxin efflux complexes at the PM, as well as the endocytic trafficking of PGP1 and PGP19 (Geisler et al., 2003). Reciprocal coimmunoprecipitation and yeast two-hybrid analyses have demonstrated that the C-termini of both PGP1 and PGP19 interact with the TWD1 protein in an NPA-sensitive fashion, and *Arabidopsis* TWD1–PGP interactions appear to be essential for proper PGP function *in planta* (Geisler et al., 2003). As with FKBP12, the PGP–TWD1 interaction appears to be independent on the PPIase activity of TWD1 (Geisler et al., 2003; Hemenway and Heitman, 1996). Further, the levels of PGPs present in endomembrane compartments are increased in *twd1* mutants, indicating that TWD1 plays a role in PGP trafficking.

The association of PGPs with a number of secretory proteins, such as ADL1A, patellin1 (PATL1), and β-adaptin (Geisler et al., 2003; Murphy et al., 2002) provides additional evidence for PGP trafficking *in planta*. The phosphotidylinositol transfer protein PATL1 and the dynamin-like protein ADL1a have been shown to function in membrane trafficking to the cell plate (Kang et al., 2003; Peterman et al., 2004), but may also have other trafficking functions. ADL1a exhibits a polar localization in nondividing transition zone root cells, where it is thought to contribute to maintenance of PM subdomains (Kang et al., 2003). The association of β-adaptin with PGPs further suggests that PGPs may be cycled in a clathrin-dependent manner similar to that observed in mammalian cells. The association of APM1 with PGP1 and PGP19 (Murphy et al., 2002) is also consistent with an IRAP-like trafficking mechanism underlying PGP localization in *Arabidopsis*.

In mammalian canalicular and multidrug-resistant cells, the presence of an external signal, such as a taurocholate or a hydrophobic drug, leads to increased cycling of PGPs to the PM and a resultant increase in transport activity (Kipp and Arias, 2002; Kipp et al., 2001). In plants, an external signal, such as light, gravity, or auxin itself, may function to increase PGP cycling to the PM and lead to an increase in auxin transport. Support for this hypothesis was provided when PGP1 expression was found to be upregulated in response to auxin treatment of *Arabidopsis* seedlings (Geisler et al., 2005; Noh et al., 2001). The recent successful construction of functional translational fusions of PGP1 and PGP19 provides an opportunity to study these mechanisms *in planta*. To date, sequestration of PGP1 in endomembrane compartments by BFA has been observed (Bandyopadhyay and Murphy, unpublished results), but the reversibility of this sequestration could not be demonstrated.

7.4
MDR/PGPs Function as ATP-Activated Hydrophobic Anion/Auxin Carriers/Transporters

Although PGPs appear to stabilize PIN proteins at the PM, they may also be capable of direct auxin transport (Sidler et al., 1998). MDR/PGP transporters in mammalian cells mediate efflux of hydrophobic cytotoxic compounds, maintenance of the blood–brain barrier, and cell membrane homeostasis (Ambudkar et al., 1999; Garrigues et al., 2002). Other substrates transported by mammalian PGPs include daunomycin, verapamil, alkaloids (such as vinblastine and vincristine), and small peptides (reviewed in Ambudkar et al., 1999). However, in plants, the substrate specificity of MDR/PGPs may be more limited, as, other than CjMDR1, an MDR/PGP that appears to mediate uptake of the alkaloid berberine in *Coptis japonica* (Yazaki et al., 2001), all other MDR/PGP activity has been linked to transport of auxin (Geisler et al., 2003, 2005; Noh et al., 2001, 2003; Terasaka et al., 2005).

Recently, studies using *Arabidopsis* protoplasts and whole plant tissues have demonstrated that AtPGP1 and AtPGP19 mediate the cellular efflux of natural and synthetic auxins, as well as oxidative auxin breakdown products in these systems (Geisler et al., 2003, 2005; Noh et al., 2001). Protoplasts from *pgp1*, *pgp19*, and *pgp1 pgp19* double mutant seedlings displayed reduced auxin transport when compared with those from wild-type controls, and PGP-mediated transport was found to be specific and sensitive to AEIs (Geisler et al., 2005). Additionally, when expressed in yeast, PGP1 mediated the transport of both natural and synthetic auxins, as well as increased resistance to the toxic auxin analog 5-fluoroindole (Geisler et al., 2005). Expression of *AtPGP1* and *AtPGP19* in the HeLa cell system commonly used to assay mammalian PGPs resulted in increased efflux of natural and synthetic auxins as well as oxidative auxin breakdown products (Geisler et al., 2005). More recently, AtPGP4 has been shown to transport auxin *in planta* and to function as an uptake transporte when expressed in mammalian cells (Terasaka et al., 2005). However, while mammalian MDR/PGPs demonstrate broad-spectrum transport of multiple hydrophobic compounds, PGP1, PGP19, and

PGP4 failed to transport common mammalian PGP substrates (Geisler et al., 2005; Terasaka et al., 2005). As such, the term MDR protein is somewhat of a misnomer in plants, where MDR/PGPs might be better characterized as ATP-activated hydrophobic anion carriers.

The observed auxin *influx* mediated by AtPGP4 may be a result of a structural divergence from the consensus MDR/PGP protein structure. Besides containing a unique 20 amino acid coiled-coil N-terminal extension, PGP4 is predicted to contain only 10 or 11 transmembrane helices (five helices at the N-terminal half of the protein, and five or six at the C-terminus), instead of the normal 12 transmembrane domains commonly found in PGP-type ABC transporters (http://aramemnon.botanik.uni-koeln.de/; http://smart.embl-heidelberg.de/). In prokaryotic systems, ABC transporters involved in efflux consistently contain 12 transmembrane helices, but that those involved in uptake contain anywhere from 8 to 20 transmembrane domains, usually as the result of coexpression of two half-transporters of 4–10 transmembrane helices each (Higgins, 1992; Locher, 2004; Locher et al., 2002). More recently, the SbtA family of *Synecosystis* cyanobacterial ABC transporters, which are involved in bicarbonate anion uptake, were found to be a single peptide containing ten transmembrane helices, with some similarity to *Arabidopsis* PGP4 (von Rozycki et al., 2004).

7.5 Members of the MDR/PGP Family Function in Auxin Transport in a Tissue-Specific Fashion *in Planta*

Expression patterns of *AtPGP1* and *AtPGP19* as well as localization of the PGP1 protein suggest a rationale for PGP function in auxin transport. PGP1 is expressed primarily in root and shoot apices, and exhibits an apolar PM localization in these regions (Geisler et al., 2005; Sidler et al., 1998). In apical regions, where auxin concentrations are high, cells are small, and rediffusion of auxin is more likely to reduce polar auxin transport, PGP1 appears to function in energy-dependent nonpolar auxin export (Geisler et al., 2005) (Fig. 2). Interestingly, in the larger, more elongated cortical cells above the root apex, PGP1 displays a polar localization, although its contribution to polar auxin transport in these tissues is unclear (Geisler et al., 2005). Like *PGP1*, *PGP19* is strongly expressed at root and shoot apices, but expression also extends throughout the entire root and, in dark-grown seedlings, throughout the hypocotyl as well (Noh et al., 2001). Additionally, *PGP19* is expressed in nonapical cell layers associated with the vascular tissue (xylem parenchyma, endodermis, and bundle sheath) where *PGP1* is not expressed, consistent with the wider role of PGP19 in auxin transport evident from mutational studies (Geisler et al., 2005, Noh et al., 2001). The 21 members of the MDR/PGP family in *Arabidopsis* exhibit overlapping expression patterns and are expected to contribute to discrete tissue-specific auxin transport streams (Fig. 1). For example, the polar local-

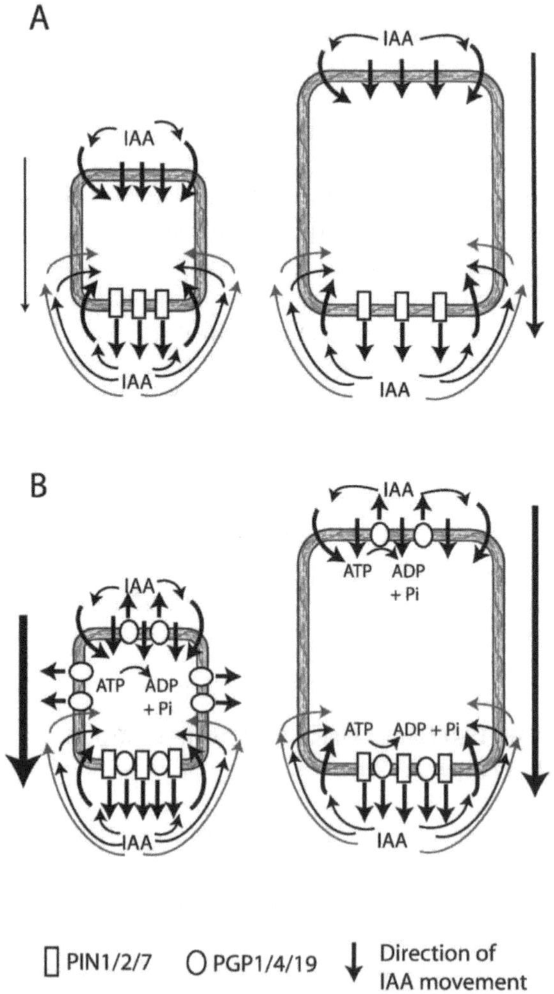

Fig. 2 Model of auxin rediffusion in smaller and larger cells. **a** In meristematic regions, cells are small, concentrations of auxin are high, and rediffusion of auxin occurs across a larger proportion of the total cell surface area, which can lead to a reduction in polar auxin transport. In maturer tissues, cells are larger and rediffusion of auxin occurs across a smaller proportion of the total cell surface area, thereby decreasing the effect of rediffusion on polar auxin transport. *Arrows* indicate the direction and magnitude of auxin flux. **b** The energy-dependent transport of auxin by MDR/PGP proteins increases the directionality of auxin transport. In meristematic regions, an energy-dependent transport mechanism is needed to overcome the localized high auxin concentration gradient. The apolar localization and nondirectional transport of MDR/PGPs overcomes the rediffusion of auxin into small cells. When MDR/PGPs and PINs colocalize subcellularly, synergistic PIN–PGP interactions lead to increased rate and specificity of auxin transport and confer a directional bias to auxin movement. In larger cells of maturer plant tissues, rediffusion of auxin is minimal, and MDR/PGP-mediated auxin transport increases the rate of auxin transport. *Arrows* indicate the direction and magnitude of auxin flux. *IAA* indole-3-acetic acid

ization of PGP4 in root epidermal cells (Terasaki et al., 2005) and PGP1 in root cortical cells suggests that, in elongated cells, PGPs may more directly contribute to polar auxin transport. Consistent with a proposed role as an auxin influx carrier, PGP4 exhibits apical and lateral localization with no signal detected at the basal ends of cells (Terasaka et al., 2005).

7.6
Interactions Between MDR/PGPs and PIN Proteins Confer Increases in Rate and Specificity of Auxin Transport

The localization of PGP and PIN proteins in DRM fractions suggests that these proteins may function together in auxin transport (Blakeslee et al., 2005). Interestingly, when either PGP1 or PGP19 was coexpressed with PIN1 in HeLa cells, a synergistic increase in auxin transport, specificity, and NPA sensitivity was observed (Blakeslee and Murphy, unpublished). An antagonistic effect was seen when the PGPs were coexpressed with PIN2 (Blakeslee and Murphy, unpublished). When the influx transporter PGP4 was coexpressed with PIN2, a synergistic increase in influx was observed (Blakeslee and Murphy, unpublished). However, although PGP4 and PIN2 are expressed in the same epidermal cell layers, it is not clear how they interact *in planta* as their subcellular localizations overlap only in the distal elogation zone. These data indicate that, while PGPs appear to mediate energy-dependent auxin transport primarily in regions of high auxin concentration, interactions between PGPs and PINs may confer increased specificity and directionality to this transport (Blakeslee et al., 2005).

8
Conclusions and Future Prospects

The recently demonstrated ability of PGPs to transport auxin, combined with the evidence for PGP cycling provide support for a model in which PGPs are cycled between the PM and an endosomal compartment, and function primarily in nonpolar, energy-dependent auxin transport in areas of high auxin concentration and perhaps in polarized transport in other tissues (Fig. 3). PGPs appear to stabilize specific PIN efflux facilitator proteins and other components of auxin efflux complexes in DRMs. Interactions between PGPs and PIN proteins appear to provide increased rate, specificity, and directionality to PGP-mediated auxin transport. The similarities between the components of the mammalian IRAP-GLUT4, APN, and MDR vesicular cycling pathways, and the *Arabidopsis* APM1, PGPs, and their associated proteins suggest that *Arabidopsis* PGPs undergo vesicular trafficking, and that a conserved mechanism functions in both plant and mammalian systems. It is unclear at this date, however, whether PGP cycling occurs in the same membrane

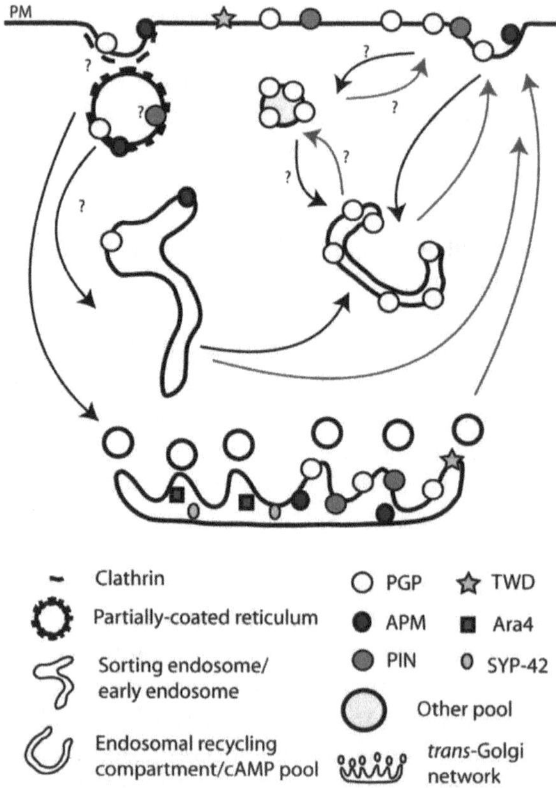

Fig. 3 Model of putative MDR/PGP cycling in *Arabidopsis*. Using mammalian models of MDR/PGP vesicular cycling (Kipp and Arias 2002), it is possible to construct a hypothetical model by which MDR/PGPs may be cycled in *Arabidopsis*. In this model, MDR/PGPs are synthesized in the endoplasmic reticulum, move into the Golgi and the trans-Golgi network (*TGN*), and are transported directly to the plasma membrane (*PM*). Once on the PM, MDR/PGPs may be recycled via an endocytic mechanism similar to that observed for PIN proteins. Although PGPs can be internalized by brefeldin A treatment, reversibility has not been demonstrated, and it is not known if PINs and PGPs cycle in the same compartments. In this model, MDR/PGPs would move from the PM into low-density endomembrane pools, where components such as APM1 and TWD1 may function to regulate MDR/PGP trafficking. MDR/PGPs present in endosomal membrane pools could be returned to the PM in response to an external stimulus (possibly in a cyclic AMP, *cAMP*, dependent fashion). Additionally, MDR/PGPs may be recycled from the PM through early endosomes and redirected to either TGN or the sorting/recycling low-density endosomal compartment for retransport to the PM. The partially coated reticulum is thought to be identical to the early endosome in plants

vesicles/compartments as PIN cycling, or whether PGPs and PINs are cycled via independent systems. More research is needed to further define the interactions between PGP and PIN trafficking, and their roles in plant growth and development.

References

Ambudkar S, Dey S, Hrycyna C, Ramachandra M, Pastan I, Gottesman M (1999) Biochemical, cellular, and pharmacological aspects of the multidrug transporter. Annu Rev Pharmacol Toxicol 39:361–398

Baluška F, Hlavacka A, Šamaj J, Palme K, Robinson DG, Matoh T, McCurdy DW, Menzel D, Volkmann D (2002) F-actin-dependent endocytosis of cell wall pectins in meristematic root cells: insights from brefeldin A-induced compartments. Plant Physiol 130:422–431

Baumann CA, Saltiel AR (2001) Spatial compartmentalization of signal transduction in insulin action. Bioessays 23:215–222

Benkova E, Michniewicz M, Sauer M, Teichmann T, Seifertova D, Jurgens G, Friml J (2003) Local, efflux-dependent auxin gradients as a common module for plant organ formation. Cell 115:591–602

Bernasconi P, Patel BC, Reagan JD, Subramanian MV (1996) The N-1-naphthylphthalamic acid-binding protein is an integral membrane protein. Plant Physiol 111:427–432

Blakeslee JJ, Peer WA, Murphy A (2005) Auxin Transport. Curr Opin Plant Biol (in press)

Blancaflor EB, Masson PH (2003) Plant gravitropism. Unraveling the ups and downs of a complex process. Plant Physiol 133:1677–1690

Blilou I, Xu J, Wildwater M, Willemsen V, Paponov I, Friml J, Heidstra R, Aida M, Palme K, Scheres B (2005) The PIN auxin efflux facilitator network controls growth and patterning in *Arabidopsis* roots. Nature 433:39–44

Borner GHH, Sherrier DJ, Weimar T, Michaelson LV, Hawkins ND, MacAskill A, Napier JA, Beale MH, Lilley KS, Dupree P (2005) Analysis of detergent-resistant membranes in *Arabidopsis*. Evidence for plasma membrane lipid rafts. Plant Physiol 137:104–116

Borst P, Zelcer N, van Helvoort A (2000) ABC transporters in lipid transport. Biochim Biophys Acta Mol Cell Biol Lipids 1486:128–144

Brown DA, London E (1998) Functions of lipid rafts in biological membranes. Annu Rev Cell Dev Biol 14:111–136

Brown DE, Rashotte AM, Murphy AS, Normanly J, Tague BW, Peer WA, Taiz L, Muday GK (2001) Flavonoids act as negative regulators of auxin transport in vivo in *Arabidopsis*. Plant Physiol 126:524–535

Chen RJ, Hilson P, Sedbrook J, Rosen E, Caspar T, Masson PH (1998) The *Arabidopsis* thaliana AGRAVITROPIC 1 gene encodes a component of the polar-auxin-transport efflux carrier. Proc Natl Acad Sci USA 95:15112–15117

Dixon M, Jacobson J, Cady C, Muday G (1996) Cytoplasmic orientation of the naphthylphthalamic acid-binding protein in zucchini plasma membrane vesicles. Plant Physiol 112:421–432

Friml J (2003) Auxin transport—shaping the plant. Curr Opin Plant Biol 6:7–12

Friml J, Palme K (2002) Polar auxin transport—old questions and new concepts? Plant Mol Biol 49:273–284

Friml J, Benkova E, Blilou I, Wisniewska J, Hamann T, Ljung K, Woody S, Sandberg G, Scheres B, Jurgens G, Palme K (2002) AtPIN4 mediates sink-driven auxin gradients and root patterning in *Arabidopsis*. Cell 108:661–673

Friml J, Wisniewska J, Benkova E, Mendgen K, Palme K (2002) Lateral relocation of auxin efflux regulator PIN3 mediates tropism in *Arabidopsis*. Nature 415:806–809

Friml J, Vieten A, Sauer M, Weijers D, Schwarz H, Hamann T, Offringa R, Jurgens G (2003) Efflux-dependent auxin gradients establish the apical-basal axis of *Arabidopsis*. Nature 426:147–153

Fruman D, Meyers R, Cantley L (1998) Phosphoinositide kinases. Annu Rev Biochem 67:481–507

Fu D, Bebawy M, Kable E, Roufogalis B (2004) Dynamic and intracellular trafficking of P-glycoprotein-EGFP fusion protein: Implications in multidrug resistance in cancer. Int J Cancer 109:174–181

Galweiler L, Guan CH, Muller A, Wisman E, Mendgen K, Yephremov A, Palme K (1998) Regulation of polar auxin transport by AtPIN1 in *Arabidopsis* vascular tissue. Science 282:2226–2230

Garrigues A, Escargueil A, Orlowski S (2002) The multidrug transporter, P-glycoprotein, actively mediates cholesterol redistribution in the cell membrane. Proc Natl Acad Sci USA 99:10347–10352

Geisler M, Kolukisaoglu HU, Bouchard R, Billion K, Berger J, Saal B, Frangne N, Koncz-Kalman Z, Koncz C, Dudler R, Blakeslee JJ, Murphy AS, Martinoia E, Schulz B (2003) TWISTED DWARF1, a unique plasma membrane-anchored immunophilin-like protein, interacts with *Arabidopsis* multidrug resistance-like transporters AtPGP1 and AtPGP19. Mol Biol Cell 14:4238–4249

Geisler M, Blakeslee JJ, Bouchard R, Lee OR, Vincenzetti V, Bandyopadhyay A, Peer WA, Bailly A, Richards EL, Edjendal KF, Smith AP, Baroux C, Grossniklaus U, Muller A, Hrycyna CA, Dudler R, Murphy AS, Martinoia E (2005) Cellular export of auxin by MDR-type ATP-binding cassette transporters of *Arabidopsis* thaliana. Plant J 44:179–194

Geissler A, Pilet P, Katekar G (1985) Growth and gravireaction of maize roots treated with a phytotropin. J Plant Physiol 119:25–34

Geldner N, Friml J, Stierhof YD, Jurgens G, Palme K (2001) Auxin transport inhibitors block PIN1 cycling and vesicle traficking. Nature 413:425–428

Geldner N, Anders N, Wolters H, Keicher J, Kornberger W, Muller P, Delbarre A, Ueda T, Nakano A, Jurgens G (2003) The *Arabidopsis* GNOM ARF-GEF mediates endosomal recycling, auxin transport, and auxin-dependent plant growth. Cell 112:219–230

Grebe M, Friml J, Swarup R, Ljung K, Sandberg G, Terlou M, Palme K, Bennett MJ, Scheres B (2002) Cell polarity signaling in *Arabidopsis* involves a BFA-sensitive auxin influx pathway. Curr Biol 12:329–334

Hemenway C, Heitman J (1996) Immunosuppressant target protein FKBP12 is required for P-glycoprotein function in yeast. J Biol Chem 271:18527–18534

Higgins CF (1992) ABC transporters – from microorganisms to man. Annu Rev Cell Biol 8:67–113

Jacobs M, Rubery PH (1988) Naturally-occurring auxin transport regulators. Science 241:346–349

Jasinski M, Ducos E, Martinoia E, Boutry M (2003) The ATP-binding cassette transporters: Structure, function, and gene family comparison between rice and *Arabidopsis*. Plant Physiol 131:1169–1177

Kang BH, Busse JS, Bednarek SY (2003) Members of the *Arabidopsis* dynamin-like gene family, ADL1, are essential for plant cytokinesis and polarized cell growth. Plant Cell 15:899–913

Katekar GF, Geissler AE (1979) Evidence of a common-mode of action for a class of auxin transport inhibitors. Plant Physiol 63:22

Kim H, Barroso M, Samanta R, Greenberger L, Sztul E (1997) Experimentally induced changes in the endocytic traffic of P-glycoprotein alter drug resistance of cancer cells. Am J Physiol Cell Physiol 42:C687–C702

Kipp H, Arias I (2002) Trafficking of canalicular ABC transporters in hepatocytes. Annu Rev Physiol 64:595–608

Kipp H, Pichetshote N, Arias IM (2001) Transporters on demand— intrahepatic pools of canalicular ATP binding cassette transporters in rat liver. J Biolo Chem 276:7218–7224

Kramer W, Girbig F, Corsiero D, Pfenninger A, Frick W, Jahne G, Rhein M, Wendler W, Lottspeich F, Hochleitner E, Orso E, Schmitz G (2005) Aminopeptidase N (CD13) is a molecular target of the cholesterol absorption inhibitor ezetimibe in the enterocyte brush border membrane. J Biol Chem 280:1306–1320

Lavie Y, Liscovitch M (2000) Changes in lipid and protein constituents of rafts and caveolae in multidrug resistant cancer cells and their functional consequences. Glycocon J 17:253–259

Lavie Y, Fiucci G, Liscovitch M (1998) Up-regulation of caveolae and caveolar constituents in multidrug-resistant cancer cells. J Biol Chem 273:32380–32383

Locher K (2004) Structure and mechanism of ABC transporters. Curr Opin Struct Biol 14:426–431

Locher K, Lee A, Rees D (2002) The E-coli BtuCD structure: A framework for ABC transporter architecture and mechanism. Science 296:1091–1098

Lomax TL, Muday GK, Rubery PH (1995) Auxin transport. In: Davies PJ (ed) Plant hormones: physiology, biochemistry, and molecular biology. Kluwer, Dordrecht, The Netherlands, pp 509–530

Loo T, Bartlett M, Clarke D (2005) The dileucine motif at the COOH terminus of human multidrug resistance P-glycoprotein is important for folding but not activity. J Biol Chem 280:2522–2528

Luker G, Pica C, Kumar A, Covey D, Piwnica-Worms D (2000) Effects of cholesterol and enantiomeric cholesterol on P-glycoprotein localization and function in low-density membrane domains. Biochemistry 39:7651–7661

Martinoia E, Klein M, Geisler M, Bovet L, Forestier C, Kolukisaoglu U, Muller-Rober B, Schultz B (2002) Multifunctionality of plant ABC transporters—more than just detoxifiers. Planta 214:345–355

Mealey K, Barhoumi R, Burghardt R, McIntyre B, Sylvester P, Hosick H, Kochevar D (1999) Immunosuppressant inhibition of P-glycoprotein function is independent of drug-induced suppression of peptide-prolyl isomerase and calcineurin activity. Cancer Chemother Pharmacal 44:152–158

Michalke W, Katekar GF, Geissler AE (1992) Phytotropin-binding sites and auxin transport in Cucurbita-pepo—evidence for 2 recognition sites. Planta 187:254–260

Mongrand S, Morel J, Laroche J, Claverol S, Carde JP, Hartmann MA, Bonneu M, Simon-Plas F, Lessire R, Bessoule JJ (2004) Lipid rafts in higher plant cells: purification and characterization of Triton X-100-insoluble microdomains from tobacco plasma membrane. J Biol Chem 279:36277–36286

Muday GK, Murphy AS (2002) An emerging model of auxin transport regulation. Plant Cell 14:293–299

Muday GK, Peer WA, Murphy AS (2003) Vesicular cycling mechanisms that control auxin transport polarity. Trends Plant Sci 8:301–304

Muller A, Guan CH, Galweiler L, Tanzler P, Huijser P, Marchant A, Parry G, Bennett M, Wisman E, Palme K (1998) AtPIN2 defines a locus of *Arabidopsis* for root gravitropism control. EMBO J 17:6903–6911

Multani DS, Briggs SP, Chamberlin MA, Blakeslee JJ, Murphy AS, Johal GS (2003) Loss of an MDR transporter in compact stalks of maize br2 and sorghum dw3 mutants. Science 302:81–84

Murphy A, Peer WA, Taiz L (2000) Regulation of auxin transport by aminopeptidases and endogenous flavonoids. Planta 211:315–324

Murphy A, Taiz L (1999) Naphthylphthalamic acid is enzymatically hydrolyzed at the hypocotyl-root transition zone and other tissues of *Arabidopsis* thaliana seedlings. Plant Physiol Biochem 37:413–430

Murphy A, Taiz L (1999) Localization and characterization of soluble and plasma membrane aminopeptidase activities in *Arabidopsis* seedlings. Plant Physiol Biochem 37:431–443

Murphy AS, Hoogner KR, Peer WA, Taiz L (2002) Identification, purification, and molecular cloning of N-1-naphthylphthalmic acid-binding plasma membrane-associated aminopeptidases from *Arabidopsis*. Plant Physiol 128:935–950

Noh B, Bandyopadhyay A, Peer WA, Spalding EP, Murphy AS (2003) Enhanced gravi- and phototropism in plant mdr mutants mislocalizing the auxin efflux protein PIN1. Nature 423:999–1002

Noh B, Murphy AS, Spalding EP (2001) Multidrug resistance-like genes of *Arabidopsis* required for auxin transport and auxin-mediated development. Plant Cell 13:2441–2454

Okada K, Ueda J, Komaki M, Bell C, Shimura Y (1991) Requirement of the auxin polar transport-system in early stages of *Arabidopsis* floral bud formation. Plant Cell 3:677–684

Palme K, Galweiler L (1999) PIN-pointing the molecular basis of auxin transport. Curr Opin Plant Biol 2:375–381

Peer WA, Brown DE, Tague BW, Muday GK, Taiz L, Murphy AS (2001) Flavonoid accumulation patterns of transparent testa mutants of *Arabidopsis*. Plant Physiol 126:536–548

Peterman T, Ohol Y, McReynolds L, Luna E (2004) Patellin1, a novel Sec14-like protein, localizes to the cell plate and binds phosphoinositides. Plant Physiol 136:3080–3094

Petrasek J, Cerna A, Schwarzerova K, Elckner M, Morris DA, Zazimalova E (2003) Do phytotropins inhibit auxin efflux by impairing vesicle traffic? Plant Physiol 131:254–263

Šamaj J, Baluška F, Voigt B, Schlicht M, Volkmann D, Menzel D (2004) Endocytosis, actin cytoskeleton and signaling. Plant Physiol 135:1150–1161

Shogomori H, Brown D (2003) Use of detergents to study membrane rafts: The good, the bad, and the ugly. Biol Chem 384:1259–1263

Sidler M, Hassa P, Hasan S, Ringli C, Dudler R (1998) Involvement of an ABC transporter in a developmental pathway regulating hypocotyl cell elongation in the light. Plant Cell 10:1623–1636

Swarup R, Bennett M (2003) Auxin transport: The fountain of life in plants? Dev Cell 5:824–826

Swarup R, Friml J, Marchant A, Ljung K, Sandberg G, Palme K, Bennett M (2001) Localization of the auxin permease AUX1 suggests two functionally distinct hormone transport pathways operate in the *Arabidopsis* root apex. Gen Dev 15:2648–2653

Terasaka K, Blakeslee JJ, Bandyopadhyay A, Peer WA, Murphy AS, Sato F, Yazaki K (2005) Involvement of AtPGP4, a P-glycoprotein-type ATP binding cassette protein, in the root basipetal auxin transport in *Arabidopsis* thaliana. Plant Cell (submitted)

Troost J, Lindenmaier H, Haefeli W, Weiss J (2004) Modulation of cellular cholesterol alters P-glycoprotein activity in multidrug-resistant cells. Mol Pharmacol 66:1332–1339

vanHelvoort A, Smith A, Sprong H, Fritzsche I, Schinkel A, Borst P, vanMeer G (1996) MDR1 P-glycoprotein is a lipid translocase of broad specificity, while MDR3 P-glycoprotein specifically translocates phosphatidylcholine. Cell 87:507–517

von Rozycki T, Schultzel M, Saier M (2004) Sequence analyses of cyanobacterial bicarbonate transporters and their homologues. J Mol Microbiol Biotech 7:102–108

Wang E, Casciano C, Clement R, Johnson W (2000) Cholesterol interaction with the daunorubicin binding site of P-glycoprotein. Biochem Biophys Res Commun 276:909–916

Yazaki K, Shitan N, Takamatsu H, Ueda K, Sato F (2001) A novel Coptis japonica multidrug-resistant protein preferentially expressed in the alkaloid-accumulating rhizome. J Exp Bot 52:877–879

Plant Cell Monogr (1)
J. Šamaj · F. Baluška · D. Menzel: Plant Endocytosis
DOI 10.1007/7089_011/Published online: 18 October 2005

Rab GTPases in Plant Endocytosis

Erik Nielsen[1,2]

[1]Donald Danforth Plant Science Center, St. Louis, Missouri 63132, USA
enielsen@danforthcenter.org

[2]Department of Biology, Washington University in St. Louis, One Brookings Dr., St. Louis, MO 63130, USA
enielsen@danforthcenter.org

Abstract The Rab family is part of the Ras superfamily of small GTPases. In eukaryotes Rab GTPases are present as members of gene families, and the different Rab GTPase isoforms are localized specific intracellular membranes, where they function as regulators of distinct steps in membrane traffic pathways. They perform these regulatory functions through the specific recruitment of cytosolic effector proteins onto membranes. This recruitment occurs when the Rab GTPase is in the GTP-bound, or active, form. Through these recruited effector proteins, Rab GTPases regulate many aspects of membrane trafficking including vesicle formation, actin- and tubulin-dependent vesicle movement, and membrane fusion. The recent sequencing of complete genomic sequences from animal, yeast, and plant organisms has revealed that a number of Rab GTPase families are conserved from yeast to animals and plants. The plant model system, *Arabidopsis thaliana,* contains 57 Rab GTPases, of which 40 distinct Rab GTPase members of four subfamilies RabA (26 members), RabC (three members), RabF (three members), and RabG (eight members) share significant similarity with Rab GTPases implicated in endocytic events in animals and yeast.

In this review we will highlight recent observations of the function of some of these plant Rab GTPases during endocytosis in plants, and discuss possible roles of plant endocytic Rab GTPases in relation to what is currently known in animal and yeast systems.

1 Introduction

In eukaryotic cells, endocytosis is an essential process that is necessary for the delivery of proteins, lipids, and extracellular components to various intracellular destinations. After internalization, selective sorting of cargo within endocytic compartments, budding of vesicle transport intermediates, and the efficient delivery and fusion of transport vesicles with their target membranes are all required for the efficient delivery of endocytic cargo to their correct subcellular destinations. As key regulators of membrane trafficking events, Rab GTPase family members specifically localize to these endocytic compartments and control aspects of these sorting, vesicle budding, and fusion events. Because of their specific subcellular distributions, Rab GTPases have also served as useful markers of the organelles present within the endocytic membrane trafficking pathways in plant, yeast, and animal systems. Further, the recent genomic sequencing projects which have led to the identification of the full complements of the Rab GTPase families for *A. thaliana* (57 members; Pereira-Leal and Seabra 2001; Vernoud et al. 2003), *S. cerevisiae* (11 members; Vernoud et al. 2003), and *H. sapiens* (~ 60 members; Stenmark and Olkkonen 2001), allows for more complete comparison and analysis of the regulation of endomembrane trafficking pathways in these three eukaryotic systems.

Much progress has been made recently in our understanding of plant endocytic membrane trafficking and the roles of various plant Rab GTPases in these pathways. However, in many cases our understanding of the functioning of these plant Rab GTPases still relies primarily upon the roles defined for their evolutionarily-related counterparts in yeast and mammalian systems. As a result, while recognizing the possibility that some aspects of endomembrane trafficking in plants are likely unique to plants, we will attempt to summarize recent progress in understanding the localization and function of plant Rab GTPases during plant endocytosis and relate this to the current understanding of the roles of Rab GTPases in regulation of endocytosis in animals and yeast.

2
The First Step: Internalization at the Plasma Membrane

In animal cells, several mechanisms have been described by which molecules at the cell surface can be internalized (Fig. 1). The best characterized of these is receptor-mediated, clathrin-dependent endocytosis, in which receptor-ligand binding stimulates receptor protein recruitment into clathrin-coated pits that subsequently invaginate and are pinched off to form clathrin-coated vesicles (CCVs). In addition to this, receptor- and/or clathrin-independent processes, broadly called fluid-phase endocytosis, internalize significant amounts of membrane and solute (Gruenberg and Maxfield 1995). Internalization of cell surface components, such as GPI-anchored proteins, enveloped viruses, and certain plasma membrane proteins can also occur via caveolae, 50–60 nm invaginations in the cell surface of some mammalian cells. This process, which is linked to the propensity of the endocytic cargo to partition into lipid rafts within the plasma membrane, also can occur in cells devoid of caveolae, although the details as yet remain murky. Finally, mammalian cells internalize large particles, such as bacteria, via an actin-dependent process called phagocytosis (Maxfield and McGraw 2004).

In mammals, Rab5 localizes to the plasma membrane and early endosomes (Gorvel et al. 1991), and blockade of Rab5 function through the over expression of dominant-negative forms of Rab5 inhibits receptor-mediated, clathrin-dependent endocytosis, but not fluid-phase endocytosis (Bucci et al. 1992). More recently, it was shown that Rab5 might also play an active role in formation and budding of clathrin-coated vesicles (McLauchlan et al. 1998).

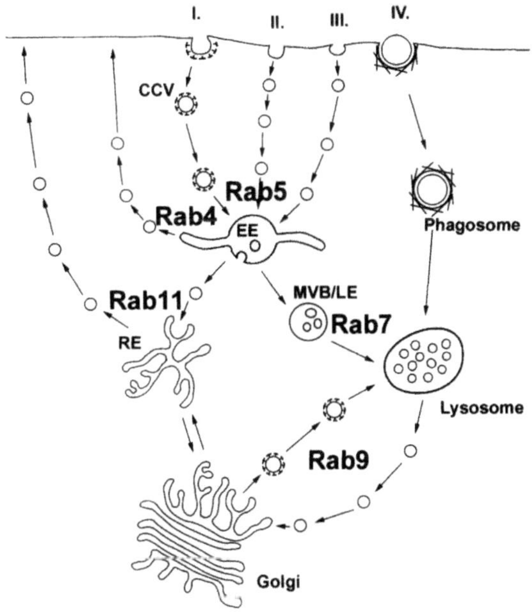

Fig. 1 Rab GTPases and the mammalian endocytic membrane trafficking pathway. Internalization from the plasma membrane occurs by several mechanisms: receptor-mediated clathrin-pit formation (I), fluid-phase endocytosis (II), caveolae-mediated uptake (III), and phagocytosis (IV). Rab5 regulates both receptor-mediated endocytosis and fusion of clathrin-coated vesicles (CCV) with early endosomes (EE). Rab4 regulates early recycling steps from the EE, while Rab11 regulates later recycling steps from the recycling endosome (RE). Newly-internalized endocytic cargo destined for degradation is sorted in the EE into multivesicular bodies (MVB). MVBs then fuse with lysosomes in a Rab7-regulated process. Recycling of lysosomal sorting receptor proteins (e.g. mannose-6-phosphate receptors) from lysosomal compartments is regulated by Rab9

In addition to the three Rab5 isoforms, mammals contain several other Rab GTPases that display high sequence conservation and segregate with the Rab5 GTPase isoforms. Interestingly, several of these Rab GTPases appear to regulate other internalization mechanisms at the plasma membrane. Rab22a appears to regulate clathrin-independent endocytosis events (Olkkonen et al. 1993; Weigert et al. 2004), while Rab17 and Rab21 appear to participate in endocytosis and recycling from early endosomes in polarized cells (Hunziker and Peters 1998; Zacchi et al. 1998; Opdam et al. 2000). Finally, Rab23, which does not co-segregate with other Rab5-subfamily members, has recently been shown to localize predominantly to the plasma membrane and negatively regulate sonic hedgehog signaling (Eggenschwiler et al. 2001; Evans et al. 2003).

While receptor-mediated endocytosis occurs in *S. cerevisiae*, and CCVs are present within the cell, there is some question as to whether clathrin-dependent endocytosis occurs in this organism (Fig. 2, Baggett and Wendland 2001). Instead, the majority of receptor-mediated endocytosis is thought to occur independently of clathrin, its associated adaptor proteins, and dynamin, in a process that depends on correct organization of the actin cytoskeleton (D'Hondt et al. 2000). Additionally, in yeast, many proteins to be endocytosed are targeted for internalization by the post-translational addition of ubiquitin (Dupre et al. 2004). *S. cerevisiae* contains three homologues of Rab5, Ypt51/52/53, and newly endocytosed cargo also traverses compartments labeled with these Rab GTPases (Singer-Kruger et al. 1995; Gerrard et al. 2000). However, it is unclear whether Ypt51 family members are directly involved in receptor clustering and vesicle budding events as described for the Rab5-like GTPases in animal cells.

While it was originally argued that turgor pressure and the presence of a rigid cell wall might preclude endocytosis in plants (Hawes 1995; Okita 1996), in recent years, a large body of evidence has emerged that, at least in some plant cells, endocytosis at the plant plasma membrane does indeed occur at significant rates (Fig. 3, Holstein 2002; Bolte et al. 2004, Šamaj et al. 2004). Internalization of plasma membrane lipids and sterols has been shown using the styryl dyes FM4-64 and FM1-43, and the sterol-binding molecule, filipin, respectively (Ueda et al. 2001; Emans et al. 2002; Grebe

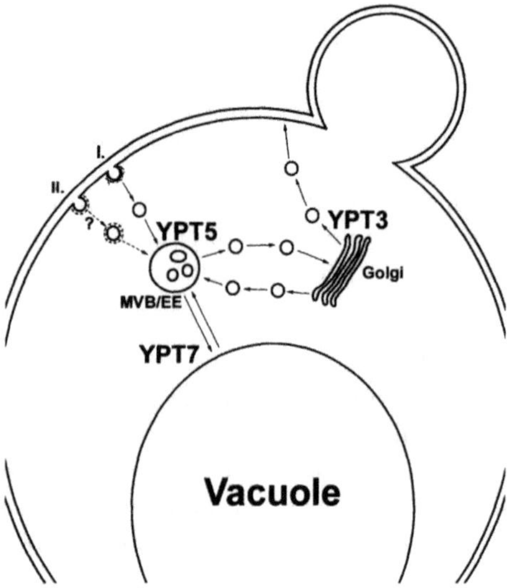

Fig. 2 Yeast endocytic Rab GTPases. In S. cerevisiae, most internalization occurs via an actin-dependent process (I). While clathrin and CCVs are present, it is not clear whether clathrin-dependent internalization from the plasma membrane occurs (II). Newly-internalized endocytic cargo first accumulates in MVBs labeled by the yeast Rab GTPase, YPT5. YPT5-labeled compartments also serve as a prevacuolar sorting compartment for trafficking of vacuolar-targeted proteins. Some internalized plasma membrane proteins are recycled to the plasma membrane in a process dependent on YPT3 Rab GTPases, which are localized to trans-Golgi cisternal elements. YPT7 labels vacuolar compartments, and regulates homotypic vacuolar fusion events

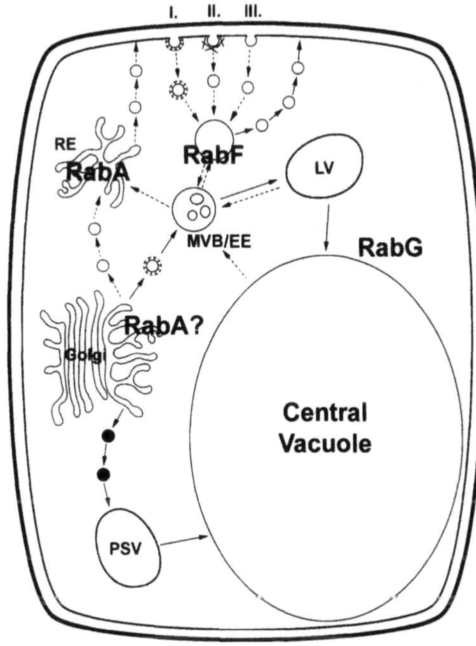

Fig. 3 Rab GTPases along the plant endocytic membrane trafficking pathways. In plants, clathrin and adaptors are present and clathrin-coated pits are observed associated with the plasma membrane (I). In addition sterols, labeled by filipin, have been observed to internalize, suggesting that lipid-raft dependent internalization occurs in plants (II). In addition, actin-dependent endocytosis from plant plasma membranes may also occur (III). Newly internalized cargo first accumulates in MVB/EE compartments labeled by members of the RabF GTPase family. Recycling of some plasma membrane proteins probably occurs from these compartments. Additionally, there appears to be distinct MVB/EE populations containing RabF1 and RabF2a/F2b in differing ratios. Additional recycling and sorting events for polarized secretion may occur in trans-Golgi-network-like compartments labeled by members of the RabA GTPase family. Vacuolar sorting receptor proteins (e.g. AtELP/BP-80) and their cognate ligands are delivered to RabF-labeled MVB/EEs en route to their delivery to plant lytic vacuoles (LV). Ultimately LVs and a separate post-Golgi compartment, the protein storage vacuoles (PSV) fuse with large central vacuoles to which members of the RabG GTPase family localize

et al. 2003; see Chap. by Šamaj, this volume, and the Chap. by Ovečka and Lichtscheidl, this volume). The concomitant internalization of soluble extracellular molecules via fluid-phase endocytosis has been demonstrated using the fluorescent dye, Lucifer Yellow (Baluška et al. 2004; see Chap. by Baluška et al., this volume), as well as by ultrastructural detection of internalized electron dense tracers (Huebner 1985; Hillmer 1986). Internalization and accumulation of several plasma membrane proteins such as, PIN1, PIN2 and the plasma membrane H^+-ATPase upon treatment with the vesicular traffic inhibitor, brefeldin A, indicates that proteins are also able to be internalized from the plasma membrane (Geldner et al. 2001; Grebe et al. 2003; see Chap. by Chen and Masson, this volume). Further, in stomatal guard cells the inward rectifying K^+-channel, KAT1, has recently been shown to continuously cycle between the plasma membrane and internal membranes (Hurst et al. 2004; Meckel et al. 2004; see Chap. by Homann, this volume). However, while these results indicate that plasma membrane localized proteins do indeed cycle via endocytic processes in plants, there is still no positive proof of receptor-mediated endocytosis in plants (Holstein 2002; see Chap. by Russinova and de Vries, this volume).

The presence of clathrin heavy and light chain proteins, of adaptor protein components in plants, as well as the visualization of clathrin coated pits on the plasma membranes of plant cells suggests that clathrin-mediated endocytosis occurs in plants (Blackbourn and Jackson 1996; Boehm and Bonifacino 2001; Scheele and Holstein 2002). In animal systems, Rab5 is known to regulate aspects of receptor-mediated clathrin-dependent endocytosis (Bucci et al. 1992; Zerial and McBride 2001). The presence of plant Rab GTPases with high sequence similarity to animal Rab5, and which are rapidly labeled upon addition of FM4-64, raises the possibility for their involvement in endocytosis from the plant plasma membrane (Ueda et al. 2001). However, as of yet there is still no direct evidence for involvement of plant Rab GTPases during endocytic internalization events at the plant plasma membrane.

3 Sorting Steps in the Early Endosome

In animals, after formation and budding from the plasma membrane, newly endocytosed CCVs shed their clathrin coat and fuse to form the first morphologically distinct endocytic compartment, the early endosome. Typically these early endosomal compartments are peripherally located and contain both tubular and vesicular elements (Gruenberg and Maxfield 1995; Maxfield and McGraw 2004). In addition to its role in CCV budding, Rab5 also regulates the initial fusion of recently uncoated clathrin-coated vesicles (CCVs) to early endosomes, homotypic fusion between early endosomes, and regulates various aspects of membrane trafficking through this early endosomal compartment (Zerial and McBride 2001). These tubular-vesicular early endosomal compartments are the first main branch point in the endocytic pathway, and it is within these compartments that newly internalized cargo is sorted and either returned to the plasma membrane, targeted to the recycling endosome for further sorting steps, or delivered to the late endosome for turnover (Maxfield and McGraw 2004). In addition to serving as a sorting compartment for newly-endocytosed cargo, the early endosome also serves as a sorting compartment for newly synthesized proteins en route to lysosomal/vacuolar membranes (Nielsen et al. 2000).

In yeast, a Rab GTPase closely related to the animal Rab5 GTPase, VPS21/YPT51, was initially identified in a screen for mutants defective in sorting of proteins to vacuolar compartments (Robinson et al. 1988). VPS21/YPT51 along with two additional closely related Rab GTPases, YPT52/53, has been demonstrated to associate with and regulate membrane trafficking through an endosomal-like, prevacuolar compartment that receives both newly endocytosed cargo (Singer-Kruger et al. 1995), as well as newly-synthesized proteins that are targeted to the yeast vacuolar compartment (Gerrard et al. 2000).

When compared with other eukaryotic Rab GTPases, the three members of the *A. thaliana* RabF sub-group show highest similarity to HsRab5 and yeast Ypt51 family members (Pereira-Leal and Seabra 2001; Vernoud et al. 2003). Subsequent experiments have demonstrated that *A. thaliana* members of this family also appear to reside on endocytic compartments. Both AtRabF2a/Rha1 and AtRabF2b/Ara7 have been shown to localize to similar internal membrane compartments that become quickly labeled by the lipophilic fluorescent styryl dye, FM 4-64 (Ueda et al. 2001). Additionally, upon BFA treatment, PIN1 and several other plasma membrane proteins and cell wall components have been shown to accumulate within RabF2b/Ara7 compartments (Geldner et al. 2001; Baluška et al. 2002). Further evidence that the RabF2a/RabF2b compartment is involved in endocytic recycling of newly internalized plasma membrane proteins comes from examination of these compartments in *gnom* mutants. Inactivation of GNOM, which is an Arf-GEF responsible for activation of Arf GTPases, was shown to disrupt the polar distribution of the PIN1 protein in *gnom* mutants (Steinmann et al. 1999). Therefore, it is intriguing that the morphology of RabF2b/Ara7-labeled compartments is altered in *gnom* mutants (Geldner et al. 2003). These observations support a model in which improper distribution of PIN1 in *gnom* mutants is due to defects in trafficking of the PIN1 protein through RabF2a/RabF2b compartments.

Interestingly, the third *A. thaliana* member of the RabF GTPase subfamily, RabF1/Ara6 appears to be a novel, plant-specific Rab GTPase. AtRabF1/Ara6 lacks the traditional carboxyl-terminal-CAAX motif for post-translational isoprenylation, and rather is myristoylated at the amino terminus. Examination of this isoform indicates that, despite these differences, AtRabF1/Ara6 partially colocalizes with AtRabF2b on plant endosomal compartments which also can be labeled with newly-internalized FM4-64 (Ueda et al. 2001; Ueda et al. 2004). Observations using the sterol-specific marker, filipin, suggest that RabF1/Ara6-labeled compartments may also function in sterol endocytosis in plants (Grebe et al. 2003).

As more and more Rab5 effector proteins have been identified in yeast and animal systems, it has become clear that one aspect of Rab5/YPT51 regulation of early endosomal trafficking is the formation of lipid subdomains upon these compartments (Sonnichsen et al. 2000; Pfeffer 2003). This occurs through the co-recruitment of phosphatidylinositol 3-OH kinases (PI-3K) and proteins containing specific phosphatidylinositol-3-phosphate (PI-3P) interaction domains such as FYVE domains, and PX domains (Song et al. 2001; Xu et al. 2001). In animal cells Rab5 has been shown to recruit two distinct PI-3Ks, the type III PI-3K, hVPS34, and a type I PI-3K, p110b (Christoforidis et al. 1999). Of these two PI-3Ks, recruitment of the type III hVPS34 is required for most of the PI-3P production upon Rab5-positive endosomal membranes. This activity is responsible for subsequent recruitment of PI-3P binding proteins, and is required for EE-EE fusion (Zerial and McBride 2001). Plant endosomal compartments labeled with members of the RabF family also appear to be preferentially labeled with PI-3P, as GFP-FYVE fusion proteins localize to these compartments (Sohn et al. 2003; Voigt et al. 2005).

4 Endosome Maturation: Formation of MVB/LE

In animal and yeast systems, once newly internalized cargo reaches the early endosomal compartments, membrane proteins and lipids targeted for degradation are further sorted into late endosomes (LEs). These are often are typified by the presence of smaller internal vesicles within their lumen. As such LEs are often also called multivesicular bodies (MVBs), (Raiborg et al. 2003). Because internal vesicles are occasionally seen within the vesicular sorting portions of the early endosomes in animal cells, current models suggest that MVBs mature from early endosomes after recycling proteins have already been sorted and removed (Katzmann et al. 2002; Raiborg et al. 2003). This model is supported by observation that trafficking from early endosomes to late endosomes is microtubule dependent, and that in the presence of microtubule-depolymerizing drugs an intermediate transport compartment, termed the endosomal carrier vesicle (ECV) accumulates. ECVs are typified by their relatively small size ($\sim 0.4\,\mu m$) and the presence of internal vesicles, and they appear to be distinct from the tubular-vesicular early endosomal compartments, as they are no longer capable of fusion with early endosomal membranes. While the presence of Rab5 on ECVs is still unknown, the inability of ECVs to fuse with Rab5-positive endosomes in vitro suggests that at some point during the formation of ECVs the localization of Rab5 to these compartments is lost (Aniento et al. 1993). Rather, in animal systems, Rab7 is the predominant Rab GTPase observed on MVB/LEs and lysosomes, and this Rab GTPase is required for delivery of endocytic cargo from early endosomes to MVB/LEs and lysosomes (Chavrier et al. 1990; Feng et al. 1995; Bucci et al. 2000).

In both yeast and plants, the distinction between early endosomal compartments and MVBs is not as clear cut as in animal systems, as compartments containing newly internalized plasma membrane proteins have as yet not been found in tubular-vesicular endosomal structures (Katzmann et al. 2002; Raiborg et al. 2003). Plant and yeast MVBs play important roles in the sorting of newly synthesized proteins to vacuolar compartments and have therefore also been referred to as prevacuolar compartments (PVCs) (Stack et al. 1995; Vitale and Raikhel 1999). In recent years, the role of the PVC compartment in plant endocytic trafficking has also become more apparent, and this compartment has been demonstrated to be labeled with members of the RabF family of Rab GTPases (Ueda et al. 2001; Tse et al. 2004). RabF2a/Rha1 and RabF2b/Ara7 have been shown to colocalize to prevacuolar/endosomal membranes containing the PVC syntaxin, SYP21, and the vacuolar sorting receptor BP-80/AtELP (Sohn et al. 2003; Tse et al. 2004). Confirmation of roles for RabF2a and RabF2b, but not RabF1, in trafficking of newly synthesized proteins to the vacuole was confirmed by demonstration that expression of dominant-negative forms of the RabF2a/Rha1 and RabF2b/Ara7 GTPases block trafficking of soluble, vacuole-targeted proteins within the prevacuolar compartments, whereas expression of dominant-negative RabF1/Ara6 did not (Sohn et al. 2003; Lee et al. 2004). On the other hand, m-Rab_{mc}, a RabF1/Ara6-like GTPase from *Mesembryanthemum*, was shown to localize to plant PVC compartments, and expression of dominant-negative form of this Rab GTPase blocked trafficking of aleurain to vacuolar compartments (Bolte et al. 2004). In addition, recent results in which RabF1/Ara6 displayed more colocalization with PVC-localized syntaxins SYP21 and SYP22, whereas RabF2a/Rha1 and RabF2b/Ara6 were instead observed with Vamp727 and were on compartments that displayed altered morphology in a *gnom* mutant background (Ueda et al. 2004). While there may be some distinction in plant endosomal compartments, it still remains unclear whether RabF1 and RabF2 GTPases perform distinct functions during trafficking of newly synthesized proteins to plant vacuolar compartments.

5
MVB/LE, Lysosomes and Vacuoles

In mammals, Rab7 localizes to MVB/LEs and lysosomes and regulates transport of cargo from early endosomes to these compartments (Feng et al. 1995; Mukhopadhyay et al. 1997). In yeast, however, Ypt7p localizes to vacuoles and is responsible for regulation of homotypic fusion between vacuolar compartments (Price et al. 2000). In plants, the RabG subfamily is significantly homologous to Rab7 and Ypt7p, and members of this family appear to localize to plant vacuolar compartments. The *A. thaliana* RabG3c localizes to at least a subset of vacuolar compartments in BY-2 cells, (E. Nielsen, unpublished results) and other RabG subfamily members from *A. thaliana* and rice have been observed on plant tonoplast membranes (Saito et al. 2002; Nahm et al. 2003).

In animals and in yeast, the primary function of vacuoles and lysosomes is as a lytic compartment for turnover of macromolecules. However, plant vacuoles may act as storage organelles in addition to having lytic functions, and single plant cells can even contain multiple vacuole types (Vitale and Raikhel 1999). In addition, the majority of the volume of a fully expanded plant cell is taken up by a large central vacuole that maintains the rigidity and determines the shape of the plant cell. As a result, in addition to lytic and storage roles, plant vacuoles also play important developmental roles, and in plant responses to the environment, such as phototropic and gravitropic responses (Surpin and Raikhel 2004). Plant vacuolar trafficking may also play important roles in plant responses to abiotic stresses as over expression of RabG3e results in increased resistance to salinity and oxidative stress (Mazel et al. 2004), and a rice RabG orthologue, OsRab7, was recently shown to be upregulated in response to cold, salt, and drought stress (Nahm et al. 2003). This increased complexity of plant vacuolar functions appears to be reflected in the large number of members in this Rab GTPase sub-group in plants. For example, *A. thaliana* contains eight distinct RabG subfamily members, as opposed to a single Rab GTPase, Ypt7p, in *S. cerevisiae*, and two, Rab7A and Rab7B, in *H. sapiens*.

As important as membrane trafficking pathways leading to lysosomal or vacuolar compartments, the specific retrieval of membranes and sorting receptor proteins delivered to these compartments is equally important for the maintenance of cellular homeostasis. In animals, newly synthesized hydrolytic enzymes are modified in late Golgi compartments by the addition of mannose-6-phosphate, and sorted away from secretory traffic by interaction with mannose-6-phosphate receptors (MPRs) in the TGN (Kornfeld 1992). These MPRs are recruited to AP-3 adaptor complexes and bud from the TGN as CCVs which subsequently fuse with maturing MVB/LE compartments (Robinson and Bonifacino 2001). MPRs are then retrieved from MVB/late endosomes to the trans-Golgi network in a process which is regulated by Rab9 (Riederer et al. 1994). While neither yeast nor plants have an obvious Rab9 homologue (Pereira-Leal and Seabra 2001; Vernoud et al. 2003), since Rab9 co-segregates with Rab7 in animals, the possibility remains that some members of the RabG subfamily in plants may regulate recycling of vacuolar sorting receptors from plant vacuolar compartments.

6
Plasma Membrane Recycling from the Early Endosome

In animal cells there are two main routes back to the plasma membrane from Rab5-positive early endosomes. The first of these involves recycling to the plasma membrane directly from the early endosomal compartment in a process that is regulated in a Rab4-dependent fashion. The second involves further sorting to internal compartments called recycling endosomes, which will be described below. The rapid recycling component of internalized transferrin receptor typifies the trafficking of proteins along this pathway (Maxfield and McGraw 2004). In this pathway, the first step in recycling occurs upon acidification (pH ~ 6.0) of the lumen of a newly formed early endosome, which results in a concomitant uncoupling of receptors from their bound ligand (Yamashiro and Maxfield 1988). Upon dissociation from their cognate receptors, free ligands accumulate along with internalized solutes in the lumen of the vesicular portion of the early endosomal compartments, while ~ 50% of the internalized membranes and receptor proteins segregate into tubular elements. Once in these tubular elements, membrane and dissociated receptor proteins are pinched off from the vesicular, sorting portion of the early endosome and subsequently fuse with the plasma membrane (Gruenberg and Maxfield 1995). Segregation of proteins into these tubules is not thought to require a specific sorting signal since the removal of the cytoplasmic domain of the transferrin receptor, while impairing receptor internalization, does not slow its recycling back to the plasma membrane from the early endosomal compartment (Jing et al. 1990). Also, while Rab4-dependent recycling from this compartment does not require clathrin, it does require β-COP, a component of the COP-I coatomer which is also utilized for retrograde trafficking steps associated with the Golgi apparatus (Daro et al. 1997; Wettey et al. 2002). This is intriguing given the fact that Rab4 shows higher similarity to, and segregates with the intra-Golgi localized Rab2, rather than with members of the Rab5 subfamily (Pereira-Leal and Seabra 2001; Stenmark and Olkkonen 2001).

Reflecting the intimate nature of the interactions of the tubular recycling and the vesicular sorting portions of these early endosomes, it is not surprising that both Rab4 and Rab5 display significant, albeit incomplete overlap within animal cells, and are observed upon contiguous membrane-bounded organelles (van der Sluijs et al. 1992). Detailed examination of these compartments *in vivo* has led to a model in which Rab4 and Rab5 GTPases are each responsible for determining the physical characteristics of the separate subdomains within early endosomes (Sonnichsen et al. 2000). Cooperation between these domains is necessary for efficient sorting of cargo between these domains, and is highlighted by the fact that these two Rab GTPases share several effector proteins (Vitale et al. 1998, de Renzis et al. 2002).

While *S. cerevisiae* may recycle some plasma membrane proteins from endocytic compartments (Wiederkehr et al. 2000; Chen et al. 2005), this recycling pathway (see below) appears to traverse TGN-like compartments prior to their return to the plasma membrane (Jedd et al. 1997; Holthuis et al. 1998; Chen et al. 2005). Again, it is worth noting that early recycling events are regulated by Rab4 in animal cells, and that this Rab GTPase shows highest similarity to members of the intra-Golgi Rab2 family. In light of this it is interesting that Rab2/4 orthologues are not present in either *S. cerevisiae* or the closely related *Candida albicans*, while they are present in several other filamentous yeast species such as *Schizosaccharomyces pombe*, *Neurospora crassa*, and *Aspergillus fumigatus*. As neither *S. cerevisiae* nor *C. albicans* exhibit tip-growth via hyphae, the lack of Rab4 orthologues in these systems has been suggested to reflect the increased requirement for efficient membrane recycling to the plasma membrane during hyphal tip-growth in filamentous fungi (Gupta and Brent Heath 2002). Given the presence of tip-growing cells such as pollen tubes and root hair cells in plants, it is somewhat surprising that Rab4 orthologues have not yet been characterized (Pereira-Leal and Seabra 2001; Vernoud et al. 2003). It should be mentioned, however, that members of the *A. thaliana* RabB family are highly similar to the Rab2 family of animals (Pereira-Leal and Seabra 2001). Because the Rab4 orthologues present in animals and filamentous yeast co-segregate with the Rab2 subfamily the possibility remains that one or more of the *A. thaliana* RabB isoforms might fulfill a Rab4-like recycling function in plants.

7 Targeting to the Recycling Endosome

In animals, only ~ 50% of newly-internalized cargo that is destined for recycling back to the plasma membrane is delivered directly from the early endosomal compartment (Maxfield and McGraw 2004). Recycling cargo not delivered directly to the plasma membrane from the early endosome enters an indirect recycling pathway regulated by members of the Rab11 subfamily of animal Rab GTPases (Ullrich et al. 1996). Rab11-regulated membrane trafficking pathways involves traversal of loosely defined compartments termed "recycling endosomes" (Gruenberg and Maxfield 1995). This pleiotropic organelle can be tightly organized around the microtubule organizing center in some mammalian cells, yet distributed peripherally in others. In addition to roles in recycling it is clear that secretion of some newly synthesized cargo to the plasma membrane involves sorting steps that occur within Rab11 positive recycling endosomal membranes (Chen et al. 1998).

Rab11A, and its close relatives Rab11B and Rab25, have been shown to regulate recycling pathways for a number of cellular processes in animal cells, including polarized epithelial transport, and the insulin-dependent delivery of the GLUT4 glucose-transporter to the plasma membranes of fat cells (Cox et al. 2000; Wang et al. 2000; Wilcke et al. 2000). In addition, in these cells many of the proteins that cycle between the plasma membrane and internal compartments accumulate in the recycling endosome due to the slow rates of sorting and exit from these compartments (Maxfield and McGraw 2004). Even in animal systems, precise identification of sorting determinants and characterization of vesicle transport intermediates leading to and from these compartments remains poorly understood. In part, the lack of understanding of the mechanisms by which cargo is sorted and recycled from this compartment reflects the complexity of sorting events, in which specific sorting determinants can differ between non-polarized and polarized cells (Gan et al. 2002).

In *S. cerevisiae*, direct recycling from early endosomal compartments to the plasma membrane have not yet been demonstrated (Dupre et al. 2004). Instead, the main mechanism for recycling of plasma membrane proteins back to the plasma membrane appears to involve transit of recycling cargo through TGN-like compartments (Gupta and Brent Heath 2002; Chen et al. 2005). While the roles of Ypt31/32 in the exit of secretory cargo from the trans-Golgi cisterna have been long recognized (Benli et al. 1996; Jedd et al. 1997), the role of these compartments in sorting and recycling of endocytic cargo has only recently been described (Chen et al. 2005).

In both yeast and animal systems the number of Rab11/YPT31/32-like Rab GTPases are relatively small in comparison to the entire Rab GTPase family complement—3 of ~60 in *H. sapiens*, and 2 of 11 in *S. cerevisiae* (Pereira-Leal and Seabra 2001; Stenmark and Olkkonen 2001). However, in plants, one of the most obvious features of the Rab GTPase family is the strikingly large number of Rab GTPases (26 of 57 in *A. thaliana*) present in the Rab11/Ypt31/32-related RabA family (Rutherford and Moore 2002; Vernoud et al. 2003). This large number of RabA GTPases seems to be a conserved feature in other plants as well (Borg et al. 1997). Whether this multitude of genes provides distinct functions, or represents functionally redundant gene families remains to be determined. However, given the roles members of this Rab GTPase subfamily play in sorting and recycling of plasma membrane proteins through the recycling endosomal membranes in both yeast and animal systems it seems likely that these pathways are present, and have been elaborated within plants (Rutherford and Moore 2002; Vernoud et al. 2003).

What role do RabA family members play in the growth and development of plants? Several lines of evidence suggest that at least some members of the RabA subfamily play important roles in secretion and/or recycling of cell wall components in plants. In *A. thaliana*, RabA4b displays polarized distribution to the tips of growing root hair cells, and this localization was absent or altered in root hair developmental mutants (Preuss et al. 2004). These results suggest a role for RabA4b in the regulation of root hair tip growth, through polarized direction of new cell wall components, or through control of polar recycling events within these tip-growing cells. Supporting a specific role in polar expansion in tip-growing cells, several other members of the RabA subfamily are dramatically upregulated in *A. thaliana* pollen (A. Szlumanski and E. Nielsen, unpublished results). Antisense inhibition of RabA subfamily GTPases in tomato results in complex developmental abnormalities and delayed fruit ripening, which could be attributed to impaired cell wall deposition (Abeliovich et al. 1999). Also, the Pea RabA homologue,

Pra2 GTPase, was shown to primarily localize to Golgi and endosomal compartments, and the closely related Pea RabA homologue, Pra3, localized to trans-Golgi network and/or prevacuolar compartments (Inaba et al. 2002). These RabA isoforms were suggested to play a role in the delivery of new cell wall components to the plasma membrane.

8 Perspectives

Clearly these last several years have resulted in important gains in our understanding of the subcellular localizations and functions of many members of the plant endocytic Rab GTPase complement. Yet significant gaps remain in our understanding of the plant endocytic membrane trafficking pathways, and the relative roles of various plant Rab GTPases within these pathways. With this in mind the following are presented as a few of a myriad of interesting, unanswered questions to be posed for future investigation:

1) What do early endosomes look like in plants? Do they resemble tubular-vesicular structures as observed in animals, or are they MVBs as in yeast?
2) Does the presence of Rab2-like RabB GTPases in plants allow early endosomal recycling as in animal and filamentous yeast systems?
3) Does recycling of newly internalized endocytic cargo occur primarily from early endosomes, as in animal systems, or does recycling occur primarily from TGN-like recycling endosomes?
4) To what extent do RabA subfamily GTPase family members regulate endocytic recycling events in plants as opposed to regulation of polarized secretion?
5) Does the larger number of RabG GTPase subfamily members represent increased control of trafficking to distinct plant vacuoles?

References

Abeliovich H, Darsow T, Emr SD (1999) Cytoplasm to vacuole trafficking of aminopeptidase I requires a t-SNARE-Sec 1 p complex composed of Tlg2p and Vps45p. EMBO J 18:6005–6016

Aniento F, Emans N, Griffiths G, Gruenberg J (1993) Cytoplasmic dynein-dependent vesicular transport from early to late endosomes. J Cell Biol 123:1373–1387

Baggett JJ, Wendland B (2001) Clathrin function in yeast endocytosis. Traffic 2:297–302

Baluška F, Šamaj J, Hlavacka A, Kendrick-Jones J, Volkmann D (2004) Actin-dependent fluid-phase endocytosis in inner cortex cells of maize root apices. J Exp Bot 55:463–473

Baluška F, Hlavacka A, Šamaj J, Palme K, Robinson DG, Matoh T, McCurdy DW, Menzel D, Volkmann D (2002) F-actin-dependent endocytosis of cell wall pectins in meristematic root cells. Insights from brefeldin A-induced compartments. Plant Physiol 130:422–431

Benli M, Doring F, Robinson DG, Yang X, Gallwitz D (1996) Two GTPase isoforms, Ypt31p and Ypt32p, are essential for Golgi function in yeast. EMBO J 15:6460–6475
Blackbourn HD, Jackson AP (1996) Plant clathrin heavy chain: sequence analysis and restricted localisation in growing pollen tubes. J Cell Sci 109:777–786
Boehm M, Bonifacino JS (2001) Adaptins: the final recount. Mol Biol Cell 12:2907–2920
Bolte S, Talbot C, Boutte Y, Catrice O, Read ND, Satiat-Jeunemaitre B (2004) FM-dyes as experimental probes for dissecting vesicle trafficking in living plant cells. J Microsc 214:159–173
Borg S, Brandstrup B, Jensen TJ, Poulsen C (1997) Identification of new protein species among 33 different small GTP-binding proteins encoded by cDNAs from Lotus japonicus, and expression of corresponding mRNAs in developing root nodules. Plant J 11:237–250
Bucci C, Thomsen P, Nicoziani P, McCarthy J, van Deurs B (2000) Rab7: a key to lysosome biogenesis. Mol Biol Cell 11:467–480
Bucci C, Parton RG, Mather IH, Stunnenberg H, Simons K, Hoflack B, Zerial M (1992) The small GTPase rab5 functions as a regulatory factor in the early endocytic pathway. Cell 70:715–728
Chavrier P, Parton RG, Hauri HP, Simons K, Zerial M (1990) Localization of low molecular weight GTP binding proteins to exocytic and endocytic compartments. Cell 62:317–329
Chen SH, Chen S, Tokarev AA, Liu F, Jedd G, Segev N (2005) Ypt31/32 GTPases and their novel F-box effector protein Rcy1 regulate protein recycling. Mol Biol Cell 16:178–192
Chen W, Feng Y, Chen D, Wandinger-Ness A (1998) Rab11 is required for trans-golgi network-to-plasma membrane transport and a preferential target for GDP dissociation inhibitor. Mol Biol Cell 9:3241–3257
Christoforidis S, Miaczynska M, Ashman K, Wilm M, Zhao L, Yip SC, Waterfield MD, Backer JM, Zerial M (1999) Phosphatidylinositol-3-OH kinases are Rab5 effectors. Nat Cell Biol 1:249–252
Cox D, Lee DJ, Dale BM, Calafat J, Greenberg S (2000) A Rab11-containing rapidly recycling compartment in macrophages that promotes phagocytosis. Proc Natl Acad Sci USA 97:680–685
Daro E, Sheff D, Gomez M, Kreis T, Mellman I (1997) Inhibition of endosome function in CHO cells bearing a temperature-sensitive defect in the coatomer (COPI) component epsilon-COP. J Cell Biol 139:1747–1759
de Renzis S, Sonnichsen B, Zerial M (2002) Divalent Rab effectors regulate the sub-compartmental organization and sorting of early endosomes. Nat Cell Biol 4:124–133
D'Hondt K, Heese-Peck A, Riezman H (2000) Protein and lipid requirements for endocytosis. Annu Rev Genet 34:255–295
Dupre S, Urban-Grimal D, Haguenauer-Tsapis R (2004) Ubiquitin and endocytic internalization in yeast and animal cells. Biochim Biophys Acta 1695:89–111
Eggenschwiler JT, Espinoza E, Anderson KV (2001) Rab23 is an essential negative regulator of the mouse Sonic hedgehog signalling pathway. Nature 412:194–198
Emans N, Zimmermann S, Fischer R (2002) Uptake of a fluorescent marker in plant cells is sensitive to brefeldin A and wortmannin. Plant Cell 14:71–86
Evans TM, Ferguson C, Wainwright BJ, Parton RG, Wicking C (2003) Rab23, a negative regulator of hedgehog signaling, localizes to the plasma membrane and the endocytic pathway. Traffic 4:869–884
Feng Y, Press B, Wandinger-Ness A (1995) Rab 7: an important regulator of late endocytic membrane traffic. J Cell Biol 131:1435–1452
Gan Y, McGraw TE, Rodriguez-Boulan E (2002) The epithelial-specific adaptor AP1B mediates post-endocytic recycling to the basolateral membrane. Nat Cell Biol 4:605–609

Geldner N, Friml J, Stierhof YD, Jurgens G, Palme K (2001) Auxin transport inhibitors block PIN1 cycling and vesicle trafficking. Nature 413:425–428
Geldner N, Anders N, Wolters H, Keicher J, Kornberger W, Muller P, Delbarre A, Ueda T, Nakano A, Jurgens G (2003) The *Arabidopsis* GNOM ARF-GEF mediates endosomal recycling, auxin transport, and auxin-dependent plant growth. Cell 112:219–230
Gerrard SR, Bryant NJ, Stevens TH (2000) VPS21 controls entry of endocytosed and biosynthetic proteins into the yeast prevacuolar compartment. Mol Biol Cell 11:613–626
Gorvel JP, Chavrier P, Zerial M, Gruenberg J (1991) rab5 controls early endosome fusion in vitro. Cell 64:915–925
Grebe M, Xu J, Mobius W, Ueda T, Nakano A, Geuze HJ, Rook MB, Scheres B (2003) *Arabidopsis* sterol endocytosis involves actin-mediated trafficking via ARA6-positive early endosomes. Curr Biol 13:1378–1387
Gruenberg J, Maxfield FR (1995) Membrane transport in the endocytic pathway. Curr Opin Cell Biol 7:552–563
Gupta GD, Brent Heath I (2002) Predicting the distribution, conservation, and functions of SNAREs and related proteins in fungi. Fungal Genet Biol 36:1–21
Hawes C, Crooks K, Coleman J, Satiat-Jeunematrie B (1995) Endocytosis in plants: fact or artefact? Plant Cell Environ 18:1245–1252
Hillmer S, Depta H, Robinson DG (1986) Confirmation of endocytosis in higher plant protoplasts using lectin-gold conjugates. Eur J Cell Biol 42:142–149
Holstein SE (2002) Clathrin and plant endocytosis. Traffic 3:614–620
Holthuis JC, Nichols BJ, Dhruvakumar S, Pelham HR (1998) Two syntaxin homologues in the TGN/endosomal system of yeast. EMBO J 17:113–126
Huebner RHD, Depta H, Robinson DG (1985) Endocytosis in maize root cap cells. Evidence obtained using heavy metal salt solutions. Protoplasma 129:214–222
Hunziker W, Peters PJ (1998) Rab17 localizes to recycling endosomes and regulates receptor-mediated transcytosis in epithelial cells. J Biol Chem 273:15734–15741
Hurst AC, Meckel T, Tayefeh S, Thiel G, Homann U (2004) Trafficking of the plant potassium inward rectifier KAT1 in guard cell protoplasts of Vicia faba. Plant J 37:391–397
Inaba T, Nagano Y, Nagasaki T, Sasaki Y (2002) Distinct localization of two closely related Ypt3/Rab11 proteins on the trafficking pathway in higher plants. J Biol Chem 277:9183–9188
Jedd G, Mulholland J, Segev N (1997) Two new Ypt GTPases are required for exit from the yeast trans-Golgi compartment. J Cell Biol 137:563–580
Jing SQ, Spencer T, Miller K, Hopkins C, Trowbridge IS (1990) Role of the human transferrin receptor cytoplasmic domain in endocytosis: localization of a specific signal sequence for internalization. J Cell Biol 110:283–294
Katzmann DJ, Odorizzi G, Emr SD (2002) Receptor downregulation and multivesicular-body sorting. Nat Rev Mol Cell Biol 3:893–905
Kornfeld S (1992) Structure and function of the mannose 6-phosphate/insulinlike growth factor II receptors. Annu Rev Biochem 61:307–330
Lee GJ, Sohn EJ, Lee MH, Hwang I (2004) The *Arabidopsis* rab5 homologs rha1 and ara7 localize to the prevacuolar compartment. Plant Cell Physiol 45:1211–1220
Maxfield FR, McGraw TE (2004) Endocytic recycling. Nat Rev Mol Cell Biol 5:121–132
Mazel A, Leshem Y, Tiwari BS, Levine A (2004) Induction of salt and osmotic stress tolerance by overexpression of an intracellular vesicle trafficking protein AtRab7 (AtRabG3e). Plant Physiol 134:118–128
McLauchlan H, Newell J, Morrice N, Osborne A, West M, Smythe E (1998) A novel role for Rab5-GDI in ligand sequestration into clathrin-coated pits. Curr Biol 8:34–45

Meckel T, Hurst AC, Thiel G, Homann U (2004) Endocytosis against high turgor: intact guard cells of Vicia faba constitutively endocytose fluorescently labelled plasma membrane and GFP-tagged K-channel KAT1. Plant J 39:182–193

Mukhopadhyay A, Funato K, Stahl PD (1997) Rab7 regulates transport from early to late endocytic compartments in Xenopus oocytes. J Biol Chem 272:13055–13059

Nahm MY, Kim SW, Yun D, Lee SY, Cho MJ, Bahk JD (2003) Molecular and biochemical analyses of OsRab7, a rice Rab7 homolog. Plant Cell Physiol 44:1341–1349

Nielsen E, Christoforidis S, Uttenweiler-Joseph S, Miaczynska M, Dewitte F, Wilm M, Hoflack B, Zerial M (2000) Rabenosyn-5, a novel Rab5 effector, is complexed with hVPS45 and recruited to endosomes through a FYVE finger domain. J Cell Biol 151:601–612

Okita TW, Rogers JC (1996) Compartmentation of proteins in the endomembrane system of plant cells. Annu Rev Plant Physiol Plant Mol Biol 47:327–350

Olkkonen VM, Dupree P, Killisch I, Lutcke A, Zerial M, Simons K (1993) Molecular cloning and subcellular localization of three GTP-binding proteins of the rab subfamily. J Cell Sci 106:1249–1261

Opdam FJ, Kamps G, Croes H, van Bokhoven H, Ginsel LA, Fransen JA (2000) Expression of Rab small GTPases in epithelial Caco-2 cells: Rab21 is an apically located GTP-binding protein in polarised intestinal epithelial cells. Eur J Cell Biol 79:308–316

Pereira-Leal JB, Seabra MC (2001) Evolution of the Rab family of small GTP-binding proteins. J Mol Biol 313:889–901

Pfeffer S (2003) Membrane domains in the secretory and endocytic pathways. Cell 112:507–517.

Preuss ML, Santos-Serna J, Falbel TG, Bednarek SY, Nielsen E (2004) The *Arabidopsis* Rab GTPase RabA4b localizes to the tips of growing root hair cells. Plant Cell 16:1589–1603

Price A, Seals D, Wickner W, Ungermann C (2000) The docking stage of yeast vacuole fusion requires the transfer of proteins from a cis-SNARE complex to a Rab/Ypt protein. J Cell Biol 148:1231–1238

Raiborg C, Rusten TE, Stenmark H (2003) Protein sorting into multivesicular endosomes. Curr Opin Cell Biol 15:446–455

Riederer MA, Soldati T, Shapiro AD, Lin J, Pfeffer SR (1994) Lysosome biogenesis requires Rab9 function and receptor recycling from endosomes to the trans-Golgi network. J Cell Biol 125:573–582

Robinson JS, Klionsky DJ, Banta LM, Emr SD (1988) Protein sorting in Saccharomyces cerevisiae: isolation of mutants defective in the delivery and processing of multiple vacuolar hydrolases. Mol Cell Biol 8:4936–4948

Robinson MS, Bonifacino JS (2001) Adaptor-related proteins. Curr Opin Cell Biol 13:444–453

Rutherford S, Moore I (2002) The *Arabidopsis* Rab GTPase family: another enigma variation. Curr Opin Plant Biol 5:518–528

Saito C, Ueda T, Abe H, Wada Y, Kuroiwa T, Hisada A, Furuya M, Nakano A (2002) A complex and mobile structure forms a distinct subregion within the continuous vacuolar membrane in young cotyledons of *Arabidopsis*. Plant J 29:245–255

Šamaj J, Baluška F, Voigt B, Schlicht M, Volkmann D, Menzel D (2004) Endocytosis, actin cytoskeleton, and signaling. Plant Physiol 135:1150–1161

Scheele U, Holstein SE (2002) Functional evidence for the identification of an *Arabidopsis* clathrin light chain polypeptide. FEBS Lett 514:355–360

Singer-Kruger B, Stenmark H, Zerial M (1995) Yeast Ypt51p and mammalian Rab5: counterparts with similar function in the early endocytic pathway. J Cell Sci 108:3509–3521

Sohn EJ, Kim ES, Zhao M, Kim H, Kim YW, Lee YJ, Hillmer S, Sohn U, Jiang L, Hwang I (2003) Rha1, an *Arabidopsis* Rab5 Homolog, plays a critical role in the vacuolar trafficking of soluble cargo proteins. Plant Cell 15:1057–1070

Song X, Xu W, Zhang A, Huang G, Liang X, Virbasius JV, Czech MP, Zhou GW (2001) Phox homology domains specifically bind phosphatidylinositol phosphates. Biochemistry 40:8940–8944

Sonnichsen B, De Renzis S, Nielsen E, Rietdorf J, Zerial M (2000) Distinct membrane domains on endosomes in the recycling pathway visualized by multicolor imaging of Rab4, Rab5, and Rab11. J Cell Biol 149:901–914

Stack JH, Horazdovsky B, Emr SD (1995) Receptor-mediated protein sorting to the vacuole in yeast: roles for a protein kinase, a lipid kinase and GTP-binding proteins. Annu Rev Cell Dev Biol 11:1–33

Steinmann T, Geldner N, Grebe M, Mangold S, Jackson CL, Paris S, Galweiler L, Palme K, Jurgens G (1999) Coordinated polar localization of auxin efflux carrier PIN1 by GNOM ARF GEF. Science 286:316–318

Stenmark H, Olkkonen VM (2001) The Rab GTPase family. Genome Biol 2: REVIEWS3007

Surpin M, Raikhel N (2004) Traffic jams affect plant development and signal transduction. Nat Rev Mol Cell Biol 5:100–109

Tse YC, Mo B, Hillmer S, Zhao M, Lo SW, Robinson DG, Jiang L (2004) Identification of multivesicular bodies as prevacuolar compartments in Nicotiana tabacum BY-2 cells. Plant Cell 16:672–693

Ueda T, Yamaguchi M, Uchimiya H, Nakano A (2001) Ara6, a plant-unique novel type Rab GTPase, functions in the endocytic pathway of *Arabidopsis* thaliana. EMBO J 20:4730–4741

Ueda T, Uemura T, Sato MH, Nakano A (2004) Functional differentiation of endosomes in *Arabidopsis* cells. Plant J 40:783–789

Ullrich O, Reinsch S, Urbe S, Zerial M, Parton RG (1996) Rab11 regulates recycling through the pericentriolar recycling endosome. J Cell Biol 135:913–924

van der Sluijs P, Hull M, Webster P, Male P, Goud B, Mellman I (1992) The small GTP-binding protein rab4 controls an early sorting event on the endocytic pathway. Cell 70:729–740

Vernoud V, Horton AC, Yang Z, Nielsen E (2003) Analysis of the small GTPase gene superfamily of *Arabidopsis*. Plant Physiol 131:1191–1208

Vitale A, Raikhel NV (1999) What do proteins need to reach different vacuoles? Trends Plant Sci 4:149–155

Vitale G, Rybin V, Christoforidis S, Thornqvist P, McCaffrey M, Stenmark H, Zerial M (1998) Distinct Rab-binding domains mediate the interaction of Rabaptin-5 with GTP-bound Rab4 and Rab5. EMBO J 17:1941–1951

Voigt B, Timmers A, Šamaj J, Hlavacka A, Ueda T, Preuss M, Nielsen E, Mathur J, Emans N, Stenmark H, Nakano A, Baluška F, Menzel D (2005) Actin-propelled endosomes accumulate at sites of actin-driven polar growth of root hairs. Eur J Cell Biol 84:609–621

Wang X, Kumar R, Navarre J, Casanova JE, Goldenring JR (2000) Regulation of vesicle trafficking in madin-darby canine kidney cells by Rab11a and Rab25. J Biol Chem 275:29138–29146

Weigert R, Yeung AC, Li J, Donaldson JG (2004) Rab22a regulates the recycling of membrane proteins internalized independently of clathrin. Mol Biol Cell 15:3758–3770

Wettey FR, Hawkins SF, Stewart A, Luzio JP, Howard JC, Jackson AP (2002) Controlled elimination of clathrin heavy-chain expression in DT40 lymphocytes. Science 297:1521–1525

Wiederkehr A, Avaro S, Prescianotto-Baschong C, Haguenauer-Tsapis R, Riezman H (2000) The F-box protein Rcy1p is involved in endocytic membrane traffic and recycling out of an early endosome in Saccharomyces cerevisiae. J Cell Biol 149:397–410
Wilcke M, Johannes L, Galli T, Mayau V, Goud B, Salamero J (2000) Rab11 regulates the compartmentalization of early endosomes required for efficient transport from early endosomes to the trans-golgi network. J Cell Biol 151:1207–1220
Xu Y, Seet LF, Hanson B, Hong W (2001) The Phox homology (PX) domain, a new player in phosphoinositide signalling. Biochem J 360:513–530
Yamashiro DJ, Maxfield FR (1988) Regulation of endocytic processes by pH. Trends Pharmacol Sci 9:190–193
Zacchi P, Stenmark H, Parton RG, Orioli D, Lim F, Giner A, Mellman I, Zerial M, Murphy C (1998) Rab17 regulates membrane trafficking through apical recycling endosomes in polarized epithelial cells. J Cell Biol 140:1039–1053
Zerial M, McBride H (2001) Rab proteins as membrane organizers. Nat Rev Mol Cell Biol 2:107–117

Plant Cell Monogr (1)
J. Šamaj · F. Baluška · D. Menzel: Plant Endocytosis
DOI 10.1007/7089_012/Published online: 27 September 2005

SNAREs in Plant Endocytosis and the Post-Golgi Traffic

Masa H. Sato[1] (✉) · Ryosuke L. Ohniwa[2] · Tomohiro Uemura[3]

[1]Faculty of Human Environmental Sciences, Kyoto Prefectural University, Shimogamo-nakaragi-cho, Sakyo-ku, 606-8522 Kyoto, Japan
mhsato@kpu.ac.jp

[2]Graduate School of Biostudies, Kyoto University, Yoshidanihontsu, Sakyo-ku, 606-8501 Kyoto, Japan

[3]Molecular Membrane Biology Laboratory, RIKEN Discovery Research Institute, 2-1 Hisosawa, Wako, 351-0198 Saitama, Japan

Abstract In eukaryotic cells, the transport vesicles carry various cargo proteins from a donor compartment to a target compartment, and discharge the cargo into the target compartment by fusing with the membrane of the target compartment. SNARE molecules have a central role for initiating membrane fusion between transport vesicles and target membranes by forming a specific trans-SNARE complex in each transport step. In higher plants, the numbers of SNARE molecules are greater than those of yeast and mammals, suggesting a higher complexity of membrane traffic in higher plant cells.

In this chapter, we will focus on the functions and subcellular localizations of plant SNARE molecules and discuss the complexity and evolution of endocytosis and the post-Golgi traffic in the higher plant cells.

1
Introduction

In eukaryotic cells, endocytosis starts from the plasma membrane to take up macromolecules, particular substances, plasma membrane surface proteins as well as lipid components of the plasma membrane by forming endocytic coated vesicles. Most of the internalized cargo molecules residing in the transport vesicles are then transported via the endocytic pathway to the lysosomes or vacuoles, where the transported proteins are finally degraded, although some molecules are recycled in the endosomes back to the cell surface.

In mammalian cells, endosomes represent organelles involved in the endocytic pathway, and they are mainly classified into four classes; early endosomes, late endosomes, recycling endosomes, and lysosomes. Early endosomes and recycling endosomes are tubulo-vesicular compartments distributed throughout the peripheral and perinuclear cytoplasm. They are regarded as sites of the recycling of internalized proteins and membranes to the plasma membrane. Late endosomes are defined as vesicular structures that accumulate internalized components after their passage through early endosomes. Late endosomes are also called multivesicular bodies (MVBs) because they often contain abundant internal membranes. Lysosomes contain hydrolytically active hydrolases and are involved in the degradative process (Mellman, 1996; Brodsky et al., 2001).

In plant cells, the occurrence of endocytosis has been in doubt for a long time because the endocytic process was regarded as unlikely to occur under high turgor pressure (Cram, 1980). However, a lot of evidence supports the occurrence of endocytosis in higher plants now (Low and Chandra, 1994; Battey et al., 1999; Holstein, 2002; Šamaj et al., 2004, 2005). For example, calculations of membrane flow during secretory vesicle fusion indicated that an endocytic pathway must exist for the retrieval of membrane material into the cell (Phillips et al., 1988). Studies using electron-dense endocytic markers applied to plant protoplasts showed that the sequential movement of these markers through the endocytic compartments was analogous to that found in animal and yeast cells (Nishizawa and Mori, 1977; Joachim and Robinson, 1984; Hillmer et al., 1986). Recently, fluorescent dyes, FM1-43 and FM4-64, were used for tracing the endocytic process in both protoplasts and intact plant cells (Ueda et al., 2001; Emans et al., 2002). These dyes initially label the plasma membrane, but are gradually internalized and then stain rapidly moving punctate organelles (Šamaj 2005, Šamaj et al., 2005). Finally, these dyes are delivered to vacuoles via ring-like organelles (Ueda et al., 2001). The endocytic process is inhibited by low temperature treatment, and inhibitors such as wortmannin, which is a phosphatidylinositol 3-kinase inhibitor, or latrunculin B, a F-actin depolymerizing drug, indicating that this process is strongly energy and actin dependent (Emans et al., 2002; Baluška et al., 2004). Furthermore, several recent studies have indicated that endocytosis plays important roles in the establishment and maintenance of the asymmetric distribution of particular membrane protein such as AtPIN1 (Geldner et al., 2001; Friml et al., 2002), and in the sequestration of receptor-like kinases (Shah et al., 2002). Thus growing evidence supports that endocytosis plays various important roles in plant physiology.

Most of the molecules involved in the endocytic process are highly conserved in plant cells (Sanderfoot and Raikhel, 1999; Sanderfoot et al., 2000; Holstein, 2002; Šamaj et al., 2004; Sanderfoot and Raikhel, 2003). In particular, SNARE molecules have a central role for initiating membrane fusion between transport vesicles and target membranes by forming specific trans-SNARE complexes throughout the fusion process.

In this chapter, we will firstly review the endocytic organelles so far identified in plant cells and then discuss the SNARE molecules that are involved in each fusion step.

2
Endocytic and the Post-Golgi Organelles in Plant Cells

Although the endomembrane system is well conserved among all eukaryotic cells, the plant endomembrane system has unique features. In mammals and yeast, each cell has only one type of lysosome or vacuole, whereas plant cells possess two types of functionally different vacuoles, a lytic vacuole and a protein storage vacuole. The lytic vacuole is comparable to the mammalian lysosomes and the yeast vacuole, whereas the protein storage vacuole is a specialized vacuole having the capacity to store proteins in seeds or vegetative cells. Therefore, protein sorting in plant cells is likely to be more complicated than in those of other eukaryotes because several separate pathways originate from the Golgi apparatus, ER or plasma membrane, respectively, and are directed towards two different vacuolar destinations (Matsuoka et al., 1995; Hohl et al., 1996; Hara-Nishimura et al., 1998; Neuhaus and Rogers, 1998; Hillmer et al., 2001; Toyooka et al., 2001).

In plant protoplasts, movement through the endocytic compartments has been shown using electron-dense markers at the electron microscopic level (Low and Chandra, 1994). Once the electron-dense markers were endocytosed at the PM into clathrin coated vesicles (CCVs), they moved to the partially coated reticulum (PCR), subsequently they appeared in the MVBs and finally in the vacuole. The plant PCR is a system of interconnected tubular membranes with clathrin-coated buds at its periphery surrounded by CCVs (Tanchak et al., 1984, 1988; Hillmer et al., 1988; Samuels and Bisalputra, 1990; Galway et al., 1993). Based on its morphological appearance, the plant PCR is considered to be the equivalent of the mammalian early endosomes (Tanchak et al., 1988; Galway et al., 1993) or TGN (Hillmer et al., 1988). MVBs are 250-700 nm organelles containing multiple 50-100 nm vesicles within their lumen (Tanchak and Fowke, 1987; Tse et al., 2004; Sheung et al., 2005). Plant MVBs contain some degradative enzymes (Halperin, 1969; Record and Griffing, 1988), indicating that they are compartments both functionally and morphologically homologous to the animal MVBs. Several recent studies have begun to shed light on the role of MVBs as a prevacuolar compartment in plants (Tse et al., 2004).

3
General Features of SNARE Molecules

In eukaryotic cells, the transport vesicles carry various cargos from a donor compartment to a target compartment, and discharge it into the target compartment by fusing with its membranes (Pryer et al., 1992; Rothman and Wieland, 1996). The specificity of membrane fusion is mediated by various molecules, such as SNAREs and Rabs (Waters and Hughson, 2000; Nebenfuhr, 2002; Ueda and Nakano, 2002). SNARE molecules contain a highly conserved coiled-coil domain, called the SNARE motif, and are anchored to the membrane by a hydrophobic transmembrane domain at the C-terminus or via post-translational lipid modification. Before membrane fusion, a trans-SNARE complex consisting of four-helical bundles of the SNARE complex from three target-SNAREs (t-SNAREs) and one vesicle-associated SNARE (v-SNARE) is formed (Ungar and Hughson, 2003). In the case of homotypic membrane fusion between identical compartments, however, one cannot discriminate between v-SNAREs and t-SNAREs. To avoid confusion, the SNARE molecules have been reclassified as "Q-SNAREs" and "R-SNAREs" according to the conserved amino acids (glutamine or arginine) in the center of the "zero-layer" of the SNARE motif. Furthermore, they are classified into five classes based on their similarities to the synaptic SNARE molecules: Qa-SNAREs (Syntaxins), Qb-SNAREs (SNAP Ns), Qc-SNAREs (SNAPCs), R-SNAREs (VAMPs) and SNAPs (SNAP Ns and Cs). A typical SNARE complex contains each copy of the Qa, Qb, Qc-, and R-SNARE motif or one copy of Qa, SNAP, and R-SNARE (Fasshauer et al., 1998). Since a particular SNARE complex is involved in each membrane fusion step at the specific vesicular transport pathway, each organelle is marked by the presence of specific resident SNARE proteins.

4
Localization and Possible Complex Formation of Plant SNAREs on the Endocytic Pathway

In the *Arabidopsis* genome, 54 SNARE and 57 Rab GTPase genes have been identified (Sanderfoot et al., 2000; Vernoud et al., 2003; Nielsen, 2005). These numbers are greater than those found in yeast and mammals (Bock et al., 2001), indicating the complexity of the plant endomembrane system. Although the localizations of few SNARE molecules have been individually determined, the subcellular localizations of most SNAREs, particularly R-SNAREs, are still unknown. Recently, the localization of SNAREs in *Arabidopsis* was systematically determined using green fluorescent protein fused to SNARE molecules (Uemura et al., 2004). The subcellular localizations of the SNARE molecules of the post-Golgi or the endocytic organelles so far identified are summarized in Fig. 1.

4.1
SNARE Molecules on the Plasma Membrane

Using the transient assay system, 18 SNARE molecules were determined to be localized on the plasma membrane of *Arabidopsis* (Uemura et al., 2004). In addition, the SNAP25 homolog, SNAP33, was reported to be in the

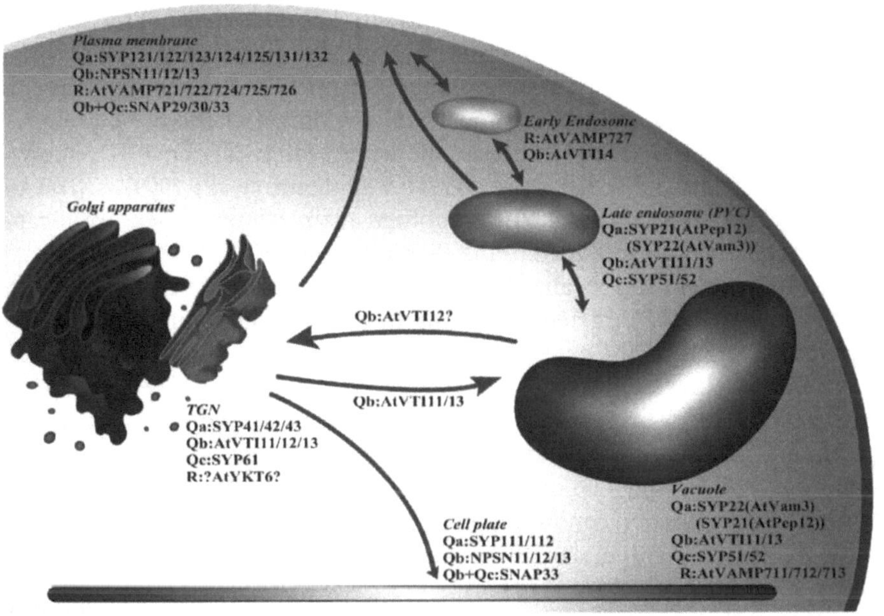

Fig. 1 Schematic model of SNARE localizations on the post-Golgi organelles. Many Qa- and R-SNARE molecules are localized to the plasma membrane, indicating the complexity of endocytic and exocytic pathways in plant cells

plasma membrane and the cell plate, forming a SNARE complex with SYP121 or KN/SYP111 (Heese et al., 2001). A plant specific Qb-SNARE molecule, NPSN11 also interacts with KN/SYP111 at the cell plate (Zheng et al., 2002). Namely, KN/SYP111 interacts with both SNAP33 and NPSN11. If NPSN11 forms a complex with KN/SYP111 and SNAP33, this SNARE complex would be unusual, because this complex contains one redundant Qb-SNARE. Alternatively, KN/SYP111 may form at least two distinct SNARE complexes with different partners. In this case, a SNARE complex containing KN/SYP111 and NPSN11 lacks appropriate Qc partners (Jürgens, 2004). A KN/SYP111 interacting SNAP33 also interacts with NtSYP121 in tobacco cells and the expression of AtSNAP33 is induced by pathogens and mechanical stimulation (Kargul et al., 2001). These data imply that SNAP33 forms several distinct SNARE complexes on the plasma membrane, and has a general role for plasma membrane traffic in plant cells.

In summary, a total of 21 SNAREs (9 Qa/syntaxin, 3 Qb, 5 R, and 3 SNAP25) were determined to reside on the plasma membrane. These proteins are not considered to be tissue-specific isoforms, because the expression of these genes was detected in almost all tissues (Uemura et al., 2004). Why are so many SNAREs expressed on the plasma membrane of *Arabidopsis* at the same time? One possibility is that SNAREs on the plasma membrane may function in distinct transport pathways to the subdomains of the plasma membrane. Some integral membrane proteins are indeed localized to special subdomains of the plasma membrane. For instance, the putative auxin influx carrier, AUX1, is localized to the apical domain of the plasma membrane (Swarup et al., 2001), whereas the auxin efflux carrier, PIN1, shows basal localization (Galweiler et al., 1998, Steinmann et al., 1999). COBRA, a GPI-anchored protein, and PIN3, another auxin efflux carrier, are localized to the lateral plasma membrane (Schindelman et al., 2001; Friml et al., 2002). PIN2 protein was localized at the apical surface of root epidermis cells (Muller et al., 1998). Targeted trafficking to plasma-membrane subdomains requires sorting from the general bulk flow, which may occur on Golgi stacks or during recycling from endosomes (Jürgens and Geldner, 2002). These proteins have to be delivered to distinct destinations on the cell surface via specific polarized vesicular transport pathways. Although the plasma membrane SNARES showed a ubiquitous distribution on the plasma membrane in the transient expression condition, these SNAREs may function in membrane traffic to specific subdomains of the plasma membrane of polarized cells in plants.

4.2 SNAREs in the TGN

SYP4 (SYP41, SYP42 and SYP43) family proteins are yeast Tlg1p homologs, and have been shown to be localized on the trans-Golgi network (TGN) (Bassham et al., 2000; Sanderfoot et al., 2001; Uemura et al., 2004) and forms complexes with SYP41/42, VTI12 and AtVPS45 in *Arabidopsis* cells (Bassham et al., 2000). SYP61 is also localized on the TGN (Sanderfoot et al., 2001; Zhu et al., 2002; Uemura et al., 2004). Interestingly, it was shown by an immunoelectron microscopic study that SYP41 and SYP42 are found on distinct domains of the TGN. The vacuolar cargo receptor, ELP, was found to preferentially localize with SYP42 at the TGN, suggesting that SYP41 and SYP 42 may form a separate SNARE complex in distinct parts of the TGN (Bassham et al., 2000).

SYP41 (a TGN marker) and SYP31 (a Golgi marker) show different localizations in the plant cells indicating that the TGN does not structurally connect to and move along with the Golgi apparatus (Uemura et al., 2004). This suggests that the TGN could be defined as a separate organelle distinct from the Golgi apparatus in *Arabidopsis* cells and is clearly not a continuation of the Golgi apparatus as has been postulated for mammalian cells.

Paris and coworkers have shown that the network structure of the TGN is not always seen at the trans-face of the Golgi apparatus, when optimal fixation procedures, such as high-pressure freezing, are used (Saint-Jore-Dupas et al., 2004). Antibodies raised against SYP41 showed strong staining of elongated, often sickle-shaped structures that appeared to be different from structures labelled with Golgi stack markers (Geldner et al., 2003). So in contrast to the original proposal by Robinson and coworkers (Hillmer et al., 1988), the PCR structures are more often observed as separate compartments not associated with the Golgi apparatus than as trans-Golgi network sensu strictu (Griffing, 1991). According to the cisternal maturation model, each Golgi cisterna matures as it migrates outwards through a stack (Mironov et al., 1997; Pelham and Rothman, 2000) and hence will contribute to the TGN at some point of maturation. The TGN may then detach itself from the Golgi apparatus and serve a function independent from the Golgi. If such a separate nature of the plant TGN holds true, we may need to redefine this intracellular compartment as a novel organelle, which might be the sorting center for the post-Golgi membrane traffic.

4.3 SNAREs in the Endosomes

In animals, the small GTPase Rab5 has been established as a marker for early endosomes (Somsel Rodman and Wandinger-Ness, 2000). In *Arabidopsis*, three Rab5 homologs, namely Ara6, Ara7 and Rha1, were identified (Nielsen, 2005). These Rab5 homologs were shown to colocalize with FM4-64 in endocytic compartments after short uptake periods (Anuntalabhochai et al., 1991; Ueda et al., 2001). Treatment with BFA resulted in the accumulation of endocytic vesicles, which were enriched with BFA-sensitive ARF-GEF GNOM, PIN1 and other plasma-membrane proteins (for recent review see Šamaj et al., 2004). GNOM also colocalized with FM4-64, Ara7 and Rha1-positive compartments, and the GNOM-residing endosomes were structurally altered in genetically engineered *gnom* mutant (insensitive to BFA) protoplasts (Geldner et al., 2003).

Since the recycling of plasma membrane proteins is thought to involve sorting in the early endosomes, this GNOM, Ara7 and Rha1 containing endocytic compartment may be defined as early endosome (Ueda et al., 2001; Geldner et al., 2003). A double labelling experiment using different color variants of fluorescent proteins revealed that VAMP727 specifically colocalizes with Ara7 and Rha1 on the GNOM-endosomes, indicating that VAMP727 is also localized on the early endosomes (Ueda et al., 2004).

The *Arabidopsis* SYP2-type syntaxins, SYP21 and SYP22, are both found on the prevacuolar compartment, although SYP22 is predominantly localized on the vacuolar membrane. Rha1 and Ara7 colocalized with SYP21 at the PVC (Kotzer et al., 2004; Lee et al., 2004). The localization of Ara7/Rha1 and Ara6 was reported to overlap partially, indicating that early endosomes and PVCs are not completely stable compartments, and gradually change their components and properties, as is the case in Golgi cisterna maturation (Ueda et al., 2004).

4.4 SNAREs on the Vacuolar Membrane

Q-SNARE proteins such as AtVam3/SYP22, SYP51, and VTI11 have been reported to be localized on the vacuolar membrane. In addition, R-SNARE proteins, VAMP711, VAMP712, and VAMP713 were also identified on the vacuolar membrane (Uemura et al., 2004). Both SYP51 and SYP61 are found on multiple organelles in the late secretory system, and interact with other syntaxins and SNAREs of the VTI1-type (Sanderfoot et al., 2001). Since both SYP21 and SYP51 interact with the SNARE VTI11 (Yano et al., 2003), it might be possible that these three proteins are part of a SNARE complex on the PVC or vacuole involved in TGN-to-PVC/vacuole trafficking.

Rojo et al., reported that *vacuoleless1* (*vcl1*), a mutant of the *Arabidopsis* homolog of yeast Vps16, is not able to form vacuoles and mislocalizes vacuolar proteins to the outside of the cell (Rojo et al., 2001). VCL1 is localized on the vacuolar membrane, and forms a complex with AtVPS11, AtVPS33, SYP21 or SYP22 as the AtC-VPS complex (Rojo et al., 2003). These data suggest that SYP2-, SYP5-, VTI11, and several VAMP71 proteins (VAMP711, VAMP712, and VAMP713) form the vacuolar-type SNARE complex, and may constitute a part of the AtC-VPS complex on the vacuolar membrane. In fact, these SNARE molecules were identified from the vacuolar membrane by analyzing the vegetative vacuole proteome of *Arabidopsis* (Carter et al., 2004).

It has been reported that there are functionally different vacuoles, and several different pathways to the two-types of vacuoles exist in the plant cells (Neuhaus and Rogers, 1998). However, we could not identify any vacuolar-type SNARE complex other than the SYP2 (Qa)/VTI11 (Qb)/SYP5 (Qc)/VAMP71 (R)-type SNARE complex so far. In fact, we could not distinguish between PVC and vacuole in terms of the SNARE localization. Why does only one-type of SNARE complex exist in the vacuolar transport pathway? One possible explanation is that the complexity of the vacuolar transport pathway may be mediated by other regulatory proteins, such as Rab family protein. In fact, there are eight different vacuole-related Rab7-type proteins, and three different endosome-related Rab5 proteins exist in the *Arabidopsis* genome (Vernoud et al., 2003; Nielsen, 2005). The complexity of the vacuolar transport pathways may be mediated by the combination of SNARE and Rab proteins.

5 SNAREs Involved in Higher Order Physiological Functions

SNAREs have been reported to be necessary for higher order physiological functions unique to plants. In pathogen resistance, the disruption of plasma membrane localized Qa-SNARE molecules, PEN1/SYP121, results in a seven-fold incidence of pathogen penetration compared with the wild type plant, and is up-regulated by pathogen infection (Collins et al., 2003). The closest homolog of SYP121, SYP122 is also responsive to the infection. *pen1* and *syp122* double mutants are both dwarfed and necrotic, suggesting that the two syntaxins have overlapping functions (Assaad et al., 2004). SYP122 was shown to be phosphorylated by stimulation of a model elicitor peptide in a calcium dependent manner. These results indicate a regulatory link between elicitor-induced calcium fluxes and the rapid phosphorylation of a plasma membrane syntaxin molecule (Nuhse et al., 2003). These data also suggest the importance of membrane trafficking in pathogen penetration resistance, possibly by delivering some enzymes to the cell wall through the plasma membrane in order to kill the pathogen. A tobacco SYP121 homolog, NtSyr1 (NtSYP121), was identified as an abscisic acid response molecule from tobacco (Leyman et al., 1999). NtSyr1 was cleaved by the Clostridium neurotoxin BotN/C, and both the neurotoxin and the cytosolic domain of NtSyr1 blocked K^+ and Cl^- channel responses to abscisic acid *in vivo* when loaded into guard cells (Leyman et al., 2000). NtSyr1 is localized to the plasma membrane, and introducing a soluble, truncated (dominant-negative) form of the protein was found to block traffic to the plasma membrane *in vivo* (Geelen et al., 2002). The expression of NtSyr1 was strongly and transiently induced in tobacco leaves by ABA, drought, salt and wounding (Leyman et al., 2000). Interestingly, a mutant *osm1*, which is a mutant allele of a Qc-SNARE, SYP61, has a similar response to ABA and salt stress. Therefore, it is tempting to speculate that a related function of these SNARE proteins is controlling guard cell ion channel activity and ABA response in vesicular trafficking.

Plants sense gravity to reorient their growth direction. Genetic analysis of shoot gravitropism (*sgr*) mutants showed that *SGR3* and *SGR4/Zig* encode the Qa-SNARE/syntaxin AtVam3/SYP22 and the Q-SNAREs AtVTI11, respectively (Kato et al., 2002; Morita et al., 2002). AtVam3/SYP22 was isolated as a protein that can complement yeast *vam3* mutant, and was shown to localize to the PVC and vacuole (Sato et al., 1997; Sanderfoot et al., 1999; Uemura et al., 2002, 2004). In *Arabidopsis*, the VTI1 family is composed of four closely related members: VTI11, VTI12, VTI13, and VTI14, although VTI1 is encoded by a single gene, and has multiple functions in yeast (Sanderfoot et al., 2000; Uemura et al., 2004). VTI11 and VTI12 are highly expressed in all organs, while VTI14 is expressed only in suspension culture cells (Zheng et al., 1999; Uemura et al., 2004).

There is evidence that VTI11 and VTI12 function in different vesicle transport pathways (Zheng et al., 1999). In the *sgr4/zig* mutant, amyloplasts do not move towards the bottom of endodermal cells, and abnormal vesicular structures are found in several tissues (Kato et al., 2002; Morita et al., 2002). Endodermis-specific expression of wild-type VTI11 in the *zig* mutant complements the shoot gravitropism defect (Morita et al., 2002; Yano et al., 2003). These results suggest that vacuole formation is important for the shoot gravitropic response, and vacuolar transport pathways contribute to shoot gravitropism. Interestingly, an overproduction or mutation (*zip1*) of *VTI12* can complement the defect of a *vti11* mutant, *zig*, and a double-mutant cross between *zig* and *vti12* is embryo lethal (Surpin et al., 2003), although VTI11 and VTI12 were suggested to function in different vesicle transport pathways (Zheng et al., 1999). An analysis of a T-DNA insertion mutant of *VTI12* showed that it has a normal phenotype under nutrient-rich growth conditions, but shows an accelerated senescence phenotype under nutrient-poor conditions, suggesting that VTI12 plays a role in the autophagy pathway in plants (Surpin et al., 2003). Furthermore, the *zip-1* mutant caused by the alternation of one amino acid of VTI12 changes the localization to the vacuoles and can form a SNARE complex with SYP22 and SYP51 (Niihama et al., 2005). These results indicate that the functions of VTI11 and VTI12 seem to be distinct, but partially overlapping in the vacuolar transport pathway. Since VTI11 and VTI12 can form different SNARE complexes with SYP2/SYP5 and SYP4/SYP6, respectively (Sanderfoot et al., 2001), these proteins may be involved in the anterograde (VTI11) and the reterograde (VTI12) pathways between TGN and PVC/vacuole.

During embryogenesis in plants, both the pattern of cell division and the cell shape are changing. A screen for mutations that affect the body organization of *Arabidopsis* seedlings yielded the seedling-lethal *knolle* mutant (*kn*) (Mayer et al., 1991). *kn* mutants have malformations in the epidermal cell layer that are due to abnormal cytokinesis and cell enlargements. The embryos are composed of multinucleate cells with incomplete cross walls. Positional cloning of the *KN* gene showed that it encodes a syntaxin related protein, which is identical with SYP111 (Lukowitz et al., 1996). KN/SYP111 is membrane associated, localized to the plane of division, and is a cytokinesis-specific syntaxin. *KN/SYP111* mRNA is most abundant in tissues with rapidly dividing cells, such as flowers and developing siliques. The *KN*/SYP111 gene is very tightly cell-cycle regulated; it must be highly transcribed during M phase to produce sufficient protein, but its mRNA must be degraded rapidly to prevent the accumulation of protein following cytokinesis (Lauber et al., 1997). Syntaxin specificity in cytokinesis was studied by expressing several syntaxins under the control of *KN/SYP111 cis*-regulatory gene sequences. Both prevacuolar SYP21 and plasma membrane-localized SYP121 failed to rescue a knolle deletion mutant. In contrast, SYP112 of the plasma-membrane group fully substituted for KN/SYP111, indicating that plasma membrane resident Qa-SNAREs have a different function in membrane traffic to the plasma membrane (Muller et al., 2003).

6
Evolution of Endocytic Compartments in Eukaryotes

Phylogenetic analysis of syntaxins across various taxas of the eukaryotic kingdom indicates that these proteins can be classified into paralog families; Syn1 (PM), Syn5 (Golgi, ER), Syn6 (TGN), Syn7 (endosomes), and Syn16 (TGN). Each family is associated with either a particular step in the transport pathway or has a specific intracellular location. The fact suggests that the duplication events of syntaxin subfamilies must have occurred prior to the divergence of that taxon (Dacks and Doolittle, 2002).

Meanwhile, R-SNARE/VAMP is composed of two family members, Brevin and Longin. The Brevin is typically quite small, having no N-terminal extension from the SNARE-domain and transmembrane helices. The Longin has SNARE helices centered on R-residues like the Brevin, however, additionally it has also a long N-terminal extension. Longins form three main subclasses (Sec22, Ykt6, and TI-VAMP/VAMP7) which are conserved across most eukaryote lineages including *Brassicaceae*. The Sec22 family is involved in ER-to-Golgi traffic. The Ykt6 family has a lipid anchor rather than a C-terminal transmembrane domain, and is involved in many different trafficking events in the cell. In mammals, TI-VAMP/VAMP7 interacts with several t-SNARE molecules, and is involved in plasmalemmal and late endosomal fusion events (Filippini et al., 2001; Gonzalez et al., 2001; Rossi et al., 2004). Interestingly, the *Arabidopsis* genome does not encode any genes for R-type Brevins. Instead, the role played by R-type Brevins in plants seems to be replaced by further members of the Longin class, especially TI-VAMP/VAMP7 (Sanderfoot et al., 2000). The TI-VAMP/VAMP7 class of *Arabidopsis* contains 11 members of related proteins which show different subcelluar distributions (Uemura et al., 2004). The *Arabidopsis* TI-VAMP/VAMP7 clade is robustly separated from the other VAMP homologs, suggesting that the gene duplication of these genes occurred after the separation of plant lineages (Fig. 2).

Since particular syntaxin molecules reside on specific organelles, the evolutionary process of syntaxin diversification is closely connected to organelle evolution. In contrast, the evolution of R-SNARE molecules can be hardly separated from the development of vesicle transport pathways, because R-SNARE molecules are attached to specific transport vesicles. Therefore, the result from the phylogenetic analyses of syntaxin and R-SNARE families may indicate, that the basic characteristic of each organelle was established prior to the divergence of plant lineages, while the transport pathways, especially those involving the post-Golgi organelles, became diversified in the course of higher plant evolution by means of gene duplication of TI-VAMP/VAMP7 family members to create specific physiological functions (Fig. 3).

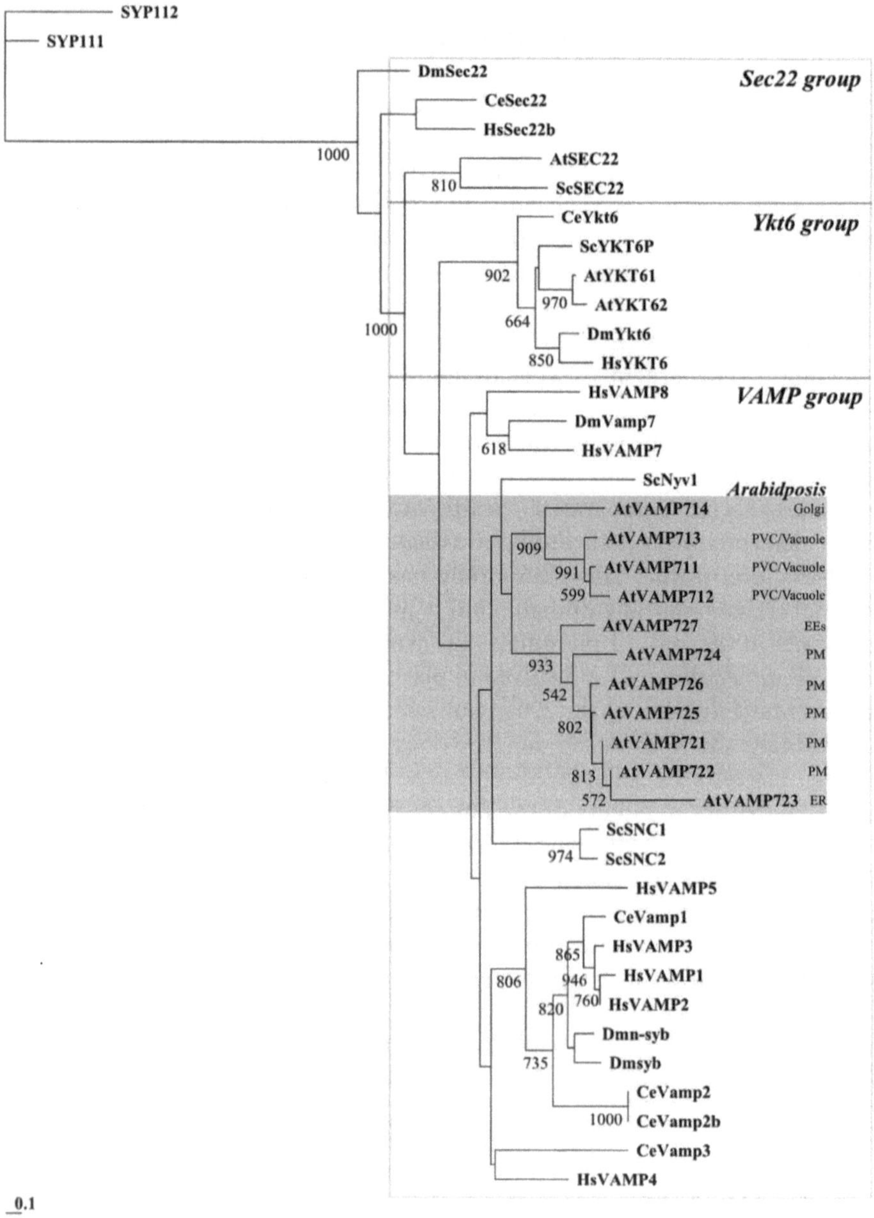

Fig. 2 A phylogenic tree of R-SNARE proteins of various eukaryotes. The phylogenic tree was constructed based on the SNARE motifs. The scale bar shows the Dayhoff distance among the SNARE molecules. The numbers on the tree are bootstrap values with 10 000 replications of SNARE motifs. Dm: *Drosophila melanogaster*, Ce: *Caenorhabditis elegans*, Hs: *Homo sapiens*, At: *Arabidopsis thaliana*, Sc: *Saccharomyces cerevisiae*

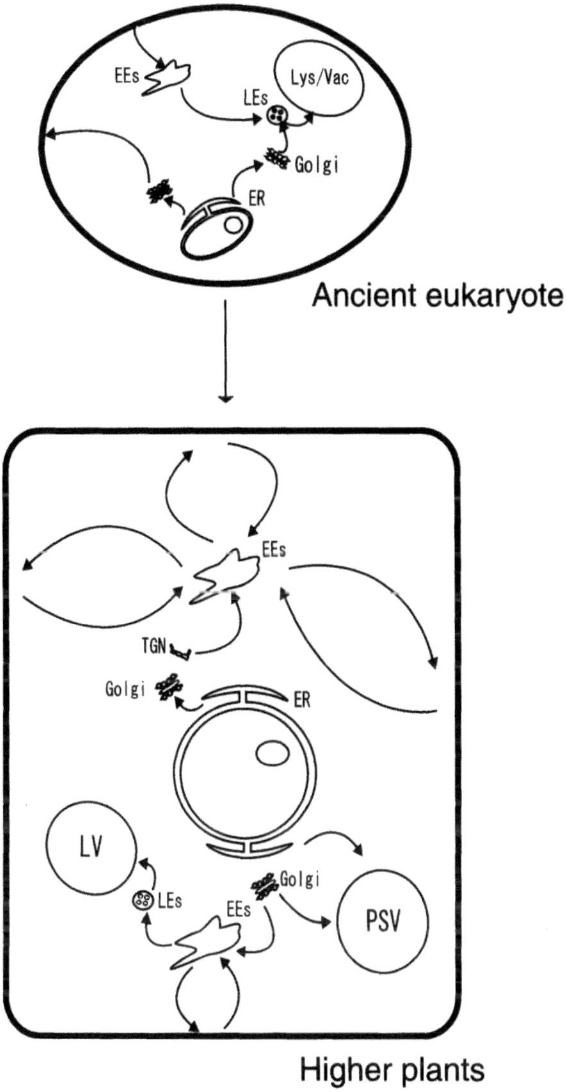

Fig. 3 A hypothesis of the establishment of organelles and transport pathways in evolution of higher plants. Phylogenic analyses of Qa/syntaxin and R-SNARE molecules of various eukaryotes indicate that a basic characteristic of each organelle is established prior to the divergence of plant lineages, while the transport pathways to the plasma membrane became diversified with the gene duplication of TI-VAMP/VAMP7 family proteins through the higher plant evolution. Lys/Vac—lasosome/vacuole, EE—early endosome, LE—late endosome, LV—lytic vacuole, PSV—protein storage vacuole, Golgi—Golgi apparatus, ER—Endoplasmic reticulum

7
Conclusions and Future Prospects

Growing evidence supporting the importance of the endocytic process for various physiological responses has been reported during the past several years in plants. At the same time, several SNARE molecules have been shown to have various functions in cellular responses that are vital for survival and growth of plants. However, little direct evidence has been published showing a relationship between the endocytic process and SNARE function due to the lack of information on the subcellular localizations of most of the SNARE molecules. Recently, the use of GFP marker technology in conjunction with transient expression assays in cultured cells gave us a clue to construct a map of the subcellular localizations of SNARE molecules in plants (Fig. 1). The map suggests that the complexity of the post-Golgi membrane traffic, particularly to the plasma membrane, has increased through plant evolution with the expansion of the numbers of SNAREs on the plasma membrane.

It is reported that the establishment of cell polarity is important for various aspects for plant growth, such as the polar tip growth of root hair or pollen tube (Voigt et al., 2005). Particular membrane proteins may recycle between specific sub-domains of the cell surface and interior of the cell by exocytosis and endocytosis in a SNARE and Rab dependent manner, thereby establishing the cell polarity that is essential for plant life. Characterization of each SNARE molecule residing in particular endocytic organelles will shed further light on the relationship between the endocytic process and SNARE function in plants.

References

Anuntalabhochai S, Terryn N, Van Montagu M, Inze D (1991) Molecular characterization of an *Arabidopsis thaliana* cDNA encoding a small GTP-binding protein, Rha1. Plant J 1:167–174

Assaad FF, Qiu JL, Youngs H, Ehrhardt D, Zimmerli L, Kalde M, Wanner G, Peck SC, Edwards H, Ramonell K, Somerville CR, Thordal-Christensen H (2004) The PEN1 syntaxin defines a novel cellular compartment upon fungal attack and is required for the timely assembly of papillae. Mol Biol Cell 15:5118–5129

Baluška F, Šamaj J, Hlavacka A, Kendrick-Jones J, Volkman D (2004) Actin-dependent fluid-phase endocytosis in inner cortex cells of maize root apices. J Exp Bot 55:463–473

Bassham DC, Sanderfoot AA, Kovaleva V, Zheng H, Raikhel NV (2000) AtVPS45 complex formation at the trans-Golgi network. Mol Biol Cell 11:2251–2265

Battey NH, James NC, Greenland AJ, Brownlee C (1999) Exocytosis and endocytosis. Plant Cell 11:643–660

Bock JB, Matern HT, Peden AA, Scheller RH (2001) A genomic perspective on membrane compartment organization. Nature 409:839–841

Brodsky FM, Chen CY, Knuehl C, Towler MC, Wakeham DE (2001) Biological basket weaving: formation and function of clathrin-coated vesicles. Annu Rev Cell Dev Biol 17:517–568

Carter C, Pan S, Zouhar J, Avila EL, Girke T, Raikhel NV (2004) The vegetative vacuole proteome of *Arabidopsis thaliana* reveals predicted and unexpected proteins. Plant Cell 16:3285–3303

Collins NC, Thordal-Christensen H, Lipka V, Bau S, Kombrink E, Qiu JL, Huckelhoven R, Stein M, Freialdenhoven A, Somerville SC, Schulze-Lefert P (2003) SNARE-protein-mediated disease resistance at the plant cell wall. Nature 425:973–977

Cram WJ (1980) Pinocytosis in plants. New Phytol 84:1–17

Dacks JB, Doolittle WF (2002) Novel syntaxin gene sequences from *Giardia*, Trypanosoma and algae: implications for the ancient evolution of the eukaryotic endomembrane system. J Cell Sci 115:1635–1642

Emans N, Zimmermann S, Fischer R (2002) Uptake of a fluorescent marker in plant cells is sensitive to brefeldin A and wortmannin. Plant Cell 14:71–86

Fasshauer D, Sutton RB, Brunger AT, Jahn R (1998) Conserved structural features of the synaptic fusion complex: SNARE proteins reclassified as Q- and R-SNAREs. Proc Natl Acad Sci USA 95:15 781–15 786

Filippini F, Rossi V, Galli T, Budillon A, D'Urso M, D'Esposito M (2001) Longins: a new evolutionary conserved VAMP family sharing a novel SNARE domain. Trends Biochem Sci 26:407–409

Friml J, Wisniewska J, Benkova E, Mendgen K, Palme K (2002) Lateral relocation of auxin efflux regulator PIN3 mediates tropism in *Arabidopsis*. Nature 415:806–809

Galway ME, Rennie PJ, Fowke LC (1993) Ultrastructure of the endocytotic pathway in glutaraldehyde-fixed and high-pressure frozen/freeze-substituted protoplasts of white spruce (*Picea glauca*). J Cell Sci 106:847–858

Galweiler L, Guan C, Muller A, Wisman E, Mendgen K, Yephremov A, Palme K (1998) Regulation of polar auxin transport by AtPIN1 in *Arabidopsis* vascular tissue. Science 282:2226–2230

Geelen D, Leyman B, Batoko H, Di Sansabastiano GP, Moore I, Blatt MR (2002) The abscisic acid-related SNARE homolog NtSyr1 contributes to secretion and growth: evidence from competition with its cytosolic domain. Plant Cell 14:387–406

Geldner N, Anders N, Wolters H, Keicher J, Kornberger W, Muller P, Delbarre A, Ueda T, Nakano A, Jurgens G (2003) The *Arabidopsis* GNOM ARF-GEF mediates endosomal recycling, auxin transport, and auxin-dependent plant growth. Cell 112:219–230

Gonzalez LC Jr, Weis W, Scheller RH (2001) A novel snare N-terminal domain revealed by the crystal structure of Sec22b. J Biol Chem 276:24 203–24 211

Griffing LR (1991) Comparisons of Golgi structure and dynamics in plant and animal cells. J Electron Microsc Tech 17:179–199

Halperin W (1969) Ultrastructural localization of acid phosphatase in cultured cells of *Daucus carota*. Planta 88:91–102

Hara-Nishimura I, Shimada T, Hatano K, Takeuchi Y, Nishimura M (1998) Transport of storage proteins to protein storage vacuoles is mediated by large precursor-accumulating vesicles. Plant Cell 10:825–836

Heese M, Gansel X, Sticher L, Wick P, Grebe M, Granier F, Jurgens G (2001) Functional characterization of the KNOLLE-interacting t-SNARE AtSNAP33 and its role in plant cytokinesis. J Cell Biol 155:239–249

Hillmer S, Depta H, Robinson DG (1986) Confirmation of endocytosis in higher plant protoplasts using lectin-gold conjugates. Eur J Cell Biol 41:142–149

Hillmer S, Freundt H, Robinson DG (1988) The partially coated reticulum and its relationship to the Golgi apparatus in higher plants. Eur J Cell Biol 47:206–212
Hillmer S, Movafeghi A, Robinson DG, Hinz G (2001) Vacuolar storage proteins are sorted in the cis-cisternae of the pea cotyledon Golgi apparatus. J Cell Biol 152:41–50
Hohl I, Robinson DG, Chrispeels MJ, Hinz G (1996) Transport of storage proteins to the vacuole is mediated by vesicles without a clathrin coat. J Cell Sci 109:2539–2550
Holstein SE (2002) Clathrin and plant endocytosis. Traffic 3:614–620
Joachim S, Robinson DG (1984) Endocytosis of cationic ferritin by bean leaf protoplasts. Eur J Cell Biol 34:212–216
Jürgens G (2004) Membrane Trafficking in Plants. Annu Rev Cell Dev Biol 20:481–504
Jürgens G, Geldner N (2002) Protein secretion in plants: from the trans-Golgi network to the outer space. Traffic 3:605–613
Kargul J, Gansel X, Tyrrell M, Sticher L, Blatt MR (2001) Protein-binding partners of the tobacco syntaxin NtSyr1. FEBS Lett 508:253–258
Kato T, Morita MT, Fukaki H, Yamauchi Y, Uehara M, Niihama M, Tasaka M (2002) SGR2, a phospholipase-like protein, and ZIG/SGR4, a SNARE, are involved in the shoot gravitropism of *Arabidopsis*. Plant Cell 14:33–46
Kotzer AM, Brandizzi F, Neumann U, Paris N, Moore I, Hawes C (2004) AtRabF2b (Ara7) acts on the vacuolar trafficking pathway in tobacco leaf epidermal cells. J Cell Sci 117:6377–6389
Lauber MH, Waizenegger I, Steinmann T, Schwarz H, Mayer U, Hwang I, Lukowitz W, Jurgens G (1997) The *Arabidopsis* KNOLLE protein is a cytokinesis-specific syntaxin. J Cell Biol 139:1485–1493
Lee GJ, Sohn EJ, Lee MH, Hwang I (2004) The *Arabidopsis* rab5 homologs rha1 and ara7 localize to the prevacuolar compartment. Plant Cell Physiol 45:1211–1220
Leyman B, Geelen D, Blatt MR (2000) Localization and control of expression of Nt-Syr1, a tobacco SNARE protein. Plant J 24:369–381
Leyman B, Geelen D, Quintero FJ, Blatt MR (1999) A tobacco syntaxin with a role in hormonal control of guard cell ion channels. Science 283:537–540
Low PS, Chandra S (1994) Endocytosis in Plants. Annu Rev Plant Physiol Plant Mol Biol 43:609–631
Lukowitz W, Mayer U, Jurgens G (1996) Cytokinesis in the *Arabidopsis* embryo involves the syntaxin-related KNOLLE gene product. Cell 84:61–71
Matsuoka K, Bassham DC, Raikhel NV, Nakamura K (1995) Different sensitivity to wortmannin of two vacuolar sorting signals indicates the presence of distinct sorting machineries in tobacco cells. J Cell Biol 130:1307–1318
Mayer U, Torres Ruiz RA, Berleth T, Misera S, Jurgens G (1991) Mutations affecting body organization in the *Arabidopsis* embryo. Nature 353:402–407
Mellman I (1996) Endocytosis and molecular sorting. Annu Rev Cell Dev Biol 12:575–625
Mironov AA, Weidman P, Luini A (1997) Variations on the intracellular transport theme: maturing cisternae and trafficking tubules. J Cell Biol 138:481–484
Morita MT, Kato T, Nagafusa K, Saito C, Ueda T, Nakano A, Tasaka M (2002) Involvement of the vacuoles of the endodermis in the early process of shoot gravitropism in *Arabidopsis*. Plant Cell 14:47–56
Muller A, Guan C, Galweiler L, Tanzler P, Huijser P, Marchant A, Parry G, Bennett M, Wisman E, Palme K (1998) AtPIN2 defines a locus of *Arabidopsis* for root gravitropism control. EMBO J 17:6903–6911
Muller I, Wagner W, Volker A, Schellmann S, Nacry P, Kuttner F, Schwarz-Sommer Z, Mayer U, Jurgens G (2003) Syntaxin specificity of cytokinesis in *Arabidopsis*. Nat Cell Biol 5:531–534

Nebenfuhr A (2002) Vesicle traffic in the endomembrane system: a tale of COPs, Rabs and SNAREs. Curr Opin Plant Biol 5:507–512

Neuhaus JM, Rogers JC (1998) Sorting of proteins to vacuoles in plant cells. Plant Mol Biol 38:127–144

Nielsen E (2005) Rab GTPases in plant endocytosis (in this volume). Springer, Berlin Heidelberg New York

Niihama M, Uemura T, Chieko S, Nakano A, Sato MH, Tasaka M, Morita MT (2005) Conversion of functional specificity in Qb-SNARE VTI11 homologues of *Arabidopsis*. Curr Biol (in press)

Nishizawa N, Mori S (1977) Invagination of plasmalemma: Its role in the absorption of macromolecules in rice roots. Plant Cell Physiol 18:767–782

Nuhse TS, Boller T, Peck SC (2003) A plasma membrane syntaxin is phosphorylated in response to the bacterial elicitor flagellin. J Biol Chem 278:45 248–45 254

Pelham HR, Rothman JE (2000) The debate about transport in the Golgi—two sides of the same coin? Cell 102:713–719

Phillips GD, Preshaw C, Steer MW (1988) Dictyosome vesicle production and plasma membrane turnover in auxin-stimulated outer epidermal cells of coleoptile segments from *Avena sativa* (L.). Protoplasma 145:59–65

Pryer NK, Wuestehube LJ, Schekman R (1992) Vesicle-mediated protein sorting. Annu Rev Biochem 61:471–516

Record RD, Griffing LR (1988) Convergence of endocytis and lysosomal pathways in soybean protoplasts. Planta 176:425–432

Rojo E, Gillmor CS, Kovaleva V, Somerville CR, Raikhel NV (2001) VACUOLELESS1 is an essential gene required for vacuole formation and morphogenesis in *Arabidopsis*. Dev Cell 1:303–310

Rojo E, Zouhar J, Kovaleva V, Hong S, Raikhel NV (2003) The AtC-VPS protein complex is localized to the tonoplast and the prevacuolar compartment in *Arabidopsis*. Mol Biol Cell 14:361–369

Rossi V, Banfield DK, Vacca M, Dietrich LE, Ungermann C, D'Esposito M, Galli T, Filippini F (2004) Longins and their longin domains: regulated SNAREs and multifunctional SNARE regulators. Trends Biochem Sci 29:682–688

Rothman JE, Wieland FT (1996) Protein sorting by transport vesicles. Science 272:227–234

Saint-Jore-Dupas C, Gomord V, Paris N (2004) Protein localization in the plant Golgi apparatus and the trans-Golgi network. Cell Mol Life Sci 61:159–171

Šamaj J (2005) Methods and molecular tools to study endocytosis in plants—an overview (in this volume). Springer, Berlin Heidelberg New York

Šamaj J, Baluška F, Voigt B, Schlicht M, Volkmann D, Menzel D (2004) Endocytosis, actin cytoskeleton and signalling. Plant Physiol 135:1150–1161

Šamaj J, Read ND, Volkmann D, Menzel D, Baluška F (2005) The endocytic network in plants. Trends Cell Biol 15:425–433

Samuels AL, Bisalputra T (1990) Endocytosis in elongating root cells of *Lobelia erinus*. J Cell Sci 97:157–165

Sanderfoot AA, Assaad FF, Raikhel NV (2000) The *Arabidopsis* genome. An abundance of soluble N-ethylmaleimide-sensitive factor adaptor protein receptors. Plant Physiol 124:1558–1569

Sanderfoot AA, Kovaleva V, Bassham DC, Raikhel NV (2001) Interactions between Syntaxins Identify at Least Five SNARE Complexes within the Golgi/Prevacuolar System of the *Arabidopsis* Cell. Mol Biol Cell 12:3733–3743

Sanderfoot AA, Kovaleva V, Zheng H, Raikhel NV (1999) The t-SNARE AtVAM3p resides on the prevacuolar compartment in *Arabidopsis* root cells. Plant Physiol 121:929–938

Sanderfoot AA, Raikhel NV (1999) The specificity of vesicle trafficking: coat proteins and SNAREs. Plant Cell 11:629–642

Sanderfoot AA, Raikhel NV (2003) In: Somerville C and Meyerowits E (eds) The secretory system of *Arabidopsis*. American Society of Plant Biologist, Rockville, MD, pp 1–24

Sato MH, Nakamura N, Ohsumi Y, Kouchi H, Kondo M, Hara-Nishimura I, Nishimura M, Wada Y (1997) The AtVAM3 encodes a syntaxin-related molecule implicated in the vacuolar assembly in *Arabidopsis thaliana*. J Biol Chem 272:24 530–24 535

Schindelman G, Morikami A, Jung J, Baskin TI, Carpita NC, Derbyshire P, McCann MC, Benfey PN (2001) COBRA encodes a putative GPI-anchored protein, which is polarly localized and necessary for oriented cell expansion in *Arabidopsis*. Genes Dev 15:1115–1127

Shah K, Russinova E, Gadella TW Jr, Willemse J, De Vries SC (2002) The *Arabidopsis* kinase-associated protein phosphatase controls internalization of the somatic embryogenesis receptor kinase 1. Genes Dev 16:1707–1720

Sheung KL, Tse YC, Jiang L, Oliviusson P, Heinzerling O, Robinson DG (2005) Plant prevacuolar compartments and endocytosis (in this volume). Springer, Berlin Heidelberg New York

Somsel Rodman J, Wandinger-Ness A (2000) Rab GTPases coordinate endocytosis. J Cell Sci 113:183–192

Steinmann T, Geldner N, Grebe M, Mangold S, Jackson CL, Paris S, Galweiler L, Palme, K, Jurgens G (1999) Coordinated polar localization of auxin efflux carrier PIN1 by GNOM ARF GEF. Science 286:316–318

Surpin M, Zheng H, Morita MT, Saito C, Avila E, Blakeslee JJ, Bandyopadhyay A, Kovaleva V, Carter D, Murphy A, Tasaka M, Raikhel N (2003) The VTI family of SNARE proteins is necessary for plant viability and mediates different protein transport pathways. Plant Cell 15:2885–2899

Swarup R, Friml J, Marchant A, Ljung K, Sandberg G, Palme K, Bennett M (2001) Localization of the auxin permease AUX1 suggests two functionally distinct hormone transport pathways operate in the *Arabidopsis* root apex. Genes Dev 15:2648–2653

Tanchak MA, Fowke LC (1987) The morphology of multivesicular bodies in soybean protoplasts and their role in endocytosis. Protoplasma 138:173–182

Tanchak MA, Griffing LR, Mersey BG, Fowke LC (1984) Endocytosis of cationized ferritin by coated vesicles of soybean protoplasts. Planta 162:481–486

Tanchak MA, Rennie PJ, Fowke LC (1988) Ultrastructure of the partially coated reticulum and dictyosomes during endocytosis by soybean protoplasts. Planta 175:433–431

Toyooka K, Okamoto T, Minamikawa T (2001) Cotyledon cells of Vigna mungo seedlings use at least two distinct autophagic machineries for degradation of starch granules and cellular components. J Cell Biol 154:973–982

Tse YC, Mo B, Hillmer S, Zhao M, Lo SW, Robinson DG, Jiang L (2004) Identification of multivesicular bodies as prevacuolar compartments in *Nicotiana tabacum* BY-2 cells. Plant Cell 16:672–693

Ueda T, Nakano A (2002) Vesicular traffic: an integral part of plant life. Curr Opin Plant Biol 5:513–517

Ueda T, Uemura T, Sato MH, Nakano A (2004) Functional differentiation of endosomes in *Arabidopsis* cells. Plant J 40:783–789

Ueda T, Yamaguchi M, Uchimiya H, Nakano A (2001) Ara6, a plant-unique novel type Rab GTPase, functions in the endocytic pathway of *Arabidopsis thaliana*. EMBO J 20:4730–4741

Uemura T, Ueda T, Ohniwa RL, Nakano A, Takeyasu K, Sato MH (2004) Systematic analysis of SNARE molecules in *Arabidopsis*: dissection of the post-Golgi network in plant cells. Cell Struct Funct 29:49–65

Uemura T, Yoshimura SH, Takeyasu K, Sato MH (2002) Vacuolar membrane dynamics revealed by GFP-AtVam3 fusion protein. Genes Cells 7:743–753.

Ungar D, Hughson FM (2003) SNARE protein structure and function. Annu Rev Cell Dev Biol 19:493–517

Vernoud V, Horton AC, Yang Z, Nielsen E (2003) Analysis of the small GTPase gene superfamily of *Arabidopsis*. Plant Physiol 131:1191–1208

Voigt B, Timmers A, Šamaj J, Hlavacka A, Ueda T, Preuss M, Nielsen E, Mathur J, Emans N, Stenmark H, Nakano A, Baluška F, Menzel D (2005) Actin-based motility of endosomes is linked to the polar tip growth of root hairs. Eur J Cell Biol 84 (in press)

Waters MG, Hughson FM (2000) Membrane tethering and fusion in the secretory and endocytic pathways. Traffic 1:588–597

Yano D, Sato M, Saito C, Sato MH, Morita MT, Tasaka M (2003) A SNARE complex containing SGR3/AtVAM3 and ZIG/VTI11 in gravity-sensing cells is important for *Arabidopsis* shoot gravitropism. Proc Natl Acad Sci USA 100:8589–8594

Zheng H, Bednarek SY, Sanderfoot AA, Alonso J, Ecker JR, Raikhel NV (2002) NPSN11 is a cell plate-associated SNARE protein that interacts with the syntaxin KNOLLE. Plant Physiol 129:530–539

Zheng H, von Mollard GF, Kovaleva V, Stevens TH, Raikhel NV (1999) The plant vesicle-associated SNARE AtVTI1a likely mediates vesicle transport from the trans-Golgi network to the prevacuolar compartment. Mol Biol Cell 10:2251–2264

Zhu J, Gong Z, Zhang C, Song CP, Damsz B, Inan G, Koiwa H, Zhu JK, Hasegawa PM, Bressan RA (2002) OSM1/SYP61: a syntaxin protein in *Arabidopsis* controls abscisic acid-mediated and non-abscisic acid-mediated responses to abiotic stress. Plant Cell 14:3009–3028

Plant Cell Monogr (1)
J. Šamaj · F. Baluška · D. Menzel: Plant Endocytosis
DOI 10.1007/7089_013/Published online: 2 September 2005

Dynamin-Related Proteins in Plant Endocytosis

D. P. S. Verma[1,2] (✉) · Z. Hong[3] · D. Menzel[4]

[1]Department of Molecular Genetics and Plant Biotechnology Center, The Ohio State University, Columbus, OH 43210, USA
verma.1@osu.edu

[2]Plant Biotechnology Center, The Ohio State University, 240 Rightmire Hall, 1060 Carmack Road, Columbus, OH 43210, USA
verma.1@osu.edu

[3]Department of Microbiology, Molecular Biology, and Biochemistry, University of Idaho, Moscow, ID 83844, USA
zhong@uidaho.edu

[4]Institut für Zelluläre und Molekulare Botanik (IZMB), Abteilung Zellbiologie der Pflanzen, Rheinische Friedrich-Wilhelms-Universität, Kirschallee 1, 53115 Bonn, Germany
dmenzel@uni-bonn.de

Abstract Over the past decade, it has become evident that multiple endocytic pathways operate in eukaryotic cells, and several of these are dependent on dynamins and dynamin-related proteins (DRPs). Many members of the DRP superfamily possess the ability to self-assemble into long spiral polymers that wrap around lipid bilayers and thus facilitate tubulation and vesicle pinching from the plasma membrane and other membrane compartments, a process that is fundamental for endocytosis. Here, we discuss the roles of dynamins and DRPs in plants. DRPs have been shown to be present at different subcellular locations in plant cells including the cell plate, plasma membrane, Golgi apparatus, vesicles, mitochondria, chloroplasts, and peroxisomes. *Arabidopsis* contains 16 DRP members that are grouped into six functional subfamilies (DRP1–6) on the basis of their phylogeny and the presence of functional motifs. Members of the DRP1 subfamily are closest to soybean phragmoplastin and mediate membrane tubulation at the cell plate. The DRP2 subfamily members represent the bona fide plant dynamins characterized by the presence of a pleckstrin homology (PH) domain in the middle of the molecule and a proline-rich (PR) motif near the C-terminus; they are involved in membrane recycling at the cell plate and the trans Golgi region. The DRP3 subfamily does not contain PH or PR motifs; their function has been implicated in the division of mitochondria and peroxisomes, whereas the DRP5 subfamily in *Arabidopsis* is likely to play a role in plastid division. A DRP5 ortholog from the red alga *Cyanidioschyzon* has recently been shown to be a component of the chloroplast outer division-ring on the cytoplasmic face of the plastid double membrane. Finally, the DRP4 subfamily contains orthologs of the animal antiviral Mx proteins, but their function has not yet been established, and the role of DRP6 is entirely unknown so far. It is obvious that plant cells employ unique DRP subfamilies to carry out the mechanochemical work required for membrane deformation and segregation in various membranous compartments. However, to understand the function of DRPs in further detail, much is yet to be learned about the proteins that apparently interact with them to regulate their activity and specify their functions.

1 Introduction

A highly dynamic state of exocytosis and endocytosis allows a eukaryotic cell to communicate with its environment, build and maintain membrane structures, and deliver cargo to different subcellular and extracellular compartments. Over the past decade, it has become evident that multiple endocytic pathways operate in animal cells, of which some are dynamin-dependent while others are dynamin-independent. Observations of fruit fly mutant *shibire* that affects endocytosis have clearly demonstrated the role of dynamin in clathrin-coated endocytosis. Plants contain a superfamily of dynamin-related proteins (DRPs, Hong et al., 2003a). Some DRP members appear to act as a "pinchase or popase" that mediates the pinching off of endocytic vesicles from the plasma membrane. Such structures are usually generated during receptor-mediated endocytosis and for membrane recycling. Other members such as DRP1 (phragmoplastin) have been proposed to function as a tubulase that wraps around homotypically fused vesicles and helps transform these round structures into tubular membrane extensions. This creates dumbbell-shaped structures which act as building blocks for the formation of cell plate during cytokinesis in plant cells (Verma, 2001). Thus, plant cells employ unique DRP subfamilies to carry out membrane vesiculation and vesicle tubulation, two basic processes required for the endocytic and exocytic pathways, respectively (Verma and Hong, 2005). Other accessory proteins must interact with DRPs since none of the DRPs is an integral membrane protein and they function at different subcellular locations.

2 Dynamin-Related Proteins in Plants

Plant DRPs constitute a superfamily of large GTPases with molecular masses of 60–110 kD (Hong et al., 2003a). They all contain the dynamin signature (L-P-[PK]-G-[STN]-[GN]-[LIVM]-V-T-R) and many of them possess the ability to self-assemble into long spiral polymers that wrap around lipid bilayers and thus facilitate membrane tubulation and vesicle pinching (Verma and Hong, 2005). The first characterized plant DRPs were *Arabidopsis* ADL1 (Dombrowski and Raikhel, 1995) and soybean phragmoplastin (Gu and Verma, 1996). These proteins have since been shown to be present in many plants at different subcellular locations including the cell plate, plasma membrane, Golgi apparatus, vesicles, mitochondria, chloroplasts, and peroxisomes. They have also been implicated in diverse subcellular processes such as cell plate formation, endocytosis, exocytosis, protein sorting to the vacuole and plasma membrane, and organelle division.

Arabidopsis contains 16 DRP members that are grouped into six functional subfamilies (DRP1–6) on the basis of their phylogeny and the presence of functional motifs (Hong et al., 2003a). The DRP1 subfamily contains five members and is closest to soybean phragmoplastin (Gu and Verma, 1996). They do not contain a pleckstrin homology (PH) domain or a proline-rich (PR) motif. DRP1A, DRP1C, and DRP1E are localized at the forming cell plate in plant cells (Gu and Verma, 1996, 1997; Hong et al., 2003b; Kang et al., 2001, 2003a, b; Lauber et al., 1997) and may also play specific roles in the plasma membrane, polar expansion of papillae cells, maturing pollen grains (Kang et al., 2003a, b) and cytoskeleton (Hong et al., 2003b). The development of exine on the pollen surface and its interconnection with the plasma membrane apparently involves significant membrane exocytosis and endocytosis activities. Extensive endocytosis of sporopollenin appears to occur at the pollen surface if the callose primary wall is not present (due to mutation in pollen-specific callose synthase, CalS5), triggering apoptotic cell death of the maturing pollen (Dong et al., 2005).

The DRP2 subfamily represents the bona fide plant dynamins as it is characterized by the presence of a PH domain in the middle of the molecule and a PR motif (RXPXXP) near the C-terminus (Fig. 1). The PH domain is believed to bind to phosphoinositides with a broad range of specificity and affinity, whereas the PR motif may interact with the SH3 domains of proteins like amphiphysins and endophilins, as shown in mammalian cells. DRP2A (also known as ADL6) has recently been shown to bind to phosphatidylinositol 3-phosphate (PI-3-P) and PI-4-P (Lee et al., 2002; Lam et al., 2002) and interacts with adaptin and an SH3 domain-containing protein in *Arabidopsis* (Lam et al., 2002). In general, DRP2 proteins may be involved in clathrin-coated vesicle trafficking between Golgi, plasma membrane including cell plate, and in vacuole membrane biogenesis (Hong et al., 2003b, Jin et al., 2001; Lam et al., 2002).

The DRP3 subfamily does not contain PH or PR motifs, and members of this subfamily have been implicated in the division of mitochondria and peroxisomes (Arimura and Tsutsumi, 2002; Nishida et al., 2003; Mano et al., 2004). The DRP4 subfamily contains orthologs of the animal antiviral Mx proteins, but their function has not been established. The DRP5 subfamily shares only 15–16% sequence homology with DRP1A (Fig. 1). *Arabidopsis* DRP5B (also known as ARC5) and its ortholog from *Cyanidioschyzon merolae* (CmDnm2) have recently been localized to the chloroplast division-ring (Gao et al., 2003; Miyagishima et al., 2003). DRP5 proteins do not contain an N-terminal signal peptide for plastid targeting. Instead, they are recruited to the cytosolic side of the chloroplast division site to form a ring in the late stage of chloroplast division. This may help in pinching off these organelles during their divisions. Finally, one gene is classified in the DRP6 subfamily which remains to be characterized.

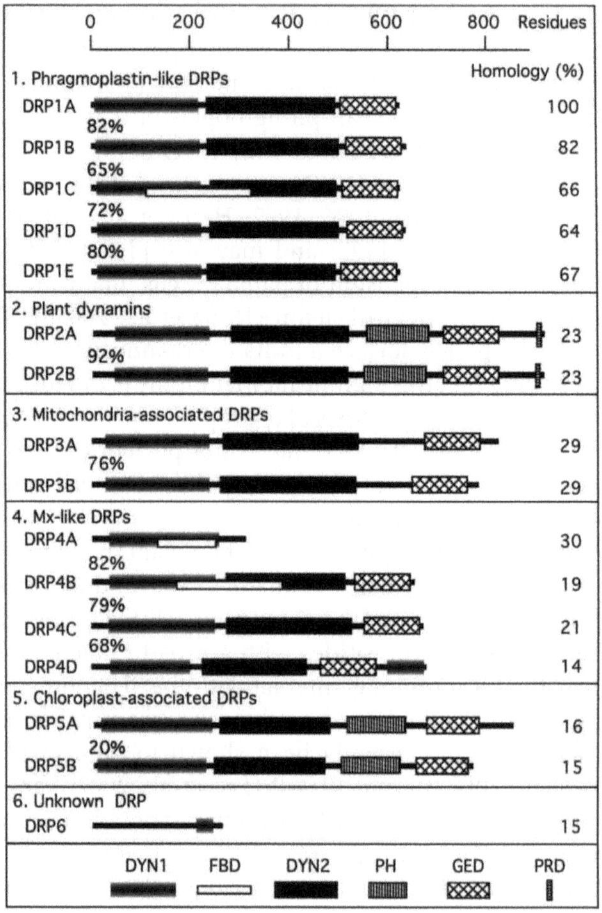

Fig. 1 Functional domains of DRPs in *Arabidopsis*. The percentage of amino acid identity with DRP1A is shown after each peptide. The identity ()% between the two adjacent peptides is shown on the left. DYN1: dynamin GTPase domain which includes a GTP-binding motif (GXXXSGKS/T) and a dynamin signature. FBD: FeoB domain which is found in the ferrous iron transport protein B. DYN2: dynamin central region. PH: pleckstrin homology domain which binds to membrane phospholipids such as PI-3-P and PI-4-P. GED: dynamin GTPase effector domain. PRD: proline-rich domain which interacts with SH3-domain proteins

3
Exocytosis and Membrane Recycling at the Cell Plate

Building the cell plate at cytokinesis in plants involves bidirectional trafficking to and from the cell plate, resembling the processes of exocytosis and endocytosis (Verma, 2001; Verma and Hong, 2005). The exocytic pathway delivers vesicles from the Golgi to the forming cell plate where they fuse with each other (homotypic fusion) or with the prototype plasma membrane of the forming cell plate (heterotypic fusion). Prototype plasma membrane refers to the membrane of the tubular network of the forming cell plate, as opposed to the “mature” and “flat” plasma membrane on the cell surface. The vesicles involved in heterotypic fusion versus homotypic fusion might differ in their membrane composition as well as the cargo they carry. At the same time, endocytosis is required to retrieve excess membrane and to recycle membrane components (phospholipids and membrane proteins) from the prototype plasma membrane of the cell plate. It has been estimated that 10–14-fold more membranes are required to build one unit of the plasma membrane, which suggests that most membrane constituents that have been delivered to the cell plate through the exocytic pathway need to be recycled back through endocytosis. The two processes (exocytosis and endocytosis) at the cell plate are apparently controlled by two different sets of DRPs.

3.1
DRP1 Mediates Membrane Tubulation at the Cell Plate

DRP1 proteins have been proposed to act as a tubulase that mediates vesicle tubulation (Verma and Hong, 2005). Tubulized membrane structures serve as building blocks for the formation of the cell plate. Such structures were observed in vivo at the cell plate in the cryofixed plant cells (Segui-Simarro et al., 2004; Samuels et al., 1995; Otegui et al., 2001; Otegui and Staehelin, 2004), in the plasma membrane of animal nerve terminals in the presence of GTP-γ-S (Takei et al., 1995), and in fruit flies expressing a clathrin antisense RNA (Iversen et al., 2003). Similar structures can also be created in vitro by purified animal dynamins acting on endocytic vesicles (Takei et al., 1996) or liposomes (Takei et al., 1998; Sweitzer and Hinshaw, 1998; Stowell, 1999). However, both plants and animals have proteins called DRP1 and dynamin-like protein DLP1, respectively, that are specifically dedicated to the formation of membrane tubes. DRP1 is believed to be responsible for the formation of the hourglass and dumbbell-shaped vesicle–tubulovesicle (VTV) structures observed in vivo at the forming cell plate (Verma, 2001; Bednarek and Falbel, 2002; Samuels et al., 1995; Otegui et al., 2001; Zhang et al., 2000), whereas mammalian DLP1 is able to tubulate membrane both in vivo and in vitro (Yoon et al., 2001). The fact that GTPase-defective DLP1 is able to tubulate membranes (Yoon et al., 2001) and dynamin treated with GTP-γ-S can form spirals tubulating membranes suggests that the control of their GTPase activity plays a critical role in the functioning of these molecules. In addition to tubulation, DRPs and DLPs may be involved in the fusion of vacuoles. Furthermore, interaction of phragmoplastin (DRP1) with the callose synthase complex (Hong et al., 2001b) suggests additional roles for this group of proteins.

3.2 DRP2 Is Involved in Membrane Recycling at the Cell Plate

DRP2 proteins are the bona fide plant dynamins which appear to function as a pinchase that hydrolyzes GTP and provides mechanical force to squeeze the vesicles off the plasma membrane (Sweitzer and Hinshaw, 1998; McNiven, 1998). A plant DRP that might act as a pinchase is present at the cell plate presumably associated with the clathrin-coated vesicles participating in endocytosis and membrane recycling (Hong et al., 2003b). Vesiculation at the Golgi might also require the participation of a bona fide plant dynamin molecule, DRP2A (i.e., ADL6; Lam et al., 2002), and may involve DRP1 since a GTPase-defective DRP1 mutant fused with GFP fails to leave the Golgi (Geisler-Lee et al., 2002; Hong et al., 2003b).

3.3 DRP1 and DRP2 Are Regulated Differently

Both DRP1 and DRP2 can form helical structures that wrap around the vesicles to create either tubes or vesicles (Sweitzer and Hinshaw, 1998). The kind of products formed appear to be determined, however, in part by GTP binding and hydrolysis activities. Both DRP1 and DRP2 are known to possess relatively high GTPase activities when they are in a dissociated state (Gu and Verma, 1996; Hinshaw, 2000), but when this activity is reduced by mutations in the GTP-binding domain, incubation with GTP-γ-S, high Ca^{2+} levels, or depletion of clathrin, dynamins can function as a tubulase generating long tubules on the plasma membrane (Takei et al., 1995; Iverson et al., 2003). Conversely, the expression of a dominant negative mutant of DRP1 causes the accumulation of tubular membrane structures throughout the cell plate (Hong et al., 2003b). This may be due to heterodimerization of the mutated and wild-type phragmoplastins. Under normal physiological conditions, the decision to act as a tubulase or a pinchase might rely either on auxiliary proteins that can regulate GTPase activities or on the presence of Ca^{2+}, which is known to affect dynamin GTPase activity (Lai et al., 1999). Both DRP1 and DRP2 contain a GTPase effector domain (GED, Hong et al., 2003a), through which the GTPase activity is regulated. Dynamins contain a PH domain for binding to phosphoinositides, and a PR motif (RXPXXP) for the interaction with the SH3 domains of proteins like amphiphysins and endophilins in animal cells (Schmid et al., 1998). Phosphorylation of the PR motif of dynamins by cyclin-dependent kinase 5 (Cdk5) blocks its interaction with amphiphysin and inhibits clathrin-mediated endocytosis of synaptic vesicles (Tomizawa et al., 2003). In contrast, animal DLP1, soybean phragmoplastins, and *Arabidopsis* DRP1 do not contain either the PH or PR domains (Hong et al., 2003a, Gu and Verma, 1996; Yoon et al., 2003). Phragmoplastins interact with a different set of proteins including a UDP-glucose transferase (Hong et al., 2001b), a subunit of the cell plate-specific callose synthase (Hong et al., 2001a, Verma and Hong, 2001), a novel RNA-binding protein, and a sumoylation enzyme (Z. Hong and D.P.S. Verma, unpublished data). Although there is less direct evidence that DRP1 and DRP2 are regulated by Ca^{2+}, the cell plate has been shown to contain high concentrations of this ion bound to the membrane (Wolniak et al., 1980; Hepler and Callaham, 1987). Calcium might regulate the GTPase activity of DRP1 and DRP2 via calcium- and calmodulin-dependent protein kinases and phosphotases such as calcineurin (Lai et al., 1999; Slepnev et al., 1998), and thus Ca^{2+} may directly control pinchase since it has been shown to inhibit dynamin GTPase activity (Cousin and Robinson, 2000).

The two processes (exocytosis and endocytosis) have to be precisely coordinated in order to build a functional cell plate compartment. We have observed that overexpression of soybean phragmoplastin in tobacco perturbs membrane recycling, resulting in the accumulation of multivesicular bodies (MVB) in the cell (Geisler-Lee et al., 2002). Accumulation of these MVBs in the path of growing ends of the plate in turn changes the direction of the cell plate to create an oblique division plane (Gu and Verma, 1997). Likewise, overexpression of phragmoplastin has been demonstrated to alter the plane of cell divisions in the tunica and carpus layers of the apical meristem, which affects cell differentiation in the meristem (Geisler-Lee et al., 2002; Wyrzykowska and Fleming, 2003).

3.4
Protein Complexes Responsible for Vesicle Docking at the Cell Plate

Two sets of protein complexes have been implicated in the control of exocytosis at the cell plate: SNARE proteins that mediate membrane fusion events, and exocyst proteins that determine the site of exocytosis. Exocyst is a complex of eight distinct protein subunits in yeast and animals (TerBush et al., 1996; Hsu et al., 2004) which, unlike t-SNAREs, is not distributed uniformly around the inner leaflet of the cytoplasmic membrane, but instead is confined to areas where Golgi-derived vesicles are destined to fuse (Hsu et al., 2004). The cytokinesis-specific SNARE complexes in *Arabidopsis* are composed of KNOLLE (a t-SNARE), Keule (a Sec1-like protein), SNAP33 (a SNAP25 homolog), and NPSN11 (a plant-specific SNARE) (Surpin and Raikhel, 2004; Assaad et al., 2001; Heese et al., 2001; Zheng et al., 2002; Lukowitz et al., 1996; Lauber et al., 1997). By comparison, little is known about the exocyst complex in plants. A search of the *Arabidopsis* database identified orthologs of all eight subunits of the exocyst complex, suggesting the conservation of this mechanism among all eukaryotic cells. GFP-tagged AtSec6 and AtExo70 are localized as particulate structures on the surface of the plasma membrane (Elias et al., 2003). Recent investigations into cell plate formation by electron tomogra-

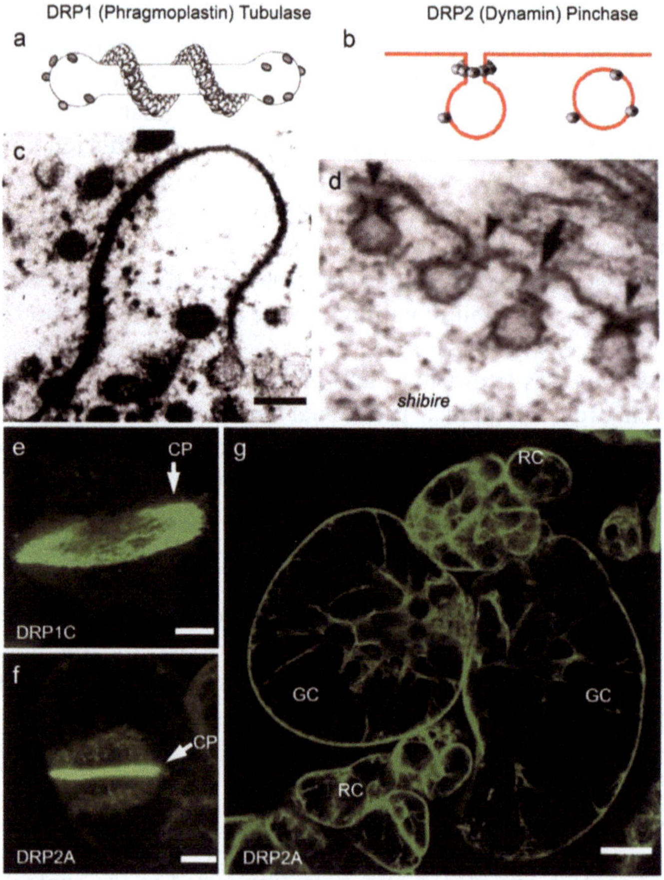

Fig. 2 DRP1 and DRP2 act as tubulase and pinchase that control exocytosis and endocytosis, respectively. **a–d** Vesicle tubulation and membrane vesiculation are mediated by structurally related DRP1 (phragmoplastins) and DRP2 (dynamins). **a** DRP1 acts as a tubulase and forms highly ordered spiral polymers on the surface of vesicles. Upon GTP hydrolysis, DRP1 provides mechanical force to convert vesicles to tubular structures. **b** DRP2, a pinchase, forms a collar ring surrounding the junction region between a budding endocytic vesicle and plasma membrane. **c** A long dumbbell-shaped tubule is observed at the nascent cell plate of tobacco BY-2 cells (Samuels et al., 1995). Note that the tubule is coated with DRP1 tubulase. **d** The pinching off of endocytic vesicles is blocked in fruit flies with a temperature-sensitive mutation at the *shibire* locus encoding dynamin (Koenig and Ikeda 1989). **e–g** Tobacco BY-2 cells expressing GFP-tagged DRP1C and DRP2A. **e** At cytokinesis, DRP1C is largely localized in the growing edges of the cell plate, forming a ringlike structure. **f** DRP2A, the plant dynamin, is also localized at the cell plate, but it might act as a pinchase that severs vesicles to retrieve excess membranes from the cell plate. **g** Approximately 5–10% of the cells expressing GFP-DRP2A are giant cells that appear to be caused by excessive endocytosis at the plasma membrane. As a consequence, vacuoles become enlarged and the cell plate formation is disrupted. These cells undergo one or two rounds of mitosis without forming a cell plate, and become multinucleate. CP, cell plate; RC, regular size cell; GC, giant cell. *Bars:* **c** 100 nm; **e–f** 10 μm; **g** 50 μm

phy have revealed the possible involvement of exocyst-like particles tethering vesicles to newly formed cell plate (Segui-Simarro et al., 2004).

Our knowledge of the proteins that participate in the control of exocytosis and endocytosis or membrane recycling at the cell plate is still quite limited. *Arabidopsis* contains two genes (DRP2A and DRP2B) for the bona fide plant dynamins that have all the functional domains present in animal dynamins (Fig. 1). By analogy to the situation in animals, these two plant dynamins are believed to participate in clathrin-coated vesicle-mediated endocytosis in plants. At cytokinesis, DRP1A is concentrated to the nascent cell plate (Hong et al., 2003b). Overexpression of GFP-DRP2A results in the formation of giant, multinucleate cells (Fig. 2g). These cells contain large vacuoles, and accumulate MVBs corresponding to late endosomes (see Sect. 3). They might be formed as a result of excessive endocytosis, leading to the expansion of endosome-like vacuoles. Since the budding of vesicles from the Golgi also requires involvement of DRPs (Hinshaw, 2000; Lam et al., 2002; Jin et al., 2001), and a dominant negative mutant of phragmoplastin (DRP1) prevents budding of such vesicles from the Golgi (Hong et al., 2003b), the role of this group of proteins in vacuolar membrane recycling is also likely (Jones et al., 1998).

4
Dynamins and Endocytosis

The role of dynamin in endocytosis was first revealed by the characterization of *Drosophila shibire* mutant followed by the cloning of this locus. The *shibire* mutation was isolated in a classical genetic screen for temperature-sensitive paralytic mutants of *Drosophila* (Grigliatti et al., 1973). When the *shibire* mutant flies are exposed to nonpermissive temperature, exocytosis at nerve terminals and in many other cell types appears to be normal, but membrane recycling/endocytosis is impaired, resulting in an accumulation of "collared pits" at the plasma membrane (Koenig and Ikeda, 1989). Cloning of the *shibire* locus (van der Bliek and Meyerowitz, 1991) revealed that this gene is homologous to the bovine brain dynamin (Obar et al., 1990), suggesting the involvement of dynamin in receptor-mediated endocytosis.

Different types of endocytosis can be distinguished according to their dependence on dynamin, clathrin, caveolin, and special lipid constituents (Fig. 3). The best characterized endocytic pathway is via clathrin-coated pits discovered 40 years ago (Roth and Porter, 1964). This pathway is the most dominant route for the efficient uptake of various lipids, proteins, and other molecules from the cell surface (Johannes and Lamaze, 2002). During the last decade, the use of protein markers and specific inhibitors has led to the identification of several clathrin-independent endocytic pathways in animals.

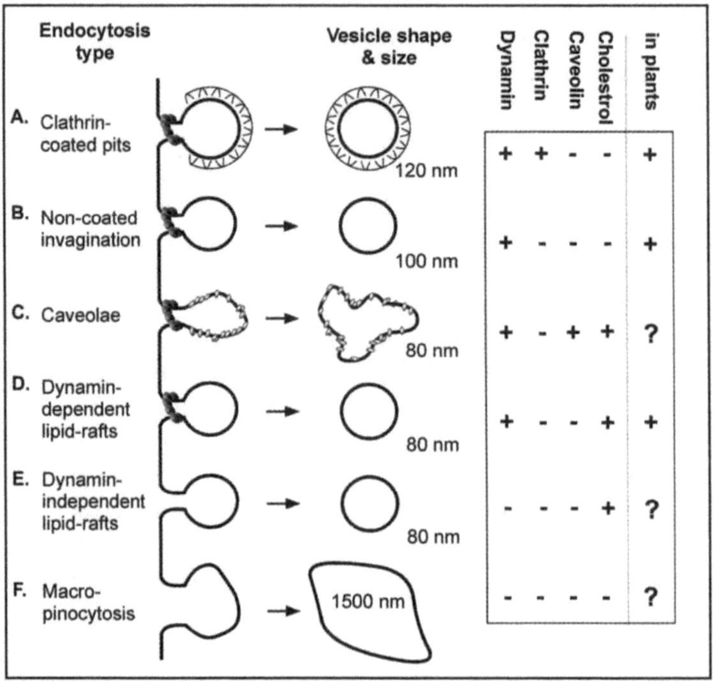

Fig. 3 Dynamin-dependent and dynamin-independent endocytic pathways and the possibility of these pathways operating in plants. Clathrin-coated pits are the best-studied endocytic pathway in both animal and plant cells. Noncoated invagination of the plasma membrane produces smooth vesicles. Caveolae which contain the characteristic marker protein caveolin are abundant in animal endothelial cells but may not exist in plant cells. Lipid rafts have recently been demonstrated in plant cells (Peskan et al., 2000; Mongrand et al., 2004). Dynamin-related proteins are present in these vesicle fractions, suggesting that the dynamin-dependent lipid-rafts route exists in plants. A dynamin-independent lipid-rafts pathway exists in animal cells but has not been documented in plants. Both caveolae and lipid rafts involve membrane microdomains that are enriched in cholesterol and glycosphingolipids. Macropinocytosis is used by animal cells in response to growth factors and mitogenic agents. It produces large vesicles and contributes to fluid-phase uptake from the medium. This pathway may not be employed by plants due to the high turgor pressure inside plant cells

Noncoated pits commonly observed at the plasma membrane require the participation of dynamin for scission of smooth vesicles. Caveolae which are abundant in endothelial cells contain an integral membrane protein, caveolin-1, which binds cholesterol and distinguishes these organelles from lipid rafts. Both caveolae and lipid rafts contain enriched cholesterol and glycosphingolipids that are present in the lateral plane of a membrane. Caveolae and lipid rafts are distinct structures at the plasma membrane and compartmentalize distinct lipids and proteins. The pinching off of vesicles in the caveolar pathway is dependent on dynamin pinchase (Henley et al., 1998; Oh et al.,

1998), while the lipid-raft pathways can be both dynamin-dependent (Lamaze et al., 2001) and dynamin-independent (Damke et al., 1994). Macropinocytosis is a result of membrane ruffling and produces large dynamic vesicles. Neither clatherin nor dynamin are involved in the process of macropinocytosis.

There have been doubts about the possibility of lipid rafts and macropinocytosis in intact plant cells because they contain high turgor pressure. However, recent evidence from cell biology and molecular biology studies in plants has favored the existence of both clathrin-dependent and clathrin-independent endocytic pathways. Abundant clathrin-coated vesicles and clathrin-coated pits have been observed in plant cells with active polarized growth, such as pollen tubes (Blackbourn et al., 1996), root hairs (Emons and Traas, 1986), and the cell plate (Samuels et al., 1995; Otegui et al., 2000, 2004; Segui-Simarro et al., 2004), suggesting a major role of the clathrin-dependent pathway in membrane recycling (see Sect. 5). On the other hand, clathrin-independent endocytic pathways also appear to have important functions in specialized cells. For instance, it has been estimated that the total content of noncoated vesicles is threefold more than that of clathrin-coated ones in guard cells (Homann and Thiel, 1999). These noncoated vesicles could be produced by noncoated invagination as well as lipid rafts. The hallmark of vesicles from lipid rafts is the presence of cholesterol-sphingolipid-enriched domains that are resistant to solubilization by nonionic detergent such as Triton X-100 at 4 °C. This detergent insolubility has allowed the isolation of lipid-rafted vesicles from tobacco leaves (Peskan et al., 2000) and cell cultures (Mongrand et al., 2004). Interestingly, proteomic study of these vesicles has identified the presence of dynamin-related proteins in lipid-rafted vesicles, suggesting a role of dynamin in the scission of rafts from the donor membrane (Mongrand et al., 2004). On the other hand, DRPs were not identified in the lipid-rafted membranes in a similar proteomic study with *Arabidopsis* (Borner et al., 2005). This result is consistent with the observation that none of the DRPs is an integral membrane protein and they only interact with membrane proteins and lipids on the surface. Most of the interacting partners of DRPs have not yet been identified, and may involve a variety of proteins since the subcellular locations and functions of DRPs are very diverse.

5
Future Perspectives

The presence of DRPs in plants with diverse functions in membrane tubulation and pinching, as well as fission of organelles, suggests that the basic property of this group of proteins, i.e., the ability to form helical structures regulated by their GTPase activity, is central to their functions. Despite a high similarity in the structure of various members of the DRP family, it is apparent that they interact with distinct proteins to perform specific roles in each of these events. A detailed dissection of various domains of these proteins and their interaction with specific partners may reveal their precise function in various cellular processes that involve membrane tubulation, pinching, and organelle divisions. Development of in vitro membrane vesiculation, tubulation, and fusion systems will further aid in dissection of the details of these mechanisms and the role of various DRPs. Finally, the function of the subfamily resembling Mx proteins remains to be elucidated, and may shed some light on the endocytosis of viruses in plant cells.

Acknowledgements This work was supported by NSF grants to D.P.S. Verma's lab. The authors would like to thank Dr. Bednarek for his critical comments on this manuscript.

References

Arimura S, Aida GP, Fujimoto M, Nakazono M, Tsutsumi N (2004) *Arabidopsis* dynamin-like protein 2a (ADL2a), like ADL2b, is involved in plant mitochondrial division. Plant Cell Physiol 45:236–242

Arimura S, Tsutsumi N (2002) A dynamin-like protein (ADL2b), rather than FtsZ, is involved in *Arabidopsis* mitochondrial division. Proc Natl Acad Sci USA 99:5727–5731

Assaad FF, Huet Y, Mayer U, Jurgens G (2001) The cytokinesis gene KEULE encodes a Secl protein that binds the syntaxin KNOLLE. J Cell Biol 152:531–543

Battey NH, James NC, Greenland AJ, Brownlee C (1999) Exocytosis and endocytosis. Plant Cell 11:643–659

Bednarek SY, Falbel TG (2002) Membrane trafficking during plant cytokinesis. Traffic 3:621–629

Blackbourn HD, Jackson AP (1996) Plant clathrin heavy chain: sequence analysis and restricted localisation in growing pollen tubes. J Cell Sci 109:777–786

Borner GH, Sherrier DJ, Weimar T, Michaelson LV, Hawkins ND, Macaskill A, Napier JA, Beale MH, Lilley KS, Dupree P (2005) Analysis of detergent-resistant membranes in *Arabidopsis*. Evidence for plasma membrane lipid rafts. Plant Physiol 137:104–116

Cousin MA, Robinson PJ (2000) Ca^{2+} influx inhibits dynamin and arrests synaptic vesicle endocytosis at the active zone. J Neurosci 20:949–957

Dombrowski JE, Raikhel NV (1995) Isolation of a cDNA encoding a novel GTP-binding protein of *Arabidopsis thaliana*. Plant Mol Biol 28:1121–1126

Damke H, Baba T, Warnock DE, Schmid SL (1994) Induction of mutant dynamin specifically blocks endocytic coated vesicle formation. J Cell Biol 127:915–934

Dong X, Hong Z, Sivaramakrishnan M, Verma DPS (2005) Callose synthase (CalS5) is required for exine formation during microgametogenesis and pollen viability in *Arabidopsis*. Plant J 42:315–328

Elias M, Drdova E, Ziak D, Bavlnka B, Hala M, Cvrckova F, Soukupova H, Zarsky V (2003) The exocyst complex in plants. Cell Biol Int 27:199–201

Emons AM, Traas JA (1986) Coated pits and coated vesicles on the plasma membrane of plant cells. Eur J Cell Biol 41:57–64

Gao H, Kadirjan-Kalbach D, Froehlich JE, Osteryoung KW (2003) ARC5, a cytosolic dynamin-like protein from plants, is part of the chloroplast division machinery. Proc Natl Acad Sci USA 100:4328–4333

Geisler-Lee J, Hong Z, Verma DPS (2002) Overexpression of the cell plate-associated dynamin-like GTPase, phragmoplastin, results in the accumulation of callose at the cell plate and arrest of plant growth. Plant Sci 163:33–42

Grigliatti TA, Hall L, Rosenbluth R, Suzuki DT (1973) Temperature-sensitive mutations in *Drosophila melanogaster*. XIV: A selection of immobile adults. Mol Gen Genet 120:107–114

Gu X, Verma DPS (1996) Phragmoplastin, a dynamin-like protein associated with cell plate formation in plants. EMBO J 15:695–704

Gu X, Verma DPS (1997) Dynamics of phragmoplastin in living cells during cell plate formation and uncoupling of cell elongation from the plane of cell division. Plant Cell 9:157–169

Heese M, Gansel X, Sticher L, Wick P, Grebe M, Granier F, Jurgens G (2001) Functional characterization of the KNOLLE-interacting t-SNARE AtSNAP33 and its role in plant cytokinesis. J Cell Biol 155:239–249

Henley JR, Krueger EW, Oswald BJ, McNiven MA (1998) Dynamin-mediated internalization of caveolae. J Cell Biol 141:85–99

Hepler PK, Callaham DA (1987) Free calcium increases during anaphase in stamen hair cells of *Tradescantia*. J Cell Biol 105:2137–2143

Hinshaw JE (2000) Dynamin and its role in membrane fission. Annu Rev Cell Dev Biol 16:483–519

Homann U, Thiel G (1999) Unitary exocytotic and endocytotic events in guard-cell protoplasts during osmotically driven volume changes. FEBS Lett 460:495–499

Hong Z, Bednarek S, Blumwald E, Hwang I, Jurgens G, Menzel D, Osteryoung K, Raikhel N, Shinozaki K, Tsutsumi N, Verma DPS (2003) A unified nomenclature for *Arabidopsis* dynamin-related large GTPases based on homology and possible functions. Plant Mol Biol 53:261–265

Hong Z, Delauney AJ, Verma DPS (2001) A cell plate-specific callose synthase and its interaction with phragmoplastin and UDP-glucose transferase. Plant Cell 13:755–768

Hong Z, Geisler-Lee J, Zhang Z, Verma DPS (2003) Phragmoplastin dynamics: multiple forms, microtubule association, and their roles in cell plate formation in plants. Plant Mol Biol 53:297–312

Hong Z, Zhang Z, Olson J, Verma DPS (2001) A novel UDP-glucose transferase interacts with callose synthase and phragmoplastin at the forming cell plate. Plant Cell 13:769–779

Hsu SC, TerBush D, Abraham M, Guo W (2004) The exocyst complex in polarized exocytosis. Int Rev Cytol 233:243–265

Iversen TG, Skretting G, van Deurs B, Sandvig K (2003) Clathrin-coated pits with long, dynamin-wrapped necks upon expression of a clathrin antisense RNA. Proc Natl Acad Sci USA 100:5175–5180

Jin JB, Kim YA, Kim SJ, Lee SH, Kim DH, Cheong GW, Hwang I (2001) A new dynamin-like protein, ADL6, is involved in trafficking from the trans-Golgi network to the central vacuole in *Arabidopsis*. Plant Cell 13:1511–1526

Johannes L, Lamaze C (2002) Clathrin-dependent or not: is it still the question? Traffic 3:443–451

Jones SM, Howell KE, Henley JR, Cao H, McNiven MA (1998) Role of dynamin in the formation of transport vesicles from the trans-Golgi network. Science 279:573–577

Jurgens G, Geldner N (2002) Protein secretion in plants: from the trans-Golgi network to the outer space. Traffic 3:605–613

Kang B, Busse JS, Dickey C, Rancour DM, Bednarek SY (2001) The *Arabidopsis* cell plate-associated dynamin-like protein, ADL1Ap, is required for multiple stages of plant growth and development. Plant Physiol 126:47–68

Kang BH, Busse JS, Bednarek SY (2003) Members of the *Arabidopsis* dynamin-like gene family, ADL1, are essential for plant cytokinesis and polarized cell growth. Plant Cell 15:899–913

Kang BH, Rancour DM, Bednarek SY (2003) The dynamin-like protein ADL1C is essential for plasma membrane maintenance during pollen maturation. Plant J 35:1–15

Kang SG, Jing JB, Hai PL, Kyeong P, Hyun JJ, Jeong L, Hwang I (1998) Molecular cloning of an *Arabidopsis* cDNA encoding a dynamin-like protein that is localized to plastids. Plant Mol Biol 38:437–447

Kim YW, Park DS, Park SC, Kim SH, Cheong GW, Hwang I (2001) *Arabidopsis* dynamin-like 2 that binds specifically to phosphatidylinositol 4-phosphate assembles into a high molecular weight complex in vivo and in vitro. Plant Physiol 127:1243–1255

Koenig JH, Ikeda K (1989) Disappearance and re-formation of synaptic vesicle membrane upon transmitter release observed under reversible blockage of membrane retrieval. J Neurosci 9:3844–3860

Lai MM, Hong JJ, Ruggiero AM, Burnett PE, Slepnev VI, De Camilli P, Snyder SH (1999) The calcineurin-dynamin 1 complex as a calcium sensor for synaptic vesicle endocytosis. J Biol Chem 274:25963–25966

Lam BC, Sage TL, Bianchi F, Blumwald E (2002) Regulation of ADL6 activity by its associated molecular network. Plant J 31:565–576

Lamaze C, Dujeancourt A, Baba T, Lo CG, Benmerah A, Dautry-Varsat A (2001) Interleukin 2 receptors and detergent-resistant membrane domains define a clathrin-independent endocytic pathway. Mol Cell 7:661–671

Lauber MH, Waizenegger I, Steinmann T, Schwarz H, Mayer U, Hwang I, Lukowitz W, Jurgens G (1997) The *Arabidopsis* KNOLLE protein is a cytokinesis-specific syntaxin. J Cell Biol 139:1485–1493

Lee SH, Jin JB, Song J, Min MK, Park DS, Kim YW, Hwang I (2002) The intermolecular interaction between the PH domain and the C-terminal domain of *Arabidopsis* dynamin-like 6 determines lipid binding specificity. J Biol Chem 277:31842–31849

Logan DC, Scott I, Tobin AK (2004) ADL2a, like ADL2b, is involved in the control of higher plant mitochondrial morphology. J Exp Bot 55:783–785

Lukowitz W, Mayer U, Jurgens G (1996) Cytokinesis in the *Arabidopsis* embryo involves the syntaxin-related KNOLLE gene product. Cell 84:61–71

Mano S, Nakamori C, Kondo M, Hayashi M, Nishimura M (2004) An *Arabidopsis* dynamin-related protein, DRP3A, controls both peroxisomal and mitochondrial division. Plant J 38:487–498

Marks B, Stowell MH, Vallis Y, Mills IG, Gibson A, Hopkins CR, McMahon HT (2001) GTPase activity of dynamin and resulting conformation change are essential for endocytosis. Nature 410:231–235

McNiven MA (1998) Dynamin: a molecular motor with pinchase action. Cell 94:151–154

Mikami K, Iuchi S, Yamaguchi-Shinozaki K, Shinozaki K (2000) A novel *Arabidopsis thaliana* dynamin-like protein containing the pleckstrin homology domain. J Exp Bot 51:317–318

Miyagishima SY, Nishida K, Mori T, Matsuzaki M, Higashiyama T, Kuroiwa H, Kuroiwa T (2003) A plant-specific dynamin-related protein forms a ring at the chloroplast division site. Plant Cell 15:655–665

Mongrand S, Morel J, Laroche J, Claverol S, Carde JP, Hartmann MA, Bonneu M, Simon-Plas F, Lessire R, Bessoule JJ (2004) Lipid rafts in higher plant cells: purification and characterization of Triton X-100-insoluble microdomains from tobacco plasma membrane. J Biol Chem 279:36277–36286

Nishida K, Takahara M, Miyagishima SY, Kuroiwa H, Matsuzaki M, Kuroiwa T (2003) Dynamic recruitment of dynamin for final mitochondrial severance in a primitive red alga. Proc Natl Acad Sci USA 100:2146–2151

Obar RA, Collins CA, Hammarback JA, Shpetner HS, Vallee RB (1990) Molecular cloning of the microtubule-associated mechanochemical enzyme dynamin reveals homology with a new family of GTP-binding proteins. Nature 347:256–261

Oh P, McIntosh DP, Schnitzer JE (1998) Dynamin at the neck of caveolae mediates their budding to form transport vesicles by GTP-driven fission from the plasma membrane of endothelium. J Cell Biol 141:101–114

Otegui MS, Staehelin LS (2004) Electron tomographic analysis of post-meiotic cytokinesis during pollen development in *Arabidopsis thaliana*. Planta 218:501–515

Otegui MS, Mastronarde DN, Kang B, Bednarek SY, Staehelin LA (2001) Three-dimensional analysis of cellularization visualized by high-resolution electron tomography. Plant Cell 13:2033–2051

Park JM, Cho JH, Kang SG, Jang HJ, Pih KT, Piao HL, Cho MJ, Hwang I (1998) A dynamin-like protein in *Arabidopsis thaliana* is involved in biogenesis of thylakoid membranes. EMBO J 17:859–867

Peskan T, Westermann M, Oelmuller R (2000) Identification of low-density Triton X-100-insoluble plasma membrane microdomains in higher plants. Eur J Biochem 267:6989–6995

Praefcke GJ, McMahon HT (2004) The dynamin superfamily: universal membrane tubulation and fission molecules? Nat Rev Mol Cell Biol 5:133–147

Roth TF, Porter KR (1964) Yolk protein uptake in the oocyte of the mosquito *Aedes aegypti*. L. J Cell Biol 20:313–332

Samuels AL, Giddings TH, Staehelin LA (1995) Cytokinesis in tobacco BY-2 and root tip cells: a new model of cell plate formation in higher plants. J Cell Biol 130:1345–1357

Schmid SL, McNiven MA, De Camilli P (1998) Dynamin and its partners: a progress report. Curr Opin Cell Biol 10:504–512

Segui-Simarro JM, Austin JR II, White EA, Staehelin LA (2004) Electron tomographic analysis of somatic cell plate formation in meristematic cells of *Arabidopsis* preserved by high-pressure freezing. Plant Cell 16:836–856

Sever S (2002) Dynamin and endocytosis. Curr Opin Cell Biol 14:463–467

Slepnev VI, Ochoa GC, Butler MH, Grabs D, De Camilli P (1998) Role of phosphorylation in regulation of the assembly of endocytic coat complexes. Science 281:821–824

Surpin M, Raikhel N (2004) Traffic jams affect plant development and signal transduction. Nat Rev Mol Cell Biol 5:1000–1009

Stowell MH (1999) Nucleotide-dependent conformational changes in dynamin: evidence for a mechanochemical molecular spring. Nat Cell Biol 1:27–32

Sweitzer SM, Hinshaw JE (1998) Dynamin undergoes a GTP-dependent conformational change causing vesiculation. Cell 93:1021–1029

Takei K, McPherson PS, Schimid S, De Camilli P (1995) Tubular membrane invaginations coated by dynamin rings are induced by GTP-γ-S in nerve terminals. Nature 374:186–190

Takei K, Mundigl O, Daniell L, De Camilli P (1996) The synaptic vesicle cycle: a single vesicle budding step involving clathrin and dynamin. J Cell Biol 133:1237–1250

Takei K, Haucke V, Slepnev V, Farsad K, Salazar M, Chen H, De Camilli P (1998) Generation of coated intermediates of clathrin-mediated endocytosis on protein-free liposomes. Cell 94:131–141

TerBush DR, Maurice T, Roth D, Novick P (1996) The exocyst is a multiprotein complex required for exocytosis in *Saccharomyces cerevisiae*. EMBO J 15:6483–6494

Tomizawa K, Sunada S, Lu YF, Oda Y, Kinuta M, Ohshima T, Saito T, Wei FY, Matsushita M, Li ST, Tsutsui K, Hisanaga S, Mikoshiba K, Takei K, Matsui H (2003) Cophosphorylation of amphiphysin I and dynamin I by Cdk5 regulates clathrin-mediated endocytosis of synaptic vesicles. J Cell Biol 163:813–824

van der Bliek AM, Meyerowitz EM (1991) Dynamin-like protein encoded by the *Drosophila shibire* gene associated with vesicular traffic. Nature 351:411–414

Verma DPS (2001) Cytokinesis and building of the cell plate in plants. Annu Rev Plant Physiol Plant Mol Biol 52:751–784

Verma DPS, Hong Z (2001) Plant callose synthase complexes. Plant Mol Biol 47:693–701

Verma DPS, Hong Z (2005) The ins and outs in membrane dynamics: tubulation and vesiculation. Trends Plant Sci 10:159–165

Volker A, Stierhof YD, Jurgens G (2001) Cell cycle-independent expression of the *Arabidopsis* cytokinesis-specific syntaxin KNOLLE results in mistargeting to the plasma membrane and is not sufficient for cytokinesis. J Cell Sci 114:3001–3012

Wolniak SM, Hepler PK, Jackson WT (1980) Detection of membrane-calcium distribution during mitosis in *Haemanthus* endosperm with chlortetracycline. J Cell Biol 87:23–32

Wyrzykowska J, Fleming A (2003) Cell division pattern influences gene expression in the shoot apical meristem. Proc Natl Acad Sci USA 100:5561–5566

Zhang Z, Hong Z, Verma DPS (2000) Phragmoplastin polymerizes into spiral coiled structures via intermolecular interaction of two self-assembly domains. J Biol Chem 275:8779–8784

Zheng H, Bednarek SY, Sanderfoot AA, Alonso J, Ecker JR, Raikhel NV (2002) NPSN11 is a cell plate-associated SNARE protein that interacts with the syntaxin KNOLLE. Plant Physiol 129:530–539

Yoon Y, Pitts KR, McNiven MA (2001) Mammalian dynamin-like protein DLP1 tubulates membranes. Mol Biol Cell 12:2894–2905

Plant Cell Monogr (1)
J. Šamaj · F. Baluška · D. Menzel: Plant Endocytosis
DOI 10.1007/7089_014/Published online: 6 September 2005

Endocytosis and Actomyosin Cytoskeleton

Jozef Šamaj[1,2] (✉) · František Baluška[1,3] · Boris Voigt[1] · Dieter Volkmann[1] · Diedrik Menzel[1]

[1]Institute of Cellular and Molecular Botany, Department of Plant Cell Biology, Rheinische Friedrich-Wilhelms-University Bonn, Kirschallee 1, 53115 Bonn, Germany
jozef.samaj@uni-bonn.de

[2]Institute of Plant Genetics and Biotechnology, Slovak Academy of Sciences, Akademicka 2, 95007 Nitra, Slovakia
jozef.samaj@uni-bonn.de

[3]Institute of Botany, Slovak Academy of Sciences, Dubravska 14, 84223 Bratislava, Slovakia

Abstract Mutual interactions between actin and endocytic assembly machineries are essential for successful clathrin-mediated endocytosis in yeast and mammals. The actin cytoskeleton is indispensable for endocytic internalization and for short-range transport of endocytic vesicles. In plants as well, actin seems to be essential for endocytic recycling of plasma membrane proteins and sterols, but surprisingly also for the turnover of cell wall pectins, which have been identified as a major cargo of endocytic vesicles. Endosomes in animal cells perform long-range movements along microtubules, whereas plant endosomes use preferentially an actin polymerization mechanism but also actin tracks for their short- and long-range movements, respectively. Thus, the actin cytoskeleton not only assists endocytic internalization and is in fact inherently associated with endosomal vesicles and endosomes, but also is responsible for their movements at the cell cortex and for their targeted delivery into the cell interior.

1 Introduction

The exact role of the actin cytoskeleton during endocytosis has been a matter of debate during the last decade (reviewed by Engqvist Goldstein and Drubin, 2003 for yeast and mammalian cells, and by Šamaj et al., 2004 for plant cells). Many possible functions have been proposed (or at least anticipated) for the actin cytoskeleton in various stages of endocytic internalization including an inhibitory effect on vesicle formation, restriction of endocytosis to distinct plasma membrane sites and invagination of plasma membrane, as well as vesicle fission, fusion and motility within the cytoplasm (reviewed by Qualmann et al., 2000; May and Machesky, 2001). It was revealed that actin polymerization and assembly is spatio-temporally coordinated with clathrin-dependent endocytosis, and many endocytic proteins were found to interact with actin-binding proteins putting them in a potentially favourite place to regulate actin dynamics (Engqvist-Goldstein and Drubin, 2003). Nevertheless, a direct functional role for actin during the formation of clathrin-coated vesicles was missing until very recently. Now some evidence has appeared supporting the notion of a functional role played by the actin cytoskeleton during endocytosis (Kaksonen et al., 2003). This new evidence suggests that filamentous actin (F-actin) dynamics is in fact a crucial and indispensable component of the endocytic machinery. Actin patches in yeast were unambiguously identified as sites of endocytic activity and their motility was characterized in more detail (Kaksonen et al., 2003; Huckaba et al., 2004). Moreover, pharmacological disruption of the actin cytoskeleton in mammalian cells with latrunculin and jasplakinolide revealed that F-actin is absolutely required for multiple stages of clathrin-dependent endocytosis including clathrin-coated pit (CCP) formation, constriction and internalization of clathrin-coated vesicles (CCVs), as well as for their lateral mobility, splitting and merging (Yarar et al., 2005). In plants, several pharmacological studies using cytochalasins, latrunculins and jasplakinolide revealed that actin is necessary for endocytic recycling of plasma membrane proteins, sterols and cell wall pectins (Geldner et al., 2001; Baluška et al., 2002; Grebe et al., 2003; Šamaj et al., 2004) and also for fluid-phase endocytosis (Baluška et al., 2004). As far as endocytosis-dependent cytomorphogenesis is concerned, it was shown that tip growth both in pollen tubes and root hairs requires a dynamic actin cytoskeleton (Gibbon et al., 1999; Baluška et al., 2000; Vidali et al., 2001; Šamaj et al., 2002; Staiger et al., 2000; Bloch et al., 2005).

2 Endocytosis and the Actin Cytoskeleton

During the last ten years, numerous studies have revealed that all forms of endocytosis in diverse eukaryotic cells ranging from yeast to plants and mammals require an intact and dynamic actin cytoskeleton, at least in some stages of endocytic internalization (reviewed by Qualmann et al., 2000; May and Machesky, 2001; Engqvist-Goldstein and Drubin, 2003; and Šamaj et al., 2004). Cell biological studies in mammalian cells demonstrated that F-actin is recruited to endocytic sites during internalization in clathrin- and caveolae-mediated endocytosis, as well as in macropinocytosis and phagocytosis. Additionally, diverse endosomes, lysosomes and vesicles use actin comet tails similar to those originally described for the pathogen *Listeria monocytogenes* for their movements via a rocketing mechanism based on localized bursts of actin polymerization (reviewed by Goldberg, 2001; Taunton, 2001; Yarar, 2003; and Engqvist-Goldstein and Drubin, 2003). In *Listeria* comet tails, WASP and ARP2/3 complexes are key components in the regulation of actin polymerization, and the same protein components are also involved in the regulation of actin dynamics in phagocytosis and macropinocytosis in lamellipodia (Insall et al., 2001; Seastone et al., 2001). Moreover, motility of newly forming pinocytic vesicles at the cell cortex of mammalian cells require actin comet tails and dynamin (Kaksonen et al., 2000; Orth et al., 2002). Association of actin comet tails and ARP2/3 complex with pinocytic vesicles and macropinosomes is transient, lasting only a few minutes (Merrifield et al., 1999; Insall et al., 2001). The actin cytoskeleton together with dynamin is also involved in caveolae-dependent endocytosis of simian virus SV40, as was demonstrated by pharmacological and microscopic studies (Pelkmans et al., 2002). This association of actin with caveolae is also transient in living cells, as revealed by simultaneous visualization of caveolin 1 tagged with CFP and actin tagged with YFP (Pelkmans et al., 2002). In spite of recent progress in the fields of caveolae/lipid raft endocytosis and pinocytosis, clathrin-dependent endocytosis represents the best studied model for actin involvement in endocytic internalization. Multiple interactions between endocytic and actin-associated proteins have been found in yeast and mammalian systems. For example, proteins such as ACK1, amphiphysin, ankyrin, proteins of the ARP2/3 complex, dynamin (see the chapter by Verma et al., this volume), HIP1 and HIP1R, profilin, WASP and synaptojanin were supposed to act as molecular linkers between endocytosis and the actin cytoskeleton (reviewed by Engqvist-Goldstein and Drubin, 2003). It was known previously that HIP1R, the protein related to mammalian Huntingtin interacting protein 1 (HIP1), interacts both with F-actin and with clathrin, and promotes clathrin assembly (Engqvist-Goldstein et al., 2001). Importantly, it was revealed in yeast that actin polymerization is required for endocytic internalization and actin patch assembly/disassembly, while mutants lacking endocytic protein Sla2p (the yeast homolog to mammalian HIP1 and HIP1R) showed defects in patch motility and extraordinarily large actin comet tails associated with endosomes (Kaksonen et al., 2003). Moreover, it was shown that endosomes covered by a network of endosomal proteins serve as nucleation sites for rapid actin assembly (Kaksonen et al., 2003). Recently, the role of HIP1R was further functionally tested in yeast cells. These data revealed that silencing of HIP1R via RNA interference (RNAi) technology resulted in impaired actin and endocytic dynamics, and in stable association between the actin cytoskeleton components and the endocytic machinery (Engqvist-Goldstein et al., 2004). Additionally, actin together with proteins promoting its polymerization, such as WASP and proteins of the ARP2/3 complex, are recruited to CCPs and CCVs (Merrifield et al., 2002, 2004). Thus, the functional association of endocytic receptors and adaptors with the network of actin-binding proteins is essential for endocytosis.

Protein kinases Prk1p and Ark1p (in yeast, homologous to GAK and AAK1 in mammals) phosphorylate Eps15-like yeast protein Pan1p, and represent key components of actin assembly since Pan1p functions as an ARP2/3 regulator during clathrin-dependent endocytosis. Recently, the molecular mechanism of these interactions was revealed. It was shown that phosphorylation of Pan1p by Prk1p prevents it from binding to F-actin and hence inhibits Arp2/3-mediated actin polymerization on endocytic vesicles, which in turn allows them to fuse with endosomes (Toshima et al., 2005). It was also convincingly shown that actin patches co-localize with FM4-64-labelled endosomes, and mutation of proteins in the ARP2/3 complex decreases cortical nonlinear movements, but not linear retrograde movements of actin patches (Huckaba et al., 2004). Moreover, actin patches also need actin cables otherwise they are not capable of linear retrograde movements (Huckaba et al., 2004). Additionally to clathrin-dependent endocytosis, actin was also proposed to play a role in compensatory endocytosis by coating and squeezing/compressing membrane-bound endocytic vesicles (Sokac et al., 2003). Altogether, these data strongly suggest that both chemical (Yarar et al., 2005) and genetic (Engqvist-Goldstein et al., 2004; Huckaba et al., 2004) disruption of the cortical actin cytoskeleton inhibit endocytosis in various systems. Thus, actin polymerization and dynamic assembly at the plasma membrane seem to control alignment and mobility of CCPs as well as internalization and lateral movements of CCVs, while the actin cytoskeleton associated with the detached CCVs and endosomes is responsible for their rapid transport from the plasma membrane into the cell interior (reviewed by Engqvist-Goldstein and Drubin, 2003; Huckaba et al., 2004; Merrifield et al., 2004).

Many important endocytic, adaptor and cytoskeletal proteins have been identified in plants (reviewed by Šamaj et al., 2004b) and filamentous fungi (see the chapter by Wendland and Walther, this volume), suggesting that actin-dependent endocytosis is operating in a very similar way in these organisms. In plants, chemical disruption of the actin cytoskeleton by actin drugs such as cytochasin D and latrunculin B resulted in the inhibition of endocytic recycling of the plasma membrane protein PIN1 (Geldner et al., 2001), as well as cell wall pectins (Baluška et al., 2002) and structural sterols of the plasma membrane (Grebe et al., 2003). Moreover, accumulation of structural sterols in endocytic brefeldin A-induced compartments is enhanced in actin2 mutants, which further supports a role for actin in sterol trafficking and recycling (Grebe et al., 2003). Additionally, recent data from our laboratory revealed that both short- and long-range movements of plant endosomes are strictly dependent on actin polymerization and actin tracks, respectively (Voigt et al., 2005b).

3
Endocytosis, Actin and Polarized Growth

In yeast and some mammalian cells, actin patches and/or foci were regularly observed at the cell cortex, especially at the sites of polarized growth during budding, shmoo protrusions formed during yeast mating or membrane movements through lamellipodia. Actin and signalling molecules are well known to regulate tip growth in several polarly growing cells, such as root hairs and pollen tubes (Baluška et al., 2000; Šamaj et al., 2002, 2004a,b, Staiger, 2000; Ketelaar et al., 2004). The function of the actin cytoskeleton was proposed to be related solely to exocytic events and delivery of secretory vesicles in these tip-growing cells, but this concept was not experimentally tested (Hepler et al., 2001). Recently, actin patches have been identified in the tip-growing fungus *Ashbya* and in plant cells, such as root hairs, and their function was related to endocytosis (Fig. 1; see also Walther and Wendland, 2004; Voigt et al., 2005a,b, and the chapter by Wendland and Walther, this volume). Endosomes were localized within growing tips of fungal hyphae and plant root hairs using either endocytic tracer FM4-64 or endosomal molecular markers such as FYVE construct, RabF2a and Ara6 (Fig. 1; see also Walther and Wendland, 2004; Voigt et al., 2005a,b, and the chapter by Wendland and Walther, this volume). Additionally, ARP3 homologues in plants were immunolocalized to tips of growing root hairs and to multivesicular bodies representing late endosomes (Van Gestel et al., 2003). Actin is an essential, structural and signalling component in tip-growing plant cells, such as root hairs and pollen tubes (Staiger, 2000; Šamaj et al., 2002, 2004a; Voigt et al., 2005a). It is believed that reorganization of the actin cytoskeleton is under the control of small GTPases belonging to the RAC/ROP subfamily of Rho GTPases in plants (e.g. Molendijk et al., 2000; Jones et al., 2002). Recently, it was reported that ectopic expression of activated AtRAC10 reorganizes the actin cytoskeleton and disrupts endocytosis and membrane recycling in

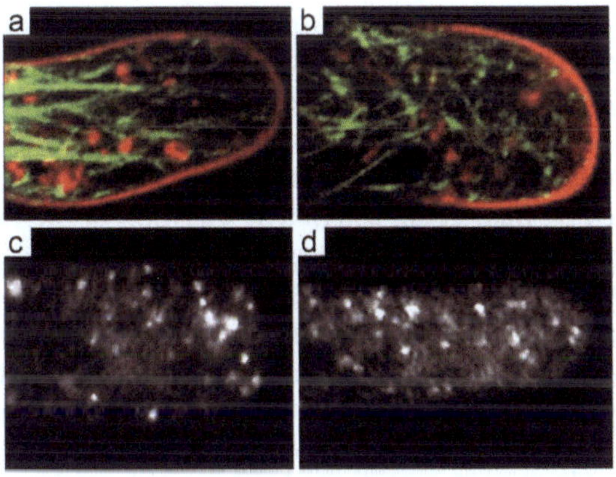

Fig. 1 **a** and **b** Simultaneous in vivo localization of F-actin with GFP-FABD construct (shown in green) and endosomes with endosomal tracer FM4-64 (shown in red) in growing root hairs of Arabidopsis. **c** and **d** In vivo localization of endosomes using molecular markers including GFP-tagged tandem FYVE domain (**c**) or YFP-tagged endosomal RabF2a (**d**) in growing root hairs of Arabidopsis. Note that actin patches and endosomes are present within expanding tips of root hair. For more details see Voigt et al. (2005b)

Arabidopsis root hairs, using a combination of genetic and pharmacological approaches (Bloch et al., 2005). In our recent study, we showed that the motility of plant endosomes at the tips of root hairs was strictly dependent on actin polymerization and tightly linked to the polarized growth of these specialized plant cells (Voigt et al., 2005b). Interestingly, endosomes and actin patches were identified as moving nonlinearly at the apex of the growing tip and did not require actin tracks (Voigt et al., 2005b), whereas their movement became linear in subapical regions and was associated with actin tracks allowing their relocation deeper into the hair tube (Voigt B., unpublished results). These recent observations raise the possibility that actin patches represent endocytic sites similar to yeast and mammalian cells, and that plant and fungal endosomes, as well as their yeast and mammalian counterparts, are also moved and propelled via transient bursts of actin polymerization, which is referred to as comet tail movement. A similar scenario is also plausible for the much faster growing pollen tubes, which harbour very tightly coupled secretory and endocytic pathways within their tips (Šamaj et al., 2005, and the chapter by Malhó et al., this volume).

4 Myosins and Endocytosis

In addition to the actin machinery, myosin VI is associated with endocytosis in polarized outgrowths known as microvilli in specialized animal cells, while plant-specific myosin VIII was localized to polarly growing plant root hairs (Baluška et al., 2000). Myosins are motor proteins which use energy released from ATP hydrolysis in order to move organelles or molecular cargo along actin tracks. In animals, myosin VI represents a special case because it moves, unlike other myosins, towards the minus (pointed) end of actin filaments (Wells et al., 1999). Recently, this myosin was implicated in endocytosis, vesicular trafficking and cellular processes such as cell migration and mitosis (reviewed by Buss et al., 2004). The question of the exact function of myosin VI during endocytosis, however, is not settled. Regarding its endocytic role, myosin VI was found to be enriched and co-localized with CCVs, and overexpression of its tail domain inhibited clathrin-mediated endocytosis of transferrin (Buss et al., 2001). Moreover, myosin VI interacts with the disabled 2-protein which binds adaptor protein AP-2 essential for clathrin-dependent endocytosis (Morris and Cooper, 2001; Morris et al., 2002). Since actin filaments are oriented with their plus (barbed) ends towards the plasma membrane, it was proposed that myosin VI may provide the necessary force for membrane invagination and for the pulling of membrane associated with CCP into the cell during vesicle formation or alternatively pushing actin filaments into the vesicle neck, thus supporting vesicle scission (Buss et al., 2001, 2004). A splice variant of myosin VI is localized to uncoated endocytic vesicles before they fuse with early endosomes, while overexpression of myosin VI tail decreased motility of endocytic vesicles through the actin-rich cell cortex (Aschenbrenner et al., 2003). Thus, another possibility would be that myosin VI may transport clathrin-coated vesicles through the meshwork of cortical actin away from the plasma membrane. Additionally, it was found that myosin VI co-assembles with adaptor protein GIPC on endocytic vesicles in rat kidney cells, which implies that it may be required for recruitment of this adaptor protein to CCPs rather than for endocytosis itself (Dance et al., 2004).

Nevertheless, Osterweil et al., (2005) have recently reported that myosin VI is highly expressed in neurons and localized to neuronal synapses. Neurons with mutated myosin VI show decreased numbers of synapses and dendritic spines associated with defective endocytic internalization of glutamate receptors. Moreover, myosin VI forms a complex with these receptors together with other adaptor and accessory proteins, such as AP-2 and SAP97, suggesting an essential role of myosin VI in synaptic functions, especially in the endocytosis of glutamate receptors (Osterweil et al., 2005).

4.1 Fluid-Phase Endocytosis in Root Apices is Driven by Plant-Specific Myosin VIII and F-actin

On the basis of the evidence that endocytosis is F-actin dependent, we have tested if the internalization of the fluid-phase marker Lucifer Yellow (LY) into cells of the inner cortex of maize root apices is sensitive to inhibitors of actin polymerization and myosin motor activities. Indeed, our recent experiments revealed that treatments with the actin and myosin inhibitors latrunculin B and 2,3-butanedione monoxime, respectively, which are both effective in maize root apices (Baluška et al., 2001; Šamaj et al., 2002) completely abolished or inhibited fluid-phase endocytosis of LY into maize root cells (Baluška et al., 2004). Moreover, the commercial availability of a specific LY antibody allowed us to perform immunogold electron microscopy (EM), which revealed that LY is taken up by the inner cortex cells of the transition zone via tubulo-vesicular invaginations found exclusively at plasmodesmata and pit fields (Baluška et al., 2004, and the chapter by Baluška et al., this volume). Interestingly, plasmodesmata domains are highly enriched with both F-actin and plant-specific myosin VIII (Reichelt et al., 1999; Baluška et al., 2004), raising the possibility that actomyosin forces are necessary for fluid-phase endocytosis at these domains. Additionally, ARP3 homologues in plants were immunolocalized to plasmodesmata domains (Van Gestel et al., 2003), suggesting that alternative mechanisms based on actin polymerization might also be involved. This link between endocytosis and plasmodesmata/pit fields gets further support from the recent studies of plant-viral movement proteins that target plasmodesmata and interact with endosomal KNOLLE, which is a plant homologue of syntaxin (Laporte et al., 2003; Uemura et al., 2004). Moreover, these proteins co-localize with Ara7 endosomal Rab GTPase within endosomes (Haupt et al., 2005) and endosomal Rab11 was reported in the plasmodesmata (Escobar et al., 2003). Obviously, plasmodesmata not only recruit vesicle trafficking pathways and the actomyosin cytoskeleton but also act as platforms for endocytosis (Baluška et al., 2004; Oparka et al., 2004).

5
Conclusions and Future Prospects

Genetic and pharmacological approaches aimed at actin disruption in both yeast and mammalian systems suggest that actin assembly is crucial for endocytosis. The same conclusion was drawn from results obtained in plant cells showing that actin disruption via cytochalasins, or more specifically by latrunculins, inhibits endocytosis. Jasplakinolide, an actin-stabilizing drug, was also recently shown to inhibit clathrin-dependent endocytosis in mammalian cells (Yarar et al., 2005). On the other hand, recent experiments with jasplakinolide performed on plants revealed that both fluid-phase endocytosis and uptake of endocytic marker FM4-64 are not affected in root cortex cells and root hairs, respectively (Baluška et al., 2004; Bloch et al., 2005).

Several lines of pharmacological, microscopic and genetic evidence unambiguously demonstrated that actin patches are, in fact, endosomal structures on which the actin cytoskeleton interacts in a multifaceted manner with the endocytic molecular machinery. Further, the actin cytoskeleton is required for vesicle abscission and the transport of vesicles away from the plasma membrane, as well as for endosomal motility. Myosin VI is necessary for clathrin-dependent receptor endocytosis in animals, while myosin VIII emerges as a putative endocytic motor for fluid-phase endocytosis in plants. However, the endocytic activities of myosin VIII remain to be documented in a functional assay as well as its mode of movement along F-actin.

Despite the recent progress, many questions remain. Which are the crucial molecular components acting at the interface between endocytic and actin machineries in plants? How is the actin cytoskeleton spatio-temporally recruited to plant endocytic sites and endosomes? How is the actin cytoskeleton regulated to become involved in the diverse modes of movements of endosomes in plant cells? We are just at the very beginning of addressing these questions and looking for answers. Future studies will have to combine genetic with cell biological approaches (Šamaj et al., 2002; Bloch et al., 2005) to shed more light on these enigmatic functions of the actin cytoskeleton related to vesicular trafficking in plant cells. On the genetic side, powerful new strategies such as silencing technology (RNAi) should be employed along with the classical genetic disruption methods (T-DNA insertions and site-directed functional mutagenesis). On the cell biological side, simultaneous in situ and live imaging of endocytic molecules and actin should be instrumental. In particular, the advanced microscopic techniques such as fluorescence resonance energy transfer (FRET) should be extremely important in revealing details of the temporal in vivo interactions between cytoskeletal/endocytic partner molecules. Functional assays will have to be performed aimed at understanding how actin binding, actin polymerization and activation/inactivation of actin-binding proteins is mediated by the group of proteins related to endocytosis in plants. And finally, as a background to these individual molecular interaction studies, it would be very helpful to have large-scale proteomics information on the expression profiles of proteins at selected developmental stages and in situations when either the cytoskeletal integrity is disrupted or endocytosis is prohibited.

Acknowledgements We thank Claudia Heym and Ursulla Mettbach for excellent technical assistance. This work was supported by the Research Trainings Network "TIPNET" from the EU, HPRN-CT-2002-00265 and by the Alexander von Humboldt foundation (Bonn, Germany). J.Š. also received support from the Slovak Grant Agency APVT (grant no. APVT-51-002302), Bratislava, Slovakia.

References

Aschenbrenner L, Lee T, Hasson T (2003) Myo6 facilitates the translocation of endocytic vesicles from cell peripheries. Mol Biol Cell 14:2728–2743

Baluška F, Salaj J, Mathur J, Braun M, Jasper F, Šamaj J, Chua N-H, Barlow PW, Volkmann D (2000) Root hair formation: F-actin-dependent tip growth is initiated by local assembly of profilin-supported F-actin meshworks accumulated within expansin-enriched bulges. Dev Biol 227:618–632

Baluška F, Hlavacka A, Šamaj J, Palme K, Robinson DG, Matoh T, McCurdy DW, Menzel D, Volkmann D (2002) F-actin-dependent endocytosis of cell wall pectins in meristematic root cells: insights from brefeldin A-induced compartments. Plant Physiol 130:422–431

Baluška F, Šamaj J, Hlavacka A, Kendrick-Jones J, Volkmann D (2004) Actin-dependent fluid-phase endocytosis in inner cortex cells of maize root apices. J Exp Bot 55:463–473

Baluška F, Jasik J, Edelmann HG, Salajova T, Volkmann D (2001) Latrunculin B-induced plant dwarfism: plant cell elongation is F-actin dependent. Dev Biol 231:113–124

Baluška F, Baroja-Fernandez E, Pozueta-Romero J, Hlavacka A, Etxeberria E, Šamaj J (2005) Endocytic uptake of nutrients, cell wall molecules, and fluidized cell wall portions into heterotrophic plant cells (in this volume). Springer, Berlin Heidelberg New York

Bloch D, Lavy M, Efrat Y, Efroni I, Bracha-Drori K, Abu-Abied M, Sadot E, Yalovsky S (2005) Ectopic expression of an activated RAC in *Arabidopsis* disrupts membrane recycling. Mol Biol Cell 16:1913–1927

Buss F, Arden SD, Lindsay M, Luzio JP, Kendrick-Jones J (2001) Myosin VI isoform localized to clathrin-coated vesicles with a role in clathrin-mediated endocytosis. EMBO J 20:3676–3684

Buss F, Spudich G, Kendrick-Jones J (2004) Myosin VI: cellular functions and motor properties. Annu Rev Cell Dev Biol 20:649–676

Dance AL, Miller M, Seragaki S, Aryal P, White B, Aschenbrenner L, Hasson T (2004) Regulation of myosin VI targeting to endocytic compartments. Traffic 5:798–813
Engqvist-Goldstein AE, Warren RA, Kessels MM, Keen JH, Heuser J, Drubin DG (2001) The actin-binding protein Hip1R associates with clathrin during early stages of endocytosis and promotes clathrin assembly in vitro. J Cell Biol 154:1209–1223
Engqvist-Goldstein AEY, Drubin DG (2003) Actin assembly and endocytosis: from yeast to mammals. Annu Rev Cell Dev Biol 19:287–332
Engqvist-Goldstein AE, Zhang CX, Carreno S, Barroso C, Heuser JE, Drubin DG (2004) RNAi-mediated Hip1R silencing results in stable association between the endocytic machinery and the actin assembly machinery. Mol Biol Cell 15:1666–1679
Escobar NM, Haupt S, Thow G, Boevink P, Chapman S, Oparka K (2003) High-throughput viral expression of cDNA-green fluorescent protein fusions reveals novel subcellular addresses and identifies unique proteins that interact with plasmodesmata. Plant Cell 15:1507–1523
Geldner N, Friml J, Stierhof Y-D, Jürgens G, Palme K (2001) Auxin-transport inhibitors block PIN1 cycling and vesicle trafficking. Nature 413:425–428
Gibbon BC, Kovar DR, Staiger CJ (1999) Latrunculin B has different effects on pollen germination and tube growth. Plant Cell 11:2349–2363
Goldberg MB (2001) Actin-based motility of intracellular microbial pathogens. Microbiol Mol Biol Rev 65:595–626
Grebe M, Xu J, Mobius W, Ueda T, Nakano A, Geuze HJ, Rook MB, Scheres B (2003) *Arabidopsis* sterol endocytosis involves actin-mediated trafficking via ARA6-positive early endosomes. Curr Biol 13:1378–1387
Haupt S, Cowan GH, Ziegler A, Roberts AG, Oparka KJ, Torrance L (2005) Two plant-viral movement proteins traffic in the endocytic recycling pathway. Plant Cell 17:164–181
Hepler PK, Vidali L, Cheung AY (2001) Polarized cell growth in higher plants. Annu Rev Cell Dev Biol 17:159–187
Huckaba TM, Gay AC, Pantalena LF, Yang HC, Pon LA (2004) Live cell imaging of the assembly, disassembly, and actin cable-dependent movement of endosomes and actin patches in the budding yeast, Saccharomyces cerevisiae. J Cell Biol 167:519–530
Insall R, Muller-Taubenberger A, Machesky L, Kohler J, Simmeth E, Atkinson SJ, Weber I, Gerisch G (2001) Dynamics of the Dicytostelium Arp2/3 complex in endocytosis, cytokinesis and chemotaxis. Cell Motil Cytoskeleton 50:115–128
Jones MA, Shen JJ, Fu Y, Li H, Yang Z, Grierson CS (2002) The *Arabidopsis* Rop2 GTPase is a positive regulator of both root hair initiation and tip growth. Plant Cell 14:763–776
Kaksonen M, Peng HB, Rauvala H (2000) Association of cortactin with dynamic actin in lamellipodia and on endosomal vesicles. J Cell Sci 113:4421–4426
Kaksonen M, Sun Y, Drubin DG (2003) A pathway for association of receptors, adaptors, and actin during endocytic internalization. Cell 115:475–487
Ketelaar T, Allwood EG, Anthony R, Voigt B, Menzel D, Hussey PJ (2004) The actin-interacting protein AIP1 is essential for actin organization and plant development. Curr Biol 14:145–149
Laporte C, Vetter G, Loudes AM, Robinson DG, Hillmer S, Stussi-Garaud C, Ritzenthaler C (2003) Involvement of the secretory pathway and the cytoskeleton in intracellular targeting and tubule assembly of grapevine fanleaf virus movement protein in tobacco BY-2 cells. Plant Cell 15:2058–2075
Malhó R, Castanho Coelho P, Pierson E, Derksen J (2005) Endocytosis and membrane recycling in pollen tubes (in this volume). Springer, Berlin Heidelberg New York
May RC, Machesky LM (2001) Phagocytosis and the actin cytoskeleton. J Cell Sci 114:1061–1077

Merrifield CJ, Moss SE, Ballestrem C, Imhof BA, Giese G, Wunderlich I, Almers W (1999) Endocytic vesicles move at the tips of actin tails in cultured mast cells. Nat Cell Biol 1:72–74

Merrifield CJ, Feldman ME, Wan L, Almers W (2002) Imaging actin and dynamin recruitment during invagination of single clathrin-coated pits. Nat Cell Biol 4:691–698

Merrifield CJ (2004) Seeing is believing: imaging actin dynamics at single sites of endocytosis. Trends Cell Biol 14:352–358

Molendijk AJ, Bischoff F, Rajendrakumar CS, Friml J, Braun M, Gilroy S, Palme K (2001) *Arabidopsis* thaliana Rop GTPases are localized to tips of root hairs and control polar growth. EMBO J 20:2779–2788

Morris SM, Cooper JA (2001) Disabled-2 colocalizes with the LDLR in clathrin-coated pits and interacts with AP-2. Traffic 2:111–123

Morris SM, Arden SD, Roberts RC, Kendrick-Jones J, Cooper JA, Luzio JP, Buss F (2002) Myosin VI binds to and localizes with Dab2, potentially linking receptor-mediated endocytosis and the actin cytoskeleton. Traffic 3:331–341

Oberholzer U, Iouk TL, Thomas DY, Whiteway M (2004) Functional characterization of myosin I tail regions in Candida albicans. Eukaryot Cell 3:1272–1286

Orth JD, Kreuger EW, Cao H, McNiven MA (2002) The large GTPase dynamin regulates actin comet formation and movement in living cells. Proc Natl Acad Sci USA 99:167–172

Osterweil E, Wells DG, Mooseker MS (2005) A role for myosin VI in postsynaptic structure and glutamate receptor endocytosis. J Cell Biol 168:329–338

Oparka KJ (2004) Getting the message across: how do plant cells exchange macromolecular complexes? Trends Plant Sci 9:33–41

Pelham RJ, Chang F (2001) Role of actin polymerization and actin cables in actin-patch movement in Schizosaccharomyces pombe. Nat Cell Biol 3:235–244

Pelkmans L, Puntener D, Helenius A (2002) Local actin polymerization and dynamin recruitment in SV40-induced internalization of caveolae. Science 296:535–539

Plastino J, Sykes C (2005) The actin slingshot. Curr Opin Cell Biol 17:62–66

Reichelt S, Knight AE, Hodge TP, Baluška F, Šamaj J, Volkmann D, Kendrick-Jones J (1999) Characterization of the unconventional myosin VIII in plant cells and its localization at the post-cytokinetic cell wall. Plant J 19:555–569

Qualmann B, Kessels MM, Kelly RB (2000) Molecular links between endocytosis and the actin cytoskeleton. J Cell Biol 159:F111–F116

Šamaj J, Peters M, Volkmann D, Baluška F (2000) Effects of myosin ATPase inhibitor 2,3-butanedione 2-monoxime on distributions of myosins, F-actin, microtubules, and cortical endoplasmic reticulum in maize root apices. Plant Cell Physiol 41:571–582

Šamaj J, Ovecka M, Hlavacka A, Lecourieux F, Meskiene I, Lichtscheidl I, Lenart P, Salaj J, Volkmann D, Bogre L, Baluška F, Hirt H (2002) Involvement of the mitogen-activated protein kinase SIMK in regulation of root hair tip growth. EMBO J 21:3296–3306

Šamaj J, Baluška F, Menzel D (2004) New signalling molecules regulating root hair tip growth. Trends Plant Sci 9:217–220

Šamaj J, Baluška F, Voigt B, Schlicht M, Volkmann D, Menzel D (2004) Endocytosis, actin cytoskeleton and signalling. Plant Physiol 135:1150–1161

Šamaj J, Read N, Baluška F (2005) Endocytosis in plants and filamentous fungi. Trends Cell Biol 15:425–433

Scott G, Leopardi S, Printup S, Madden BC (2002) Filopodia are conduits for melanosome transfer to keratinocytes. J Cell Sci 115:1441–1451

Seastone DJ, Harris E, Temesvari LA, Bear JE, Saxe CL, Cardelli J (2001) The WASP-like protein scar regulates macropinocytosis, phagocytosis and endosomal membrane flow in Dicytostelium. J Cell Sci 114:2673–2683

Sokac AM, Co C, Taunton J, Bement W (2003) Cdc42-dependent actin polymerization during compensatory endocytosis in Xenopus eggs. Nat Cell Biol 5:727–732

Taunton J (2001) Actin filament nucleation by endosomes, lysosomes and secretory vesicles. Curr Opin Cell Biol 13:85–91

Toshima J, Toshima JY, Martin AC, Drubin DG (2005) Phosphoregulation of Arp2/3-dependent actin assembly during receptor-mediated endocytosis. Nat Cell Biol 7:246–254

Van Gestel K, Slegers H, Von Witsch M, Šamaj J, Baluška F, Verbelen JP (2003) Immunological evidence for the presence of plant homologues of the actin-related protein Arp3 in tobacco and maize: subcellular localization to actin-enriched pit fields and emerging root hairs. Protoplasma 222:45–52

Verma DPS, Hong Z, Menzel D (2005) Dynamin-related proteins in plant endocytosis (in this volume). Springer, Berlin Heidelberg New York

Vidali L, McKenna ST, Hepler PK (2001) Actin polymerization is essential for pollen tube growth. Mol Biol Cell 12:2534–2545

Voigt B, Timmers ACJ, Šamaj J, Hlavacka A, Ueda T, Preuss M, Nielsen E, Mathur J, Emans N, Stenmark H, Nakano A, Baluška F, Menzel D (2005a) Actin-propelled motility of endosomes is tightly linked to polar tip growth of root hairs. Eur J Cell Biol (in press)

Voigt B, Timmers T, Šamaj J, Müller J, Baluška F, Menzel D (2005b) GFP-FABD2 fusion construct allows in vivo visualization of the dynamic actin cytoskeleton in all cells of *Arabidopsis* seedlings. Eur J Cell Biol (in press)

Walther A, Wendland J (2004) Apical localization of actin patches and vacuolar dynamics in Ashbya gossypii depend on the WASP homolog Wal1p. J Cell Sci 117:4947–4958

Weiner OD, Servant G, Welch MD, Mitchison TJ, Sedat JW, Bourne HR (1999) Spatial control of actin polymerization during neutrophil chemotaxis. Nat Cell Biol 1:75–81

Wells AL, Lin AW, Chen LQ, Safer D, Cain SM, Hasson T, Carragher BO, Milligan RA, Sweeney HL (1999) Myosin VI is an actin-based motor that moves backwards. Nature 401:505–508

Wendland J, Walther A (2005) Tip growth and endocytosis in fungi (in this volume). Springer, Berlin Heidelberg New York

Yarar D (2003) Cortical patches on the move. Cell 115:475–487

Yarar D, Waterman-Storer CM, Schmid SL (2005) A dynamic actin cytoskeleton functions at multiple stages of clathrin-mediated endocytosis. Mol Biol Cell 16:964–975

Plant Cell Monogr (1)
J. Šamaj · F. Baluška · D. Menzel: Plant Endocytosis
DOI 10.1007/7089_015/Published online: 6 September 2005

Endocytosis and Endosymbiosis

Antonius C. J. Timmers[1] · Marcelle Holsters[2] (✉) · Sofie Goormachtig[2]

[1]Laboratoire Interactions Plantes Micro-organismes, Unité Mixte de Recherche 441–2594, B.P. 52627, Chemin de Borde Rouge-Auzeville, 31326 Castanet Tolosan, France
ton.timmers@toulouse.inra.fr

[2]Department of Plant Systems Biology, Flanders Interuniversity Institute for Biotechnology (VIB), Ghent University, Technologiepark 927, 9052 Gent, Belgium
marcelle.holsters@psb.ugent.be, sofie.goormachtig@psb.ugent.be

Abstract Symbioses are widespread in nature and occur between organisms that belong to a large variety of taxonomic divisions (Hentschel et al. 2000). Most often, only two partners are involved and the outcome may be either beneficial to both, i.e. mutualism, or detrimental to one of them, i.e. parasitism. Mutualism varies from simple protection against a hostile environment to an intimate cohabitation with exchange of essential nutrients. Important and well-studied examples are the symbiosis between nitrogen-fixing bacteria and plants of the Leguminosae family (approximately 750 genera and 20 000 species) and the arbuscular mycorrhizal interactions that involve more than 80% of land plants with fungi of the Glomeromycota. In the first case, plants profit through the supply of a nitrogen source, and in the second, through an uptake of phosphate. The microsymbionts benefit through the acquisition of carbon sources in a specific and exclusive ecological niche. In both types of interactions, the microsymbionts invade the plant host and the nutrient exchange takes place inside specialised plant cells. The establishment of the symbiosis is a complex process that requires the coordinated action of both symbionts and most probably the involvement of endocytosis in a number of critical events. In this chapter, we will describe both types of endosymbiosis in view of endocytosis and endocytosis-like processes.

1 Legume–Rhizobia Interactions

Even though molecular nitrogen makes up 78% of the atmosphere, the availability of fixed nitrogen is often a limiting factor for plant growth. Most organisms rely on the capacity of a restricted number of prokaryotes to reduce atmospheric nitrogen to ammonia that can be assimilated into organic molecules. The nitrogenase enzyme complex catalyses the process of biological nitrogen fixation, which accounts for approximately half of all the fixed nitrogen annually, the other half being provided by industrial production of ammonium. The rhizobium–legume symbiosis contributes up to 30% of the total biological nitrogen fixation and provides fixed nitrogen directly to a number of important agricultural crops (Vitousek et al. 1997).

Because nitrogenase is irreversibly inactivated by oxygen, biological nitrogen fixation requires anoxic or microaerobic conditions. During legume symbiosis, these conditions are created in newly formed root organs, the nodules, in which the bacteria reside (Schultze and Kondorosi 1998). Legume-nodulating bacteria are referred to as rhizobia and include taxonomically diverse genera, such as *Allorhizobium*, *Azorhizobium*, *Bradyrhizobium*, *Mesorhizobium*, *Rhizobium*, *Sinorhizobium*, and the recently described *Methylobacterium*, *Burkholderia*, and *Ralstonia* sp. All these bacteria have in common the capacity to produce nodulation (Nod) factors, which are signalling molecules that play an essential role in initiation of nodule development and bacterial invasion (Oldroyd and Downie 2004). Nod factors are lipochitooligosaccharides and consist of an oligomeric backbone of β-1,4-linked *N*-acetyl-D-glucosaminyl residues, N-acylated at the nonreducing end, and decorated with species-/strain-specific substituents (D'Haeze and Holsters 2002). Bacterial *nod*, *nol*, and *noe* genes code for the enzymes of the Nod factor biosynthesis and secretion pathway (Batut et al. 2004). Individual rhizobium strains may synthesise Nod factors of two to approximately 60 different types and host specificity is determined, at least in part, by qualitative and quantitative aspects of the Nod factor population. Also bacterial surface polysaccharides, such as exopolysaccharides (EPS), lipopolysaccharides (LPS), and succinoglycan are important in determining the host range and the host specificity of interactions (Fraysse et al. 2003).

In general, two types of nodules are distinguished, determinate and indeterminate (Hirsch and LaRue 1997). Indeterminate nodules are elongated and have a persistent meristem that continually gives rise to new nodule cells. Representative legumes with indeterminate nodules are *Medicago sativa* (alfalfa), *M. truncatula*, *Pisum sativum* (pea), *Vicia* sp. (vetch), and *Trifolium* sp. (clover). Determinate nodules are round and lack a persistent meristem; they occur on *Lotus japonicus* and on tropical plants, such as *Glycine max* (soybean) and *Vicia faba* (broad bean). In both types, nodules are formed through a number of sequential developmental stages. We will describe the infection process of the indeterminate nodules, which are presumably the ancestral type (Sprent 2002) and include many processes that also take place during determinate nodule development.

Before their actual physical contact, the plant and the bacteria communicate through the exchange of signal molecules. Roots of host plants secrete flavonoids and betaines, which are sensed by the bacteria and induce the *nod* genes (Dénarié et al. 1996). The synthesis and export of Nod factors, in turn, trigger early nodulation responses in the host, including root hair membrane depolarisation, intracellular calcium oscillations, root hair deformation, and the initiation of cell divisions in the root cortex (Downie and Walker 1999; Miklashevichs et al. 2001). Rhizobial attachment to young growing root hairs is mediated by host lectins and bacterial adhesins (Kijne 1992).

Subsequently, the root hair curls around the bacteria, thereby isolating them from the external environment. Within the curl, local degradation of the cell wall (Hubbell 1981) allows the bacteria to penetrate the cell and initiate infection threads (ITs), tubular structures that result from an inward polar growth process and that morphologically resemble tip growth of root hairs and pollen tubes. In the ITs, the bacteria are embedded in a matrix of plant glycoproteins from which they are shielded by a layer of surface polysaccharides. Surface polysaccharides, the matrix, and the cell wall isolate the bacteria from contact with the plant plasma membrane; hence, the root hair ingests the expanding IT matrix rather than the rhizobia (Kijne 1992). Bacterial division and local cell wall and matrix deposition probably drive progression of the ITs. Solidification of the matrix under plant control may contribute to the mechano-physical forces needed to direct inward growth (Brewin 2004).

Initiation of ITs coincides with an arrest in tip growth of the root hairs. The nucleus is uncoupled from the root hair tip and recoupled to the advancing tip of the IT. Possibly, the plant cytoskeleton plays an important role in this process, because a dense array of microtubules is observed between the nucleus and the IT tip (Timmers et al. 1999). Rhizobia inside the IT grow and divide exclusively near the extending tip without ever reaching the tip itself (Gage 2004). At the junction between the epidermis and the underlying cortical cell, the plant prepares the passage of the bacteria by local remodelling of the cell wall before arrival of the IT (van Spronsen et al. 1994). Upon exit of the epidermal cell, the IT fuses with the plasma membrane and bacteria pass through the intercellular space; invagination and tip growth, similar to those at the beginning of the IT growth, occur in the underlying cortical cell. In outer cortex cells, a column of cytoplasm, the preinfection thread (PIT), marks the path of the growing IT (Kijne 1992). While the IT propagates further towards the root interior, it branches and enters the previously formed nodule primordium. In summary, the initiation of indeterminate nodules involves two series of events that take place concurrently: one starts in the root epidermis with root hair activation and curling, IT initiation, IT growth, and the reentry of outer cortical cells into the cell cycle, leading to the formation of PITs (Yang et al. 1994); the other begins in pericycle and inner cortex cells in a region opposite to the protoxylem poles, where dedifferentiation followed by cell divisions result in a cell mass that constitutes the nodule primordium (Timmers et al. 1999).

When the ITs are about to reach the nodule primordium, some centrally located noninvaded cells organise into the nodule meristem, which ensures outward growth of the nodule towards the root surface. The IT network is now located proximal to the meristem. As the nodule grows, a typical organisation of central and peripheral tissues develops (Vasse et al. 1990). Within the central tissue, most distally located is the nodule meristem, which is composed of a few layers of actively dividing cells, followed by the infec-

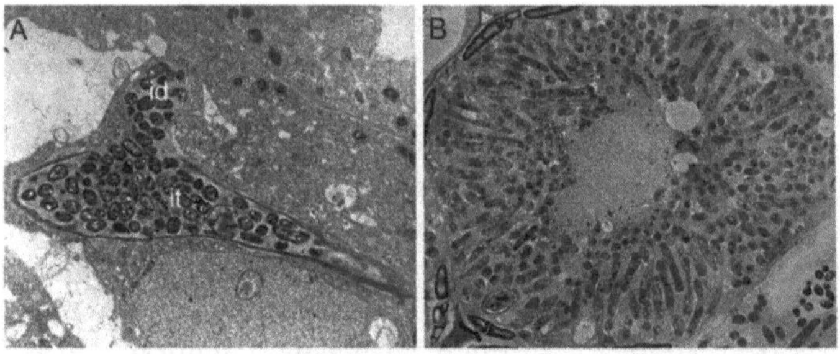

Fig. 1 Transmission electron microscopical micrographs of nodular development in *Medicago truncatula*. **a** Section through an infection thread (*it*) on which an infection droplet (*id*) has formed to release the bacteria into the plant cell cytoplasm. **b** Section through an infected cell. The cytoplasm is filled with symbiosomes. (Courtesy of Françoise de Billy, Laboratoire Interactions Plantes Micro-organismes, Castanet-Tolosan, France)

tion zone where the infection network extends towards the outward-moving meristem. In this zone, bacteria are liberated from unwalled outgrowths of ITs, so-called infection droplets (Brewin 2004) (Fig. 1a), by a process that has often been compared with phagocytosis in animal cells (Parniske 2000). During release, the rhizobia, now called bacteroids (Oke and Long 1999a), remain surrounded by a plant membrane, the peribacteroid or symbiosome membrane, with features of the host plasma membrane, the endoplasmic reticulum, and de novo delivered components (Roth and Stacey 1989). Uptake coincides with changes in the surface coat of bacteria and a physical interaction between peribacteroid/symbiosome and bacteroid membranes is suggested by in vitro experiments (Brewin 2004). Bacteroids within symbiosomes divide once or a few times; depending on the nodule type, the symbiosome either divides or does not, resulting in a symbiosome with one bacteroid or several bacteroids, respectively. Finally, the infected cells become completely filled with symbiosomes (Fig. 1b). Nitrogen-fixing bacteroids in the central zone of the nodule resemble plant organelles with a specific function, such as mitochondria and plastids. Plant cells invaded with bacteroids become round, they increase in size, and undergo an endoreduplication process, resulting in cells with a DNA content up to 32N (Cebolla et al. 1999). Nodules also contain a network of noninvaded cells, which are smaller in size, do not endoreduplicate, and probably have a transport function. Proximal to the fixation zone is a zone of senescence where first the bacteroids and later the complete content of the infected cells are resorbed (W. Van de Velde and M. Holsters, unpublished data). Peripheral tissues surrounding the central tissue include the nodule parenchyma, a vascular system, the nodule endodermis, and an outer cortex. Typically, nodules lack a surrounding epidermis.

2 Endocytosis in the Rhizobium–Legume Interaction

In a number of steps of the rhizobium–legume interaction, endocytosis-like processes play a role, for instance in Nod factor signal perception and transduction, root hair curling, IT initiation and progression, release of bacteria from ITs in the form of infection droplets, and subsequent degradation of bacteria in the senescent zone. In the last zone, the molecules from decaying bacteroids are remobilised through the peribacteroid membrane, a process that might display features of endocytic uptake of extracellular material (Baluška et al. 2002, and in this volume). However, alternative mechanisms might operate at diverse developmental stages of the symbiosis. Overall, the IT growth and bacterial uptake are more appropriately described as regulated exocytic processes, because the invaded cell massively synthesises membranes either coupled with or uncoupled from wall synthesis.

2.1 Nod Factor Perception and Signal Transduction

Nod factors are biologically active at very low concentrations, down to picomolar levels, implying their perception by plant receptors that initiate a signal amplification pathway with morphological responses and nodulation-specific gene expression in diverse root cell types, as a consequence. A number of candidate receptors have been identified from *M. truncatula*, *L. japonicus*, and pea (Riely et al. 2004) and include receptor-like kinases that have LysM domains, such as NFR1, NFR5/NFP, and LYK3 and LYK4. Prokaryotic LysM domains bind peptidoglycans, molecules that are structurally similar to *N*-acetyl glucosamines, which are the basic structure of the Nod factor, suggesting that they might be responsible for Nod factor binding. The exact nature of the Nod factor binding activity and the molecular mechanism of Nod factor recognition are unknown, but bacterial mutant analysis suggests that receptor endocytosis might be important for proper signalling. Interactions between the *nodFL* double mutant of *Sinorhizobium meliloti* and wild-type *M. truncatula* (Ardourel et al. 1994) are characterised by continued root hair deformation/curling and the development of microcolonies within root hair curls. Upon Nod factor binding, the receptor might fail to be internalised and sorted to lysosomal compartments for its degradation as observed for receptor–ligand interactions in animal cells. In the wild-type situation, the activated receptor and its ligand, the Nod factor, would be removed from the cell surface by endocytosis to attenuate the receptor-mediated signalling. The *nodFL* double mutant produces non-O-acetylated Nod factors in which unsaturated C16 fatty acids are replaced by vaccenic acid and a C20:1 fatty acid. These modifications do not prevent binding of the Nod factor to the putative receptor, because the plant still responds with root hair deformations and cortical cell divisions, but might interfere with its proper internalisation and signal downregulation.

Two studies provide more direct indications for the internalisation of Nod factors as a part of their signalling mechanism. By using biologically active fluorescent analogues of Nod factors and Nod factor-specific antibodies, the appearance of fluorescence inside root hairs and the presence of the Nod factor in invaded cells in the infection zone of nodules of *M. sativa* were shown (Philip-Hollingsworth et al. 1997; Timmers et al. 1998), respectively. The intracellular presence of Nod factors was linked with microtubular changes and plant cell differentiation. However, the internalisation of fluorescent Nod factors was not confirmed (Goedhart et al. 2000). Instead, Nod factors were immobilised in the cell wall of root hairs.

2.2 Root Hair Curling

According to recent models, root hair curling implies the continuous shift of the root hair growth axis towards the bacteria representing a point source of Nod factors (Esseling et al. 2003). As a consequence, the root hair extends more on one side and less on the other until a curl is formed that entraps rhizobia. Probably, concomitant with asymmetric growth, cell wall material and plasma membrane have to be retrieved from the inside part of the curl to accommodate the bacterial colony. Uptake of cell wall material has been shown to take place in meristematic cells of roots in maize and wheat (Baluška et al. 2002). This uptake is sensitive to brefeldin A, depends on F-actin, and is inhibited by short-term boron deprivation (Baluška et al. 2002; Yu et al. 2002; Šamaj et al. 2004). The sensitivity of root hair curling to brefeldin A has not been studied, but F-actin is most probably involved, and interestingly, the absence of borate was observed to impair IT development (Bolaños et al. 1996).

2.3 Infection Thread Initiation and Progression

Numerous bacterial mutants with altered Nod factor production (such as the previously mentioned *nodFL* mutant of *S. meliloti*, but also a *nodFELMNTO* mutant of *Rhizobium leguminosarum* bv. *viciae* [Walker and Downie 2000]) induce the curling of root hairs upon inoculation, but fail to initiate an IT. Instead, large aggregates of bacteria accumulate in the root hair curl. Apparently, cell wall degradation necessary for bacterial entry and subsequent IT formation does not take place: either receptor-mediated endocytosis might be involved through a defect in the attenuation of the Nod factor signal (Sect. 2.1) or, alternatively, intracellular signal propagation and amplification might be hampered at the level where the Nod factor is targeted to an intracellular site. Examples of endocytosis involved in signalling between the plasma membrane and the cell nucleus have recently been identified in animal systems (Miaczynska et al. 2004). Signalling endosomes with Nod factors might be implicated in the generation of calcium spiking that occurs around the nucleus of root hairs approximately 10 min after Nod factor treatment and plays a role in Nod-factor-induced plant gene expression. This observation would explain the presence of fluorescent analogues of the Nod factor in the vicinity of the root hair nucleus reported by Philip-Hollingsworth et al. (1997). However, it should be emphasised that the internalisation of fluorescent Nod factors has not been confirmed by other studies yet.

Also bacterial surface polysaccharides like EPS, LPS, and succinoglycan may play a signalling role during nodulation, for instance, for suppression of defence responses and for proper progression of nodule development (Mithöfer 2002; Mathis et al. 2005). Hence, recognition by specific receptors on the plant plasma membrane and receptor-mediated endocytosis of rhizobial elicitors and suppressors could potentially intervene in these processes. Bacterial EPS and LPS have been reported to be endocytosed by plant cells (Romanenko et al. 2002; Gross et al. 2005). For a detailed description of the role of surface polysaccharides during rhizobium–legume interactions, the reader is referred to reviews by Fraysse et al. (2003) and D'Haeze and Holsters (2004).

Several mutants have been identified in *P. sativum*, *M. truncatula*, and *L. japonicus* in which IT initiation or progression is blocked. These include *sym7*, *sym34*, *sym35*, *sym37*, and *sym38* of *P. sativum* (Tsyganov et al. 2002), *hcl* (Catoira et al. 2001), *lyk3* (Limpens et al. 2003), *rit1*, and *bit1* of *M. truncatula* (Mitra and Long 2004), and *sym79* (*crinkle*; Tansengco et al. 2003) and *sym4* of *L. japonicus* (Bonfante et al. 2000). From these mutants, only *lyk3* has been cloned and codes for a receptor-like kinase as a putative Nod factor entry receptor (Limpens et al. 2003).

IT initiation requires both local cell wall degradation and plasma membrane invagination. Microscopical studies show that the root hair engulfs the thread matrix rather than rhizobia, because the IT tip, which is in contact with the plasma membrane, is devoid of bacteria (Kijne 1992). Although invaginated growth might involve a process with characteristics of receptor-mediated phagocytosis (Conner and Schmid 2003), we propose an alternative mechanism. In our model, the first step to IT initiation is local cell wall degradation within the root hair curl that houses a bacterial microcolony. Both plant and bacterial enzymes might play a role (Hubbell 1981). As a result, cell wall degradation products accumulate within the space between the plasma membrane and the bacterial colony, locally increasing osmolarity. Water uptake into the infection pocket with bacteria builds up a local pressure (swelling), forcing the invagination of the plasma membrane and extracellular matrix together with the residing rhizobia into the root hair. Concomitantly, the invaded plant cell synthesises a new plasma membrane to accommodate the enlarged surface area. Directly behind the IT tip, new cell wall material is laid down and the IT matrix solidifies, probably by the action of specific peroxidases (Wisniewski et al. 2000). IT progression is maintained by bacterial cell division and by the constant production of polysaccharides and glycoproteins of both plant and bacterial origin, resulting in an inward-oriented pushing force of the IT against the plasma membrane. Paradoxically, in this model, although the IT is internalised by the plant cell, IT initiation and progression are dependent on processes mechanistically coupled to localised secretion.

In the outer cortex, the pathway for the progression of ITs is predetermined by the formation of cytoplasmic bridges or PITs, which are formed after the cell cycle is activated in cells arrested in G2 (Yang et al. 1994). As during cell division, the nucleus migrates towards the centre of the cell and a cytoplasmic structure containing dense arrays of microtubules that connect both periclinal walls is assembled. Most likely, the cellular machinery for phragmoplast formation uncoupled from mitosis and cytokinesis is recruited at this stage for membrane and wall synthesis of the passing IT.

2.4 Bacterial Release from Infection Threads

When the IT reaches the nodule primordium in the inner cortex, bacterial liberation begins. The process of bacterial uptake has often been compared with endocytosis or phagocytosis from animal cells (Brewin 1998, 2004; Parniske 2000). Arguments put forward are based on microscopical observations (Brewin 1998), the acidic nature of the peribacteroid space (Whitehead and Day 1997), expression inside *L. japonicus* nodules of Rab5 (Borg et al. 1997), a protein associated with compartments of the endocytic pathway in animal cells, and the possible participation of proteins involved in vesicle transport, such as Rab1p or Rab7p (Cheon et al. 1993).

The presence of a large number of bacterial and plant mutants in which either the uptake of the rhizobia or the maintenance of the symbiotic compartment are altered suggests that both partners are actively involved in the process. Bacterial mutants, which fail to be engulfed by the host plant, include mutant *T8-1* of *Bradyrhizobium japonicum* (Morrison and Verma 1987), *Rm6963* of *S. meliloti* (Niehaus et al. 1998), and *pssD133* of *R. leguminosarum* (Król et al. 1998). Also auxotrophic mutants, such as the *S. meliloti hemA* mutant that controls the first step of heme biosynthesis (Dickstein et al. 1991), as well as rhizobial mutants defective in the synthesis of signal peptides (Müller et al. 1995) produce nodules without rhizobial uptake. EPS may play a role in this process, besides their role (Sect. 2.3) during IT initiation and progression (Fraysse et al. 2003).

The study of the role of Nod factors in the late process of bacterial uptake is difficult, because Nod factors are evidently required for the very early initiation of the interaction. Genes responsible for Nod factor production are transcribed in bacteria present in ITs, but usually not in bacteroids. Moreover, Nod factors have been localised in the nodule infection zone (Timmers et al. 1998). However, evidence is accumulating that Nod factors are not needed for the bacterial uptake itself but play a role in guidance of IT growth. A few cases have been described during which Nod-factor-deficient strains could reach the inside of a nodule and be internalised to fix nitrogen as long as Nod factors were added into the root medium (Relić et al. 1994; D'Haeze et al. 1998; J. Den Herder and M. Holsters, unpublished data). The Nod-factor-deficient strains NGR234 of *Rhizobium* sp. and ESDA110 of *B. japonicum* were able to invade and nodulate *Vigna unguiculata* (cowpea) and *G. max* when NGR234 Nod factors were added at a concentration of 10^{-7}–10^{-6} M (Relić et al. 1994). Because it is very improbable that externally applied Nod factors diffuse to the inside of the plant tissue (Goedhart et al. 2000), these experiments suggest that Nod factors are locally unimportant for bacterial release into the plant cells. An analogous observation has been made for the lateral root base nodulation on hydroponic roots of *S. rostrata* by coinoculating a bacterial mutant with altered surface polysaccharides (ORS571-X15) but normal Nod factors and a mutant without Nod factor production (ORS571-V44; D'Haeze et al. 1998). *Azorhizobium caulinodans* can invade its host *S. rostrata* by a different mechanism depending on the physiological growth conditions of the plant (Goormachtig et al. 2004). In aerated soils, root hair curl invasion is used; however, upon submergence, the azorhizobia invade intercellularly through cracks at the lateral root bases. The bacteria directly invade the root cortex and form big intercellular infection pockets from which ITs grow towards the nodule primordium. Crack-entry invasion at lateral root bases depends on Nod factors, although the structural requirements are less stringent than those for root hair curl invasion (Goormachtig et al. 2004). The ORS571-V44 strain cannot enter the host for lack of Nod factors and does not provoke any nodulation-related effect on *S. rostrata* (Goethals et al. 1989). ORS571-X15 induces nodule primordia at the bases of lateral roots but cannot invade the host. Nodule primordia are arrested and the bacteria stay in superficially located infection pockets. However, ORS571-X15 can complement ORS571-V44 for nodule invasion and functional nodules are formed, which are exclusively occupied by ORS571-V44 bacteria (D'Haeze et al. 1998). Again, bacteria unable to produce Nod factors can be internalised as long as Nod factors signal from at a distance. Presumably, Nod factors trigger developmental gradients that allow bacterial invasion and cell division. Nod-factor-dependent gene expression was turned off all along the ORS571-V44 invasion track (J. Den Herder and M. Holsters, unpublished data). Because ORS571-V44 is taken up by plant cells and functional symbiosomes are formed, Nod factors are probably not needed locally to enable this process. Using a very similar approach, Mathis et al. (2005) have recently shown that mutant rhizobia with totally aberrant LPS can be internalised in *S. rostrata* root cells upon intercellular complementation or in the presence of exogenously supplied purified wild-type LPS. These observations led to the conclusion that LPS are signalling molecules for progression of nodule development and are presumably not required as a bacterial ligand for uptake in this system.

A number of fix⁻ mutants of *P. sativum* have been identified in which bacteria are not liberated, but accumulate in hypertrophied droplets of ITs with relatively thick walls, including the mutations in *sym33* and *sym40* (Tsyganov et al. 1994, 1998), and the mutant *RisFixV* (Morzhina et al. 2000). Similar phenotypes have been observed using *M. truncatula nip* (Veereshlingam et al., 2004) and the *L. japonicus alb1* (*LjSym74*) mutants (Imaizumi-Anraku et al. 1997).

Also SymRK, a receptor-like kinase that is important for Nod factor-dependent signal transduction in the epidermis (Stracke et al. 2002) might play a role during bacterial uptake. This second function for SymRK has been observed during the lateral root-based nodulation on *S. rostrata*, a nodulation system in which the epidermis is omitted for bacterial invasion. By using RNA interference technology allowing knock-down analysis, with only 7% of the total SymRK RNA left, the bacteria could invade the cortex, make ITs, albeit disturbed ones, but were hampered at the level of uptake. Ultrastructural analysis showed that the infection droplets were irregularly shaped and that the few symbiosomes within the cytoplasm had an anomalous form (Capoen et al. 2005). Analogous observations were made with *M. truncatula* plants that were partially knocked down (Limpens et al. 2005), illustrating that the SymRK might function at the heart of the symbiosis, namely at the uptake of rhizobia. Which role SymRK plays is still unknown, but the nonsymbiotic phenotype of SymRK mutants might give a hint. An epidermal phenotype has been shown to be caused by an elevated touch sensitivity (Esseling et al. 2004). When care is taken during manipulations, root hairs curl upon rhizobial treatment and stop when the tip touches the root hair shank. Therefore, touch-mediated responses might be involved during rhizobial uptake.

The maintenance processes of the symbiosome and bacterial differentiation also appear to be delicate. A large number of bacterial genes are expressed within nodules and either are required for the physiological adaptation to the new environment within the nodule or provide essential functions for bacteroid formation or maintenance (Oke and Long 1999b). In most fix⁻ mutants, rhizobia are liberated from ITs but the symbiosome rapidly degrades. The same occurs with the *SipS* mutant of *B. japonicum* (Müller et al. 1995), and *lpsB* (Campbell et al. 2002) and *bacA* (Ferguson et al. 2004) mutants of *S. meliloti*. Plant mutants with a similar phenotype include *P. sativum*, with a mutation in *sym13* (Kneen et al. 1990), *sym26* (Morzhina et al. 2000), *Rab1* (Cheon et al. 1993) of *G. max*, and *M. truncatula*, with a mutation in *MtSym1* (Bénaben et al. 1995).

Although during symbiosome formation bacteria are taken up by the plant cell in a process that at first sight resembles the uptake of bacteria through phagocytosis or induced internalisation by animal cells, some differences are apparent. During phagocytosis, the animal cell actively surrounds the bacterium with membrane protrusions, a process whereby the actin cytoskeleton plays a pivotal role. In contrast, during rhizobial uptake the bacteria appear to protrude into the plant cell by some kind of external mechanical force that could be linked to the cytoskeleton. The mechanism for bacterial uptake within nodules might partly overlap with that proposed for IT initiation and progression (Sect. 2.3), and depends on localised IT cell wall degradation, accumulation of osmotically active compounds, swelling of the IT droplet, and local plasmolysis of the host cell concomitant with an increase of the plant membrane surface area that is uncoupled from wall synthesis. Interesting in this respect is the high expression level of nodulin 26, a putative aquaporin present in the symbiosome membrane, within nodules of *G. max* (Morrison and Verma 1987; Mitra et al. 2004) and *M. truncatula* (El Yahyaoui et al. 2004). Following this reasoning, symbiosome formation involves active exocytosis to accommodate for massive membrane proliferation (Roth and Stacey 1989) and increase in cell volume rather than a process with similarities to endocytic or phagocytic uptake in animal cells.

Liberation of endosomes and phagosomes in animal cells requires the activity of a number of cytoskeletal proteins of which the GTPase dynamin is of crucial importance (Conner and Schmid 2003). Besides proteins implicated in vesicle transport (Borg et al. 1997; Son et al. 2003; Schiene et al. 2004), only one (*LjRab5*; Borg et al. 1997) orthologue of proteins of the endocytic or phagocytic machinery in animal cells has been identified in nodules so far by large-scale gene expression analysis. Likewise, the properties of a number of genes postulated to play a role in bacterial uptake within nodules, such as *MtLEC4* (Mitra and Long 2004), *MtN12*, *MtENOD11*, *MtENOD12* (Journet et al. 1994, 2001), *MtENOD16*, *MtENOD20* (Greene et al. 1998), *PsRNE1* (Rathbun et al. 2002), and *PsENOD5* (Scheres et al. 1990) do not put forward an endocytic nature of bacterial uptake. Interestingly, several genes mediating exocytosis appear in the nodule-specific expressed sequence tag datasets, such as syntaxin and other *N*-ethylmaleimide-sensitive factor attachment protein receptors (SNAREs) (W. Capoen and M. Holsters, unpublished results). Taken together, the mode of bacterial uptake in nodules probably involves a plant-specific process, related to the walled structure of plant cells, with an important role for exocytosis and localised secretion.

3
Endocytosis in Arbuscular Mycorrhiza

To adopt a land lifestyle approximately four hundred and fifty million years ago, plants established a symbiosis with fungi belonging to the Glomeromycota (Remy et al. 1994; Schüßler et al. 2001) to create an intimate interaction, which is called arbuscular mycorrhiza (AM). This ancient symbiosis that still holds today enables plants to gain access to nutrients, such as phosphorus, copper, and zinc, and provides protection against various biotic and abiotic stresses (Strack et al. 2003). Approximately 80% of land plants engage in this beneficial interaction and, in contrast to the rhizobium–legume interaction, only a low degree of specificity has been found between the fungal and plant partner (Sanders 2003). Nutrients are exchanged mainly in root cortical cells that harbour arbuscules, i.e. extensive ramifications of the fungal hyphae within the plant cell. The fungus is always surrounded by a periarbuscular membrane that is continuous with the host plasma membrane, but that has several special functions (Karandashov and Bucher 2005). Recently, both mycorrhizal and plant phosphate transporters have been identified that might function in the arbuscules (Rausch et al. 2001; Harrison et al. 2002; Karandashov and Bucher 2005). Between the fungal and the plant membrane, an apoplastic compartment is present that contains primary cell wall components, such as pectins, xyloglucans, nonesterified polygalacturonans, and hydroxyproline-rich glycoproteins (Balestrini et al. 1994; Perotto et al. 1994; Bonfante and Perotto 1995).

The arbuscule is not the only stage during which the fungus becomes intracellular. Two types of AM exist, "Arum" and "Paris", named after the first plant species in which they were characterised. Both types start with spore germination, extensive hyphal branching, and appressorium formation. A plant signal is needed to initiate extensive hyphal branching from germinating spores of *Gigaspora* and *Glomus* spp. (Giovannetti et al. 1993; Buee et al. 2000); in return, a fungal factor activates gene expression in the host (Chabaud et al. 2002; Kosuta et al. 2003; Weidmann et al. 2004). The nature of the branching factor has recently been identified (Akiyama et al. 2005). After appressorium formation, a mycelium penetrates into the root. Most of the time, the Paris-type fungi enter the epidermis intracellularly and slowly proceed intracellularly, whereas in the Arum-type fungi, rapid intercellular growth is observed. In reality, each particular interaction has to be carefully analysed and most interactions are mixtures between both types (Strack et al. 2003). Recently, the symbiosis between the model legume *L. japonicus* and *Glomus intraradices* has been studied in detail (Demchenko et al. 2004; Parniske 2004). The fungus enters the epidermis intercellularly and passes intracellularly through the exodermis and the first cortical cells. However, intracellular passage through an epidermal cell prior to exodermis invasion has also been reported (Novero et al. 2002). During intracellular passage, a plant-derived membrane surrounds the fungus. Next, the fungus leaves the outer cortex and spreads between midcortical cells. In the inner cortex, arbuscules are formed, which is the second intracellular stage of the fungus.

Not much is known about the mechanism by which the fungal hyphae are internalised. The microtubule organisation of the plant cell changes when arbuscules are formed, indicating an active role of the cytoskeleton in uptake

(Bonfante et al. 1996; Genre and Bonfante 1997; Blancaflor et al. 2001). Extensive gene sequencing projects led to an array of fungal and plant genes that are upregulated during the successive stages of the interaction (Manthey et al. 2004; Weidmann et al. 2004), but functional analyses are scarce. Through mutant analysis, a number of candidate plant genes have been identified that are involved in mycorrhization, some of which were detected in tomato (*Lycopersicon esculentum*) but most in legumes, such as *M. truncatula* and *L. japonicus* (Marsh and Schultze 2001). Because many of the legume mutants are also defective in nodulation, nodulation must have recruited functions from the much more ancient AM symbiosis (Kistner and Parniske 2002). Detailed microscopic analysis of the mutant phenotypes and discovery of the gene functions might give a hint as to which plant genes are involved in the symbiont uptake.

Three genes have been clearly correlated with the intracellular stages of the fungus, *LjSymRK*, *LjSym4*, and *LjSym15* (Novero et al. 2002; Demchenko et al. 2004). *LjSymRK*, which codes for a receptor-like kinase, is needed for intracellular passage through the exodermis/outer cortex. Inoculation of the mutant resulted in epidermal intercellular passage, but blocked the intracellular route (Demchenko et al. 2004). At this stage, *LjSym4* is needed as well (Novero et al. 2002). By studying weak alleles that allow the first intracellular passage, *LjSym4* was also found to be necessary at the second intracellular stage, during the formation of arbuscules (Novero et al. 2002). For *LjSymRK*, the situation is less clear. No weak alleles were analysed; however, in meristematically arrested roots, arbuscules were formed in the *LjSymRK* but not in the *LjSym15* mutant, indicating that *LjSym15* is essential, whereas *LjSymRK* might be conditional for arbuscule formation (Demchenko et al. 2004). The nature of the genes that are mutated in *LjSym4* and *LjSym15* is unfortunately not known yet.

Symbiososmes, arbuscules, and haustoria have much in common. In all structures, the symbiont is surrounded by plant-derived membranes through which exchange or export of nutrients happens (Parniske 2000).

Recently, the involvement of SNARE-mediated exocytosis in resistance against biotrophic fungi has been evidenced (Collins et al. 2003). An innate defence system might be misused by pathogens or symbionts to allow uptake (Schultze-Lefert 2004) and disequilibrium between exocytosis and endocytosis would engulf the symbiont. In contrast, pure endocytic uptake seems improbable, taking into account the size of the rhizobia, the partial uptake of the fungi, and the turgor pressure of plant cells. Although the interface membranes (peribacteroid/symbiosome, periarbuscular, and extrahaustorial membranes) are derived from and, in the case of the fungi, continuous with the plant plasma membrane, they are specialised and have specific functions, indicating that the plant secretory system can specifically recognise them. The same target membrane SNARE, AtSNAP33, is induced by pathogens and upon mechanical stimulation (Wick et al. 2003). Specific SNARE complexes, which are part of the "cell wall integrity control system" might be involved in the uptake of microbial symbionts (Schulze-Lefert 2004). We speculate that SymRK might function within this system. As previously mentioned, part of the nodulation phenotye of *SymRK* mutants has been attributed to an elevated touch sensitivity that might reflect an imbalance in this cell wall integrity survey system. Moreover, at the stages in which SymRK is involved, such as root hair curling, intracellular passage of AM fungi, and microbial uptake, the local cell wall changes make it very plausible that SymRK plays a crucial role in allowing touch-mediated intracellular penetration of microbes.

4 Conclusions and Future Prospects

Exchange of signals and uptake of extracellular material together with microbes by plant cells are pivotal during plant–microorganism interactions. However, it remains to be demonstrated whether these processes involve secretory endocytosis-like or exocytosis-like mechanisms or both of them.

A first indication for the involvement of endocytosis will come from localisation studies using specific endosomal markers of both animal and plant origin. A number of fluorescent protein-based markers have recently become available and the easy transformation of *M. truncatula* (Boisson-Dernier et al. 2001) makes this approach feasible. Examples of such markers are the FYVE domain of the mouse Hrs protein, which binds to PI(3)P as well as endosomal Rab GTPases RabF2a and ARA6 from *Arabidopsis thaliana* (Voigt et al. 2005). These markers have been shown to be valid endosome markers in plant cells in double-labelling experiments including the fluorescent endocytic tracer FM4-64. By studying the distribution of these markers during nodulation and the mycorrhizal interaction, insight will be gained into the implication of endocytosis during all stages, from the first contact to the formation of nodules and arbuscules.

An important plant component in endocytosis is the cytoskeleton (Šamaj et al. 2004, and in this volume). Detailed information about cytoskeletal structure during nodulation and mycorrhizal interaction has been obtained only from cytochemical approaches on tissue sections so far (Genre and Bonfante 1997; Timmers et al. 1998, 1999; Davidson and Newcomb 2001a, b). In the future, both live cell imaging and electron microscopical methods should be employed. The use of markers based on fluorescent proteins to visualise microtubules, actin filaments and the endoplasmic reticulum has already revealed the importance of these structures during the first stages of the interaction between the endomycorrhizal fungus *Gigaspora gigantea* and *M. truncatula* (A. Genre, M. Chabaud, A. Timmers, P. Bonfante, and D. Barker, unpublished results). Valuable information will be gained from studies using specific inhibitors of cytoskeletal function, such as oryzalin for microtubules, latrunculin for actin filaments, and 2,3-butanedione monoxime for myosin (reviewed by Šamaj et al. 2004). Especially, the first stages of nodulation that take place in surface-exposed root hairs are amenable to such studies. Because the effects of cytoskeletal inhibitors on bacterial uptake within nodules and arbuscle formation during the mycorrhizal interaction will be difficult to analyse, alternative techniques have to be developed, for instance, by using controlled ectopic expression of proteins with a stabilising or depolymerising activity on either microtubules or actin microfilaments. Candidate proteins for such an approach are the microtubule-severing protein katanin and actin-depolymerising factor.

The possible role of plant orthologues of animal proteins implicated in endocytosis and/or phagocytosis should be studied during plant symbiosis. Of notable interest is the GTPase dynamin. This protein is thought to play an important role during bacterial uptake in animal cells and many isoforms of dynamin-like proteins are present in the *Arabidopsis* genome (Verma et al., this volume). The identification of mutant forms with a dominant negative effect on endocytosis would make this protein a valuable tool for the study of the role of endocytosis during plant–microorganism interactions.

In spite of the availability of a number of plant and bacterial mutants defective in bacterial uptake within nodules (Tsyganov et al. 2002), almost none of the genes mutated have been analysed. A notable exception are bacterial mutants affected in the production of surface polysaccharides that provided evidence of similar requirements for prolonged survival within host cells for both the plant symbiont *S. meliloti* and the mammalian pathogen *Brucella abortus* (LeVier et al. 2000). However, other genes have not been identified yet. A thorough exploitation of this valuable mutant collection should provide important insight into the mechanism of microbial uptake into plant cells.

References

Akiyama K, Matsuzaki K-I, Hayashi H (2005) Plant sesquiterpenes induce hyphal branching in arbuscular mycorrhizal fungi. Nature 432:824–827

Ardourel M, Demont N, Debellé F, Maillet F, de Billy F, Promé J-C, Dénarié J, Truchet G (1994) *Rhizobium meliloti* lipooligosaccharide nodulation factors: different structural requirements for bacterial entry into target root hair cells and induction of plant symbiotic developmental responses. Plant Cell 6:1357–1374

Balestrini R, Romera C, Puigdomènech P, Bonfante P (1994) Location of cell-wall hydroxyproline-rich glycoprotein, cellulose and β-1,3-glucans in apical and differentiated regions of maize mycorrhizal roots. Planta 195:201–209

Baluška F, Hlavacka A, Šamaj J, Palme K, Robinson DG, Matoh T, McCurdy DW, Menzel D, Volkmann D (2002) F-actin-dependent endocytosis of cell wall pectins in meristematic root cells. Insights from brefeldin A-induced compartments. Plant Physiol 130:422–431

Batut J, Andersson SGE, O'Callaghan D (2004) The evolution of chronic infection strategies in the α-proteobacteria. Nat Rev Microbiol 2:933–945

Bénaben V, Duc G, Lefebvre V, Huguet T (1995) TE7, an inefficient symbiotic mutant of *Medicago truncatula* Gaertn. cv Jemalong. Plant Physiol 107:53–62

Blancaflor EB, Zhao L, Harrison MJ (2001) Microtubule organization in root cells of *Medicago truncatula* during development of an arbuscular mycorrhizal symbiosis with *Glomus versiforme*. Protoplasma 217:154–165

Boisson-Dernier A, Chabaud M, Garcia F, Bécard G, Rosenberg G, Barker DG (2001) *Agrobacterium rhizogenes*-transformed roots of *Medicago truncatula* for the study of nitrogen-fixing and endomycorrhizal symbiotic associations. Mol Plant-Microbe Interact 14:695–700

Bolaños L, Brewin NJ, Bonilla I (1996) Effects of boron on *Rhizobium*–legume cell-surface interactions and nodule development. Plant Physiol 110:1249–1256

Bonfante P, Perotto S (1995) Strategies of arbuscular mycorrhizal fungi when infecting host plants. New Phytol 130:3–21

Bonfante P, Bergero R, Uribe X, Romera C, Rigau J, Puigdomenech P (1996) Transcriptional activation of a maize α-tubulin gene in mycorrhizal maize and transgenic tobacco plants. Plant J 9:737–743

Bonfante P, Genre A, Faccio A, Martini I, Schauser L, Stougaard J, Webb J, Parniske M (2000) The *Lotus japonicus LjSym4* gene is required for the successful symbiotic infection of root epidermal cells. Mol Plant-Microbe Interact 13:1109–1120

Borg S, Brandstrup B, Jensen TJ, Poulsen C (1997) Identification of new protein species among 33 different small GTP-binding proteins encoded by cDNAs from *Lotus japonicus*, and expression of corresponding mRNAs in developing root nodules. Plant J 11:237–250

Brewin NJ (1998) Tissue and cell invasion by Rhizobium: The structure and development of infection threads and symbiosomes. In: Spaink HP, Kondorosi A, Hooykaas PJJ (eds) The *Rhizobiaceae*. Molecular biology of model plant-associated bacteria. Kluwer, Dordrecht, The Netherlands, pp 417–429

Brewin NJ (2004) Plant cell wall remodelling in the rhizobium–legume symbiosis. Crit Rev Plant Sci 23:293–316

Buee M, Rossignol M, Jauneau A, Ranjeva R, Bécard G (2000) The pre-symbiotic growth of arbuscular mycorrhizal fungi is induced by a branching factor partially purified from plant root exudates. Mol Plant-Microbe Interact 13:693–698

Campbell GRO, Reuhs BL, Walker GC (2002) Chronic intracellular infection of alfalfa nodules by *Sinorhizobium meliloti* requires correct lipopolysaccharide core. Proc Natl Acad Sci USA 99:3938–3943

Capoen W, Goormachtig S, Schroeyers K, Holsters M (2005) SrSymRK, a plant receptor essential for symbiosome formation. Proc Natl Acad Sci USA 102:10369–10374

Catoira R, Timmers ACJ, Maillet F, Galera C, Penmetsa RV, Cook D, Dénarié J, Gough C (2001) The *HCL* gene of *Medicago truncatula* controls *Rhizobium*-induced root hair curling. Development 128:1507–1518

Cebolla A, Vinardell JM, Kiss E, Oláh B, Roudier F, Kondorosi A, Kondorosi E (1999) The mitotic inhibitor *ccs52* is required for endoreduplication and ploidy-dependent cell enlargement in plants. EMBO J 18:4476–4484

Chabaud M, Venard C, Defaux-Petras A, Bécard G, Barker DG (2002) Targeted inoculation of *Medicago truncatula in vitro* root cultures reveals *MtENOD11* expression during early stages of infection by arbuscular mycorrhizal fungi. New Phytol 156:265–273

Cheon C-I, Lee N-G, Siddique A-BM, Bal AK, Verma DPS (1993) Roles of plant homologs of Rab1p and Rab7p in the biogenesis of the peribacteroid membrane, a subcellular compartment formed *de novo* during root nodule symbiosis. EMBO J 12:4125–4135

Collins NC, Thordal-Christensen H, Lipka V, Bau S, Kombrink E, Qiu J-L, Hückelhoven R, Stein M, Freialdenhoven A, Somerville SC, Schulze-Lefert P (2003) SNARE-protein-mediated disease resistance at the plant cell wall. Nature 425:973–977

Conner SD, Schmid SL (2003) Regulated portals of entry into the cell. Nature 422:37–44

D'Haeze W, Holsters M (2002) Nod factor structures, responses, and perception during initiation of nodule development. Glycobiology 12:79R–105R

D'Haeze W, Gao M, De Rycke R, Van Montagu M, Engler G, Holsters M (1998) Roles for azorhizobial Nod factors and surface polysaccharides in intercellular invasion and nodule penetration, respectively. Mol Plant-Microbe Interact 11:999–1008

Davidson AL, Newcomb W (2001) Changes in actin microfilament arrays in developing pea nodule cells. Can J Bot 79:767–776

Davidson AL, Newcomb W (2001) Organization of microtubules in developing pea root nodule cells. Can J Bot 79:777–786

Demchenko K, Winzer T, Stougaard J, Parniske M, Pawlowski K (2004) Distinct roles of *Lotus japonicus SYMRK* and *SYM15* in root colonization and arbuscule formation. New Phytol 163:381–392

Dénarié J, Debellé F, Promé J-C (1996) Rhizobium lipo-chitooligosaccharide nodulation factors: signaling molecules mediating recognition and morphogenesis. Annu Rev Biochem 65:503–535

Dickstein R, Scheirer DC, Fowle WH, Ausubel FM (1991) Nodules elicited by *Rhizobium meliloti* heme mutants are arrested at an early stage of development. Mol Gen Genet 230:423–432

Downie JA, Walker SA (1999) Plant responses to nodulation factors. Curr Opin Plant Biol 2:483–489

El Yahyaoui F, Küster H, Ben Amor B, Hohnjec N, Pühler A, Becker A, Gouzy J, Vernié T, Gough C, Niebel A, Godiard L, Gamas P (2004) Expression profiling in *Medicago truncatula* identifies more than 750 genes differentially expressed during nodulation, including many potential regulators of the symbiotic program. Plant Physiol 136:3159–3176

Esseling JJ, Lhuissier FGP, Emons AMC (2003) Nod factor-induced root hair curling: continuous polar growth towards the point of Nod factor application. Plant Physiol 132:1982–1988

Esseling JJ, Lhuissier FGP, Emons AMC (2004) A nonsymbiotic root hair tip growth phenotype in *NORK*-mutated legumes: implications for nodulation factor-induced signaling and formation of a multifaceted root hair pocket for bacteria. Plant Cell 16:933–944

Ferguson GP, Datta A, Baumgartner J, Roop RM II, Carlson RW, Walker GC (2004) Similarity to peroxisomal-membrane protein family reveals that *Sinorhizobium* and *Brucella* BacA affect lipid-A fatty acids. Proc Natl Acad Sci USA 101:5012–5017

Fraysse N, Couderc F, Poinsot V (2003) Surface polysaccharide involvement in establishing the rhizobium–legume symbiosis. Eur J Biochem 270:1365–1380

Gage DJ (2004) Infection and invasion of roots by symbiotic, nitrogen-fixing rhizobia during nodulation of temperate legumes. Microbiol Mol Biol Rev 68:280–300

Genre A, Bonfante P (1997) A mycorrhizal fungus changes microtubule orientation in tobacco root cells. Protoplasma 199:30–38

Giovannetti M, Sbrana C, Citernesi AS, Logi C (1993) Differential hyphal morphogenesis in arbuscular mycorrhizal fungi during pre-infection stages. New Phytol 125:587–593

Goethals K, Gao M, Tomekpe K, Van Montagu M, Holsters M (1989) Common *nodABC* genes in *Nod* locus 1 of *Azorhizobium caulinodans*: nucleotide sequence and plant-inducible expression. Mol Gen Genet 219:289–298

Goedhart J, Hink MA, Visser AJWG, Bisseling T, Gadella TWJ Jr (2000) *In vivo* fluorescence correlation microscopy (FCM) reveals accumulation and immobilization of Nod factors in root hair cell walls. Plant J 21:109–119

Goormachtig S, Capoen W, James EK, Holsters M (2004) Switch from intracellular to intercellular invasion during water stress-tolerant legume nodulation. Proc Natl Acad Sci USA 101:6303–6308

Greene EA, Erard M, Dedieu A, Barker DG (1998) MtENOD16 and 20 are members of a family of phytocyanin-related early nodulins. Plant Mol Biol 36:775–783

Gross A, Kapp D, Nielsen T, Niehaus K (2005) Endocytosis of *Xanthomonas campestris* pathovar *campestris* lipopolysaccharides in non-host plant cells of *Nicotiana tabacum*. New Phytol 165:215–226

Harrison MJ, Dewbre GR, Liu J (2002) A phosphate transporter from *Medicago truncatula* involved in the acquisition of phosphate released by arbuscular mycorrhizal fungi. Plant Cell 14:2413–2429

Hentschel U, Steinert M, Hacker J (2000) Common molecular mechanisms of symbiosis and pathogenesis. Trends Microbiol 8:226–230

Hirsch AM, LaRue TA (1997) Is the legume nodule a modified root or stem or an organ *sui generis*? Crit Rev Plant Sci 16:361–392

Hubbell DH (1981) Legume infection by *Rhizobium*: a conceptual approach. BioScience 31:832–837

Imaizumi-Anraku H, Kawaguchi M, Koiwa H, Akao S, Syōno K (1997) Two ineffective-nodulating mutants of *Lotus japonicus*—different phenotypes caused by the blockage of endocytic bacterial release a•l nodule maturation. Plant Cell Physiol 38:871–881

Journet EP, Pichon M, Dedieu A, de Billy F, Truchet G, Barker DG (1994) *Rhizobium meliloti* Nod factors elicit cell-specific transcription of the *ENOD12* gene in transgenic alfalfa. Plant J 6:241–249

Journet E-P, El-Gachtouli N, Vernoud V, de Billy F, Pichon M, Dedieu A, Arnould C, Morandi D, Barker DG, Gianinazzi-Pearson V (2001) *Medicago truncatula ENOD11*: a novel RPRP-encoding early nodulin gene expressed during mycorrhization in arbuscule-containing cells. Mol Plant-Microbe Interact 14:737–748

Karandashov V, Bucher M (2005) Symbiotic phosphate transport in arbuscular mycorrhizas. Trends Plant Sci 10:22–29

Kijne JW (1992) The Rhizobium infection process. In: Stacey G, Burris RH, Evans HJ (eds) Biological nitrogen fixation. Chapman & Hall, New York, pp 349–398

Kistner C, Parniske M (2002) Evolution of signal transduction in intracellular symbiosis. Trends Plant Sci 7:511–518

Kneen BE, LaRue TA, Hirsch AM, Smith CA, Weeden NF (1990) *sym13*—a gene conditioning ineffective nodulation in *Pisum sativum*. Plant Physiol 94:899–905

Kosuta S, Chabaud M, Lougnon G, Gough C, Dénarié J, Barker DG, Bécard G (2003) A diffusible factor from arbuscular mycorrhizal fungi induces symbiosis-specific *MtENOD11* expression in roots of *Medicago truncatula*. Plant Physiol 131:952–962

Król J, Wielbo J, Mazur A, Kopcińska J, Łotocka B, Golinowski W, Skorupska A (1998) Molecular characterization of *pssCDE* genes of *Rhizobium leguminosarum* bv. *trifolii* strain *TA1: pssD* mutant is affected in exopolysaccharide synthesis and endocytosis of bacteria. Mol Plant-Microbe Interact 11:1142–1148

LeVier K, Phillips RW, Grippe VK, Roop RM II, Walker GC (2000) Similar requirements of a plant symbiont and a mammalian pathogen for prolonged intracellular survival. Science 287:2492–2493

Limpens E, Franken C, Smit P, Willemse J, Bisseling T, Geurts R (2003) LysM domain receptor kinases regulating rhizobial Nod factor-induced infection. Science 302:630–633

Limpens E, Mirabella R, Federova E, Franken C, Franssen H, Bisseling T, Geurts R (2005) Formation of organelle-like N_2-fixing symbiosomes in legume root nodules is controlled by *DIM2*. Proc Natl Acad Sci USA 102:10375–10380

Manthey K, Krajinski F, Hohnjec N, Firnhaber C, Pühler A, Perlick AM, Küster H (2004) Transcriptome profiling in root nodules and arbuscular mycorrhiza identifies a collection of novel genes induced during *Medicago truncatula* root endosymbioses. Mol Plant-Microbe Interact 17:1063–1077

Marsh JF, Schultze M (2001) Analysis of arbuscular mycorrhizas using symbiosis-defective plant mutants. New Phytol 150:525–532

Mathis R, Van Gijsegem F, De Rycke R, D'Haeze W, Van Maelsaeke E, Anthonio E, Van Montagu M, Holsters M, Vereecke D (2005) Lipopolysaccharides as a communication signal for progression of legume endosymbiosis. Proc Natl Acad Sci USA 102:2655–2660

Miaczynska M, Pelkmans L, Zerial M (2004) Not just a sink: endosomes in control of signal transduction. Curr Opin Cell Biol 16:400–406

Miklashevichs E, Röhrig H, Schell J, Schmidt J (2001) Perception and signal transduction of rhizobial NOD factors. Crit Rev Plant Sci 20:373–394

Mithöfer A (2002) Suppression of plant defence in rhizobia-legume symbiosis. Trends Plant Sci 7:440–444

Mitra RM, Long SR (2004) Plant and bacterial symbiotic mutants define three transcriptionally distinct stages in the development of the *Medicago truncatula*/*Sinorhizobium meliloti* symbiosis. Plant Physiol 134:595–604

Mitra RM, Shaw SL, Long SR (2004) Six nonnodulating plant mutants defective for Nod factor-induced transcriptional changes associated with the legume-rhizobia symbiosis. Proc Natl Acad Sci USA 101:10217–10222

Morrison N, Verma DPS (1987) A block in the endocytosis of *Rhizobium* allows cellular differentiation in nodules but affects the expression of some peribacteroid membrane nodulins. Plant Mol Biol 9:185–196

Morzhina EV, Tsyganov VE, Borisov AY, Lebsky VK, Tikhonovich IA (2000) Four developmental stages identified by genetic dissection of pea (*Pisum sativum L.*) root nodule morphogenesis. Plant Sci 155:75–83

Müller P, Ahrens K, Keller T, Klaucke A (1995) A Tn*phoA* insertion with the *Bradyrhizobium japonicum sipS* gene, homologous to prokaryotic signal peptidases, results in extensive changes in the expression of PBM-specific nodulins of infected soybean (*Glycine max*) cells. Mol Microbiol 18:831–840

Niehaus K, Lagares A, Pühler A (1998) A *Sinorhizobium meliloti* lipopolysaccharide mutant induces effective nodules on the host plant *Medicago sativa* (alfalfa) but fails to establish a symbiosis with *Medicago truncatula*. Mol Plant-Microbe Interact 11:906–914

Novero M, Faccio A, Genre A, Stougaard J, Webb KJ, Mulder L, Parniske M, Bonfante P (2002) Dual requirement of the *LjSym4* gene for mycorrhizal development in epidermal and cortical cells of *Lotus japonicus*. New Phytol 154:741–749

Oke V, Long SR (1999) Bacteroid formation in the *Rhizobium*-legume symbiosis. Curr Opin Microbiol 2:641–646

Oke V, Long SR (1999) Bacterial genes induced within the nodule during the *Rhizobium*-legume symbiosis. Mol Microbiol 32:837–849

Oldroyd GED, Downie JA (2004) Calcium, kinases and nodulation signalling in legumes. Nat Rev Mol Cell Biol 5:566–576

Parniske M (2000) Intracellular accommodation of microbes by plants: a common developmental program for symbiosis and disease? Curr Opin Plant Biol 3:320–328

Parniske M (2004) Molecular genetics of the arbuscular mycorrhizal symbiosis. Curr Opin Plant Biol 7:414–421

Perotto S, Brewin NJ, Bonfante P (1994) Colonization of pea roots by the mycorrhizal fungus *Glomus versiforme* and by *Rhizobium* bacteria: immunological comparison using monoclonal antibodies as probes for plant cell surface components. Mol Plant-Microbe Interact 7:91–98

Philip-Hollingsworth S, Dazzo FB, Hollingsworth RI (1997) Structural requirements of *Rhizobium* chitolipooligosaccharides for uptake and bioactivity in legume roots as revealed by synthetic analogs and fluorescent probes. J Lipid Res 38:1229–1241

Rathbun EA, Naldrett MJ, Brewin NJ (2002) Identification of a family of extensin-like glycoproteins in the lumen of *Rhizobium*-induced infection threads in pea root nodules. Mol Plant-Microbe Interact 15:350–359

Rausch C, Daram P, Brunner S, Jansa J, Laloi M, Leggewie G, Amrhein N, Bucher M (2001) A phosphate transporter expressed in arbuscule-containing cells in potato. Nature 414:462–466

Relić B, Perret X, Estrada-García MT, Kopcinska J, Golinowski W, Krishnan HB, Pueppke SG, Broughton WJ (1994) Nod factors of *Rhizobium* are a key to the legume door. Mol Microbiol 13:171–178

Remy W, Taylor TN, Hass H, Kerp H (1994) Four hundred-million-year-old vesicular arbuscular mycorrhizae. Proc Natl Acad Sci USA 91:11841–11843

Riely BK, Ané J-M, Penmetsa RV, Cook DR (2004) Genetic and genomic analysis in model legumes bring Nod-factor signaling to center stage. Curr Opin Plant Biol 7:408–413

Romanenko AS, Rifel' AA, Salyaev RK (2002) Endocytosis of exopolysaccharides of the potato ring rot causal agent by host-plant cells. Dokl Biol Sci 386:451–453

Roth LE, Stacey G (1989) Bacterium release into host cells of nitrogen-fixing soybean nodules: the symbiosome membrane comes from three sources. Eur J Cell Biol 49:13–23

Šamaj J, Baluška F, Voigt B, Schlicht M, Volkmann D, Menzel D (2004) Endocytosis, actin cytoskeleton and signaling. Plant Physiol 135:1150–1161

Sanders IR (2003) Preference, specificity and cheating in the arbuscular mycorrhizal symbiosis. Trends Plant Sci 8:143–145

Scheres B, van Engelen F, van der Knaap E, van de Wiel C, van Kammen A, Bisseling T (1990) Sequential induction of nodulin gene expression in the developing pea nodule. Plant Cell 2:687–700

Schiene K, Donath S, Brecht M, Pühler A, Niehaus K (2004) A Rab-related small GTP binding protein is predominantly expressed in root nodules of *Medicago sativa*. Mol Gen Genomics 272:57–66

Schultze M, Kondorosi A (1998) Regulation of symbiotic root nodule development. Annu Rev Genet 32:33–57

Schulze-Lefert P (2004) Knocking on the heaven's wall: pathogenesis of and resistance to biotrophic fungi at the cell wall. Curr Opin Plant Biol 7:377–383

Schüßler A, Schwarzott D, Walker C (2001) A new fungal phylum, the *Glomeromycota*: phylogeny and evolution. Mycol Res 105:1413–1421

Son O, Yang H-S, Lee H-J, Lee M-Y, Shin K-H, Jeon S-L, Lee M-S, Choi S-Y, Chun J-Y, Kim H, An C-S, Hong S-K, Kim N-S, Koh S-K, Cho MJ, Kim S, Verma DPS, Cheon C-I (2003) Expression of *srab7* and *SCaM* genes required for endocytosis of *Rhizobium* in root nodules. Plant Sci 165:1239-1244

Sprent JI (2002) Nodulation in legumes. Royal Botanical Gardens, Kew

Strack D, Fester T, Hause B, Schliemann W, Walter MH (2003) Arbuscular mycorrhiza: biological, chemical, and molecular aspects. J Chem Ecol 29:1955–1979

Stracke S, Kistner C, Yoshida S, Mulder L, Sato S, Kaneko T, Tabata S, Sandal N, Stougaard J, Szczyglowski K, Parniske M (2002) A plant receptor-like kinase required for both bacterial and fungal symbiosis. Nature 417:959–962

Tansengco ML, Hayashi M, Kawaguchi M, Imaizumi-Anraku H, Murooka Y (2003) *crinkle*, a novel symbiotic mutant that affects the infection thread growth and alters the root hair trichome and seed development in *Lotus japonicus*. Plant Physiol 131:1054–1063

Timmers ACJ, Auriac M-C, de Billy F, Truchet G (1998) Nod factor internalization and microtubular cytoskeleton changes occur concomitantly during nodule differentiation in alfalfa. Development 125:339–349

Timmers ACJ, Auriac M-C, Truchet G (1999) Refined analysis of early symbiotic steps of the *Rhizobium-Medicago* interaction in relationship with microtubular cytoskeleton rearrangements. Development 126:3617–3628

Tsyganov VE, Borisov AY, Rozov SM, Tikhonovich IA (1994) New symbiotic mutants of pea obtained after mutagenesis of laboratory line SGE. Pisum Genet 26:36–37

Tsyganov VE, Morzhina EV, Stefanov SY, Borisov AY, Lebsky VK, Tikhonovich IA (1998) The pea (*Pisum sativum L.*) genes *sym33* and *sym40* control infection thread formation and root nodule function. Mol Gen Genet 259:491–503

Tsyganov VE, Voroshilova VA, Priefer UB, Borisov AY, Tikhonovich IA (2002) Genetic dissection of the initiation of the infection process and nodule tissue development in the *Rhizobium*-pea (*Pisum sativum L.*) symbiosis. Ann Bot 89:357–366

van Spronsen PC, Bakhuizen R, van Brussel AAN, Kijne JW (1994) Cell wall degradation during infection thread formation by the root nodule bacterium *Rhizobium leguminosarum* is a two-step process. Eur J Cell Biol 64:88–94

Vasse J, de Billy F, Camut S, Truchet G (1990) Correlation between ultrastructural differentiation of bacteroids and nitrogen fixation in alfalfa nodules. J Bacteriol 172:4295–4306

Veereshlingam H, Haynes JG, Penmetsa RV, Cook DR, Sherrier DJ, Dickstein R (2004) *nip*, a symbiotic mutant that forms root nodules with aberrant infection threads and plant defense-like response. Plant Physiol 136:3692–3702

Vitousek PM, Aber JD, Howarth RW, Likens GE, Matson PA, Schindler DW, Schlesinger WH, Tilman DG (1997) Human alteration of the global nitrogen cycle: sources and consequences. Ecol Appl 7:737–750

Voigt B, Timmers ACJ, Šamaj J, Hlavacka A, Ueda T, Preuss M, Nielsen E, Mathur J, Emans N, Stenmark H, Nakano A, Baluška F, Menzel D (2005) Actin-based motility of endosomes is linked to the polar tip growth of root hairs. Eur J Cell Biol 84:609-621

Walker SA, Downie JA (2000) Entry of *Rhizobium leguminosarum* bv. *viciae* into root hairs requires minimal Nod factor specificity, but subsequent infection thread growth requires *nodO* or *nodE*. Mol Plant-Microbe Interact 13:754–762

Weidmann S, Sanchez L, Descombin J, Chatagnier O, Gianinazzi S, Gianinazzi-Pearson V (2004) Fungal elicitation of signal transduction-related plant genes precedes mycorrhiza establishment and requires the *dmi3* gene in *Medicago truncatula*. Mol Plant-Microbe Interact 17:1385–1393

Whitehead LF, Day DA (1997) The peribacteroid membrane. Physiol Plant 100:30–44

Wick P, Gansel X, Oulevey C, Page V, Studer I, Dürst M, Sticher L (2003) The expression of the t-SNARE AtSNAP33 is induced by pathogens and mechanical stimulation. Plant Physiol 132:343–351

Wisniewski J-P, Rathbun EA, Knox JP, Brewin NJ (2000) Involvement of diamine oxidase and peroxidase in insolubilization of the extracellular matrix: implications for pea nodule initiation by *Rhizobium leguminosarum*. Mol Plant-Microbe Interact 13:413–420

Yang W-C, de Blank C, Meskiene I, Hirt H, Bakker J, van Kammen A, Franssen H, Bisseling T (1994) *Rhizobium* Nod factors reactivate the cell cycle during infection and nodule primordium formation, but the cycle is only completed in primordium formation. Plant Cell 6:1415–1426

Yu Q, Hlavacka A, Matoh T, Volkmann D, Menzel D, Goldbach HE, Baluška F (2002) Short-term boron deprivation inhibits endocytosis of cell wall pectins in meristematic cells of maize and wheat root apices. Plant Physiol 130:415–421

Plant Cell Monogr (1)
J. Šamaj · F. Baluška · D. Menzel: Plant Endocytosis
DOI 10.1007/7089_016/Published online: 10 September 2005

Endocytosis in Guard Cells

Ulrike Homann

Institute of Botany, Darmstadt University of Technology, 64287 Darmstadt, Germany
homann-u@bio.tu-darmstadt.de

Abstract Stomatal movement requires large and repetitive changes to cell volume and consequently surface area. These alterations in surface area are accomplished by addition and removal of plasma membrane material. Recent studies of membrane turnover in guard cell protoplasts using electrophysiology and fluorescence imaging techniques implicate that exocytosis and endocytosis are sensitive to changes in membrane tension. This may provide a regulatory mechanism for the adaptation of surface area to osmotically driven changes in cell volume in guard cell protoplasts as well as turgid guard cells. In addition guard cells also exhibit constitutive membrane turnover. Constitutive and tension-driven membrane turnover were found to be associated with addition and removal of K^+ channels. This implies that some of the exocytosis and endocytic vesicles carry K^+ channels. Together the results demonstrate that exocytosis and endocytosis are essential for stomatal movement and thus gas exchange in plants.

1
Introduction

Guard cells mediate opening and closing of the stomatal pores which regulate gas exchange in plants. Accumulation of K^+ salts and subsequent water influx lead to swelling of guard cells and opening of the stomatal pore. The reverse process closes the pore. Thus, during stomatal movement, guard cells undergo large osmotically driven changes in cell volume and consequently surface area over a period of minutes. These large changes in surface area of up to 40% (Raschke 1979) cannot result from stretching of the existing membrane as the maximum possible stretching of membranes is limited to about 2% (Wolfe et al. 1986). In addition, the large turgor pressure of up to 5 MPa (Franks et al. 1998) prevents the guard cell plasma membrane from maintaining infoldings that could provide excess surface area. Therefore, alterations in surface area must be accomplished by addition and removal of membrane material to and from the plasma membrane, respectively. However, the mechanisms underlying these osmotically induced changes in surface area are largely unknown. Recently, the application of new cell biology techniques, namely patch-clamp capacitance measurements and microscopical imaging of membranes stained with fluorescent styryl dyes, has led to a more detailed understanding of the processes occurring during opening and closing of the stomatal pore. Results from these studies are summarised here.

2 Osmotically induced Exocytosis and Endocytosis in Guard Cell Protoplasts

2.1 Investigation of Exocytosis and Endocytosis by Patch-clamp Capacitance Measurements

Osmotically driven and pressure-driven changes in surface area of guard cell protoplasts have been investigated extensively by patch-clamp capacitance measurements. This technique allows the examination of exocytosis and endocytosis in single living protoplasts. Recordings can be performed with a resolution that is high enough to detect fusion and fission of vesicles with a diameter as low as 60 nm and a temporal resolution on the order of some 10 ms (Neher and Marty 1982; Kreft and Zorec 1997). Measurements of exocytosis and endocytosis via capacitance recordings are based on the fact that a biological membrane can be viewed as a capacitor. The capacitance of this capacitor depends on its surface area. For a number of plant protoplasts, including guard cells, such a linear relationship between membrane capacitance and surface area has been demonstrated, and a specific capacitance (capacitance per unit surface area of membrane) between 7.5 and 8.1 mF m^{-2} has been calculated (Zorec and Tester 1992; Thiel et al. 1994; Carroll et al. 1998; Homann 1998). Under the valid assumption that the specific capacitance remains constant during the time of observation, the changes in plasma membrane surface area resulting from exocytic and endocytic activity can be monitored by measuring membrane capacitance. In principle, the membrane capacitance of a cell is determined from the current measured in response to a voltage command which is applied to the cell (Gillis 1995; Homann and Tester 1998; Thiel et al. 2001). When measurements are carried out in the so-called whole-cell configuration, the cytoplasm is rapidly dialysed by the pipette solution. This allows control of cytoplasmic composition and the introduction of potential regulators of exocytosis and endocytosis.

The high temporal resolution and the potential of manipulating the cytosolic composition via the patch pipette make patch-clamp capacitance measurements a powerful tool for studying exocytosis and endocytosis. The main limitations of patch-clamp capacitance measurements are the general limitations of patch-clamp measurements: the requirement of an accessible membrane (measurements are generally carried out on protoplasts) and possible loss of endogenous substances that affect exocytosis and endocytosis during cell dialysis.

2.2
Membrane Tension as a Stimulus for Exocytosis and Endocytosis in Guard Cell Protoplasts

Patch-clamp capacitance measurements have been used to study osmotically induced surface area changes of guard cell protoplasts. Results from these measurements demonstrated that osmotically induced swelling and shrinking of guard cell protoplasts are associated with incorporation and removal of membrane material into and out of the plasma membrane, respectively (Homann 1998). High-resolution capacitance measurements which allow the detection of single exocytic and endocytic events revealed fusion and fission of single vesicles with a median diameter of 300 nm during osmotically induced changes in surface area (Homann and Thiel 1999). However, for most of the recordings the change in surface area occurred without resolvable exocytic or endocytic events. This was most likely due to fusion and fission of vesicles below the resolution limit (less than 200 nm). Hence, vesicles with a diameter below 300 nm almost certainly also contribute significantly to the increase and reduction of the surface area. The vesicular retrieval of plasma membrane material during osmotically induced shrinking of guard cell protoplasts was confirmed by imaging of guard cell protoplasts stained with the fluorescent membrane probe FM1-43 (Kubitscheck et al. 2000). Confocal images of protoplasts incubated with FM1-43 at constant ambient osmotic pressure revealed a slow internalisation of FM1-43 labelled membrane into the cytoplasm without changes in cell perimeter. This indicated the occurrence of constitutive endocytosis (Šamaj, this volume). Hyperosmotic treatment of protoplasts led to a rapid internalisation of FM1-43 fluorescence into the cytoplasm and a corresponding decrease in cell perimeter. Only occasionally was the shrinking of protoplasts associated with the internalisation of large vesicles (median diameter 2.7 μm). Most hyperosmotically treated protoplasts showed a diffuse distribution of the FM1-43 label throughout the cytoplasm without any resolvable vesicular structures. This led to the conclusion that endocytosis of small vesicles below the resolution limit accommodates for the osmotically induced decrease in surface area (Kubitscheck et al. 2000).

Osmotically induced fusion and fission of plasma membrane material was not affected by changes in intracellular or extracellular Ca^{2+} concentration (Homann 1998); however, the rate of change in surface area was dependent on the size of the difference in the osmotic potential applied. The larger the osmotic difference the faster the change in surface area (Homann 1998). This strongly indicated that changes in membrane tension resulting from osmotically induced water influx or efflux can modulate exocytosis and endocytosis in guard cell protoplasts. Exocytic and endocytic activity of guard cell protoplasts could also be modulated by application of hydrostatic pressure, confirming the hypothesis of tension-sensitive surface area regulation (Fig. 1; Bick et al. 2001).

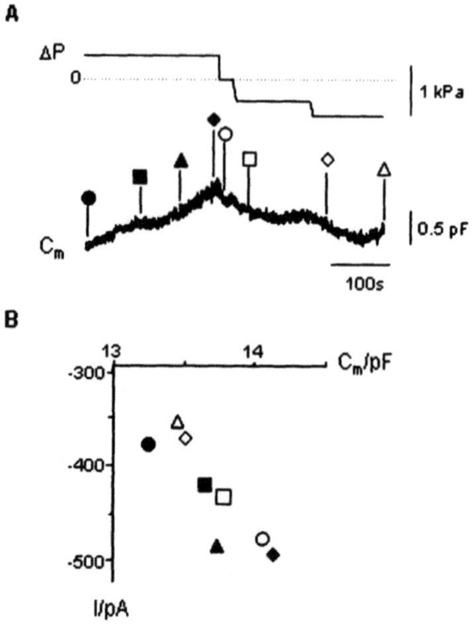

Fig. 1 Insertion and retrieval of K^+ channels during pressure-stimulated exocytosis and endocytosis in a guard cell protoplast **A** Application of a hydrostatic pressure (ΔP) via the patch-pipette resulted in an increase and a decrease in membrane capacitance (C_m), which corresponds to a change in surface area via exocytosis and endocytosis. At the time points indicated by different *symbols* a voltage pulse of – 140 mV was applied and the resulting current passing through the K^+ inward rectifier channel was measured. **B** Linear correlation between changes in the current passing through the K^+ inward rectifier channel (I) and the membrane capacitance (C_m). The current was measured at the time points indicated in **A**

Tension-sensitive exocytosis and endocytosis have been implicated to be important components of surface area regulation not only in plant cells but also in animal cells (Morris and Homann 2001). Cells seem to detect and respond to deviations from a membrane tension set point. An increase in membrane tension above this set point results in addition of membrane material to the plasma membrane until the membrane tension set point is restored. Upon a decrease in membrane tension, excess plasma membrane material is retrieved to reestablish the resting tension. The mechanisms by which cells sense changes in membrane tension are not yet known. Neither have the signal transduction pathways been identified which link changes in membrane tension to changes in the rate of exocytosis or endocytosis.

Other important but yet unresolved questions in tension-modulated surface area changes are the origin and quality of the membrane material which is added and removed in the course of this process. In guard cells the addition of membrane material could often be detected immediately after application of hydrostatic pressure (Fig. 1 in Bick et al. 2001). This indicates the existence of an intracellular reservoir of membrane material which is instantly available for incorporation into the plasma membrane. Guard cell protoplasts can undergo several cycles of swelling and shrinking. It is therefore most likely that the membrane material that is retrieved from the plasma membrane during surface area decrease is reused in subsequent cell swelling.

2.3 Insertion and Retrieval of Plasma Membrane K^+ Channels

Two types of plasma membrane K^+ channels play a central role in the accumulation and loss of K^+ during opening and closing of the stomatal pore. One, a K^+ inward rectifier, conducts K^+ uptake, the other, a K^+ outward rectifier, mediates K^+ discharge from guard cells. The density of these channels in the plasma membrane is an important factor for determining the transport rate across the membrane and thus for the cell function. The channel density can be modulated by exocytotic insertion and endocytic retrieval of ion channels, which in turn alters the membrane conductance.

Parallel measurements of membrane conductance and membrane capacitance provide a valuable tool to study the insertion and retrieval of ion channels. While the membrane capacitance is proportional to the cell surface area and monitors excursions in the plasma membrane area the membrane conductance provides information on the activity of plasma membrane ion channels.

When guard cell protoplasts are subject to swelling the membrane capacitance and the current passing through the K^+ inward rectifier increase nearly in parallel (Fig. 1). A decrease in surface area is accompanied by a corresponding reduction in K^+ inward current (Fig. 1). This implies that the vesicular membrane which is inserted and retrieved during pressure-driven changes in surface area carries active K^+ channels. Detailed measurements of membrane capacitance and conductance in guard cell protoplasts demonstrated that osmotically driven and pressure-driven changes in surface area of guard cell protoplasts are associated with insertion and removal of K^+ inward and K^+ outward rectifiers (Homann and Thiel 2002; Hurst et al. 2004).

From the parallel measurements of changes in conductance and capacitance the number of K^+ channels added for a given increase in surface area can be estimated. This led to the conclusion that only about one of nine vesicles, which fuse with the plasma membrane, contains a K^+ channel (Homann and Thiel 2002). Similar results were obtained for endocytic vesicles.

The observation that the increase in surface area and the incorporation of K^+ channels occurs immediately (i.e. within seconds) upon pressure stimulation (Fig. 1; Homann and Thiel 2002) means that the vesicles containing the active K^+ channels are already present in the cell, probably in a pool close to the plasma membrane. Incorporation and retrieval of K^+ channels can be observed even after several cycles of swelling and shrinking. It therefore seems likely that endocytosed K^+ channels are retrieved back into the pool of "ready-to-fuse" vesicles.

During swelling of guard cell protoplasts an increase in plasma membrane channel density occurred (Homann and Thiel 2002). This increase was reversed during subsequent shrinking of protoplasts (Hurst et al. 2004). The results implied that the density of the K^+ channels in the membrane of exocytic and endocytic vesicles is higher (by a factor of about 10) compared with the channel density of the plasma membrane. (Homann and Thiel 2002; Hurst et al. 2004). Since it is unlikely that channels are first concentrated in small areas before they are retrieved by endocytosis, this strongly suggests that channels form stable clusters in the plasma membrane and remain in clusters during exocytosis and endocytosis. The formation of channel clusters is also supported by analysis of fluorescence images of turgid guard cells expressing the K^+ inward rectifier KAT1 fused to green fluorescent protein (GFP) (see later; Meckel et al. 2004).

The available data also have some implications for the physiology of guard cell movement. Modulators of ion channel activity in guard cells typically have opposite effects on the K^+ outward and K^+ inward rectifiers owing to their different role in the regulation of the stomatal pore. They promote either uptake (K^+ inward rectifier) or release (K^+ outward rectifier) of K^+, resulting in opening or closing of the pore, respectively (Blatt 2000). In the case of vesicle-mediated insertion or retrieval of K^+ channels, however, the inward and outward rectifiers change in parallel. This implies that this process has nothing to do with the physiological regulation of K^+ channel activity in the context of guard cell function. The observation that the channel density changes during swelling and shrinking of guard cell protoplasts suggests that osmotically induced insertion and retrieval of ion channels act as a mechanism for a reversible increase of the ion channel density without discriminating between K^+ inward and outward rectifiers. However, in the physiological situation of stomatal movement the process of vesicle delivery and retrieval to and from the plasma membrane may be more complex, involving distinct mechanisms of vesicle sorting. In this context it is feasible to speculate that a more balanced insertion of K^+ channels into the plasma membrane may occur. This can then serve as a mechanism for homeostasis of channel density during variations in surface area.

3
Endocytosis in Turgid Guard Cells

Endocytosis in plants has long been questioned because of the turgor pressure which acts against invaginations of the plasma membrane. This is especially true for guard cells where the turgor pressure in open stomata of *Vicia faba* can increase to up to 5 MPa (Franks et al. 1998). However, recent investigations of living turgid guard cells clearly demonstrated that endocytosis in guard cells occurs constitutively and in the process of stomatal closure.

3.1 Osmotically Induced Endocytosis

Osmotically driven changes in volume and surface area are the underlying mechanism of opening and closing of the stomatal pore. Osmotically driven exocytosis and endocytosis should therefore not only occur in guard cell protoplasts but also in turgid guard cells. However, previous attempts to demonstrate internalisation of plasma membrane material during closing of stomata have failed. Using the membrane impermeable fluorescent dye Lucifer Yellow (LY), Diekmann et al. (1993) found an uptake of the dye in hyperosmotically treated guard cell protoplasts. In contrast, they could not detect any internalisation of LY in turgid guard cells even after plasmolysis of the cells (Diekmann et al. 1993). The failure to visualise internalisation of membrane material may be due to insufficient incorporation of the dye into endocytic vesicles rather than lack of endocyosis. Considering that the diameter of endocytic vesicles is supposed to be around 70–90 nm (Barth and Holstein 2004; Holstein 2002, and in this volume) the amount of dye endocytosed may not be detected with the fluorescence microscope. Recent investigations of endocytosis in turgid guard cells using FM dyes demonstrated that a decrease in surface area under hyperosmotic conditions correlated with the internalisation of the membrane marker FM4-64 (Shope et al. 2003). The internalisation of the dye was fully reversible upon hypoosmotic treatment; however, endocytic vesicles could not be identified.

In contrast to these studies, Meckel et al. (2005) identified FM4-64 stained structures of variable size ranging from greater than 1 μm down to 270 nm or less (diffraction-limited) in the cytoplasm of hyperosmotically treated turgid guard cells. The authors suggest that the size of structures internalised upon hyperosmotic treatment is correlated with the osmotic conditions (Meckel et al. 2005). In vivo, the osmotic potential of guard cells increases gradually over several minutes via accumulation of K^+ salts. Such small and gradual changes in osmotic potential difference are most likely accomplished via retrieval of small vesicles. Small vesicles, which are below the detection limit of previous studies using the fluid-phase marker LY (Diekmann et al. 1993) or styryl dyes (Shope et al. 2003), are therefore most likely responsible for plasma membrane retrieval during stomatal closure.

Tension-sensitive exocytosis and endocytosis may represent a regulatory mechanism underlying the reversible internalisation of plasma membrane material in both guard cell protoplasts and turgid guard cells. Even though the cell wall prevents a large expansion or even rupturing of the plasma membrane the rather flexible wall of guard cells (Wilmer and Fricker 1996) may still allow for stretching of the plasma membrane. During opening of the stomatal pore, both osmotically driven water influx and subsequent increase in cell volume may result in sufficient stretching of the plasma membrane in order to stimulate tension-driven exocytosis. In the process of stomatal closure, loss of water and a decrease in cell volume would be accompanied by a decrease in membrane tension, which would stimulate endocytosis.

3.2 Constitutive Endocytosis

The plasma membrane of cells is subject to constant turnover via constitutive exocytosis and endocytosis. Recent investigations in turgid guard cells demonstrated that endocytosis in these cells not only occurs during closing of the stomatal pore but also occurs in the absence of changes in cell volume. Using the fluorescent membrane marker FM4-64 Meckel et al. (2004) demonstrated that the dye was taken up into small structures of diffraction-limited size within the cortical cytoplasm. The size distribution of these structures was very similar to that of endocytic vesicles obtained from patch-clamp capacitance recordings. They were therefore identified as endocytic vesicles. Investigation of guard cells transfected with the K^+ inward rectifier KAT1 fused to GFP (KAT1::GFP) revealed that FM-stained endocytic vesicles were frequently colabelled with KAT1::GFP. This implied that turgid guard cells undergo constitutive endocytosis and retrieve membrane material including the K^+ channel KAT1 via endocytic vesicles. Besides the endocytic vesicles, the FM dyes also specifically labelled structures of intermediate sizes (approximately 470 nm). Circumstantial evidence suggested that they are prevacuolar or endosomal-like compartments and have therefore been referred to as endosomes (Meckel et al. 2004).

Analysis of another plasma membrane protein, namely a fusion of GFP with the 23 amino acid long transmembrane domain of the human lysosomal protein LAMP1 (GFP::TM23), revealed no colocalisation of GFP::TM23 and FM-stained endocytic vesicles (Meckel et al. 2005). This lack of internalisation may result from the absence of an endocytic signal motif in TM23. Such a signal should consequently exist in the K^+ channel KAT1 but has not been identified yet.

4 Conclusion and Future Prospects

Modern cell biology techniques now allow us to study membrane turnover in living cells. These investigations have clearly demonstrated that exocytosis and endocytosis are essential for stomatal movement. Membrane tension has been found to be an important regulator of exocytic and endocytic activity in guard cells. It allows the cell to adapt its surface area to changes in cell volume. In addition, there is clear evidence for constitutive endocytosis even in turgid guard cells. However, the underlying mechanisms of tension-driven and constitutive exocytosis and endocytosis are not yet understood. In particular, it is still not clear how endocytosis can occur against the large turgor pressure in plants. Further identification of molecular components of the endocytosis machinery in plants in combination with modern cell biology techniques will certainly increase our understanding of this important process.

Acknowledgements I thank Annette C. Hurst (Darmstadt) for critical reading of the manuscript.

References

Barth M, Holstein SEH (2004) Identification and functional characterization of *Arabidopsis* AP180, a binding partner of plant αC-adaptin. J Cell Sci 117:2051–2062
Bick I, Thiel G, Homann U (2001) Cytochalasin D attenuates the desensitisation of pressure-stimulated vesicle fusion in guard cell protoplasts. Eur J Cell Biol 80:521–526
Blatt MR (2000) Cellular signaling and volume control in stomatal movements in plants. Annu Rev Cell Dev Biol 16:221–241
Carroll AD, Moyen C, Van Kesteren P, Tooke F, Battey NH, Brownlee C (1998) Ca^{2+}, annexins, and GTP modulate exocytosis from maize root cap protoplasts. Plant Cell 10:1267–1276
Diekmann W, Hedrich R, Raschke K, Robinson DG (1993) Osmocytosis and vacuolar fragmentation in guard cell protoplasts: their relevance to osmotically-induced volume changes in guard cells. J Exp Bot 44:1569–1577
Franks PJ, Cowan IR, Faquhar GD (1998) A study of stomatal mechanics using the cell pressure probe. Plant Cell Env 21:94–100
Gillis KD (1995) Techniques for membrane capacitance measurements. In: Sakmann B, Neher E (eds) Single-channel recording. Plenum, New York, pp 155–197
Holstein SEH (2002) Clathrin and plant endocytosis. Traffic 3:614–620
Homann U (1998) Fusion and fission of plasma-membrane material accommodates for osmotically induced changes in the surface area of guard-cell protoplasts. Planta 206:329–333
Homann U, Tester M (1998) Patch clamp measurements of capacitance to study exocytosis and endocytosis in plants. Trends Plant Sci 3:110–114
Homann U, Thiel G (1999) Unitary exocytotic and endocytic events in guard-cell protoplasts during osmotic-driven volume changes. FEBS Lett 460:495–499
Homann U, Thiel G (2002) Alteration of K^+-channel numbers in the plasma membrane of guard cell protoplasts during surface area changes. Proc Natl Acad Sci USA 99:10215–10220
Hurst AC, Meckel T, Tayefeh S, Thiel G, Homann U (2004) Trafficking of the plant potassium inward rectifier KAT1 in guard cell protoplasts of Vicia faba. Plant J 37:391–397
Kreft M, Zorec R (1997) Cell-attached measurements of attofarad capacitance steps in rat melanotrophs. Pfluegers Arch Eur J Physiol 434:212–214
Kubitscheck U, Homann U, Thiel G (2000) Osmotic evoked shrinking of guard cell protoplasts causes retrieval of plasma membrane into the cytoplasm. Planta 210:423–431

Meckel T, Hurst AC, Thiel G, Homann U (2004) Endocytosis against high turgor: intact guard cells of Vicia faba constitutively endocytose fluorescently labelled plasma membrane and GFP-tagged K^+-channel KAT1. Plant J 39:182–193

Meckel T, Hurst AC, Thiel G, Homann U (2005) Guard cells undergo constitutive and pressure driven membrane turnover. Protoplasma (in press)

Morris CE, Homann U (2001) Cell surface area regulation and membrane tension. J Membr Biol 179:79–102

Neher E, Marty M (1982) Discrete changes of cell membrane capacitance observed under conditions of enhanced secretion in bovine adrenal chromaffin cells. Proc Natl Acd Sci USA 21:6712–6716

Raschke K (1979) Movements of stomata. In: Haupt W, Feinleb ME (eds) Encyclopedia of plant physiology, vol 7. Physiology of movements. Springer, Berlin Heidelberg New York, pp 382–441

Shope JC, DeWald DB, Mott KA (2003) Changes in surface area of intact guard cells are correlated with membrane internalization. Plant Physiol 133:1314–21

Thiel G, Rupnik M, Zorec R (1994) Raising the cytosolic Ca^{2+} concentration increases the membrane capacitance of maize coleoptile protoplasts: evidence for Ca^{2+}-stimulated exocytosis. Planta 195:305–308

Thiel G, Sutter J-U, Homann U (2001) Electrophysiological methods: monitoring exo- and endocytosis in real time. In: Hawes C, Satiat-Jeunemaitre B (eds) Plant cell biology. Oxford University Press, Oxford, pp 171–187

Willmer CM, Fricker M (1996) Stomata, 2nd edn. Chapman and Hall, London

Wolfe J, Dowgert MF, Steponkus PL (1986) Mechanical study of the deformation of the plasma membranes of protoplasts during osmotic expansions. J Membr Biol 93:63–74

Zorec R, Tester M (1992) Cytosolic calcium stimulates exocytosis in plant secretory cells. Biophys J 63:864–867

Plant Cell Monogr (1)
J. Šamaj · F. Baluška · D. Menzel: Plant Endocytosis
DOI 10.1007/7089_017/Published online: 27 September 2005

Endocytosis and Membrane Recycling in Pollen Tubes

Rui Malhó[1] (✉) · Pedro Castanho Coelho[1] · Elizabeth Pierson[2] · Jan Derksen[2]

[1]Universidade de Lisboa, Faculdade de Ciências de Lisboa, ICAT, 1749-016 Lisbon, Portugal
r.malho@fc.ul.pt

[2]Department of Plant Cell Research, Institute for Wetland and Water Research, Faculty of Science, Radboud University Nijmegen, Toernooiveld 1, 6525 ED Nijmegen, The Netherlands

Abstract In plants, tip-growing cells are an ideal system to investigate signal transduction mechanisms and, among these, pollen tubes are one of the favourite models. Many signalling pathways have been identified during germination and tip growth and, not surprisingly, the apical secretory machinery, essential for tip growth, seems to be an intersection point for all these pathways. Here we review previous data on the pollen tube endocytic machinery and its coupling to the exocytic delivery of new cell wall material. Additionally, we discuss new methodologies and how these are shaping our current working hypothesis to explain endocytosis in pollen tubes.

1 Introduction

Pollen tubes, the active male gametophytes of seed plants, are the vectors carrying the male sperm cells to the egg cell of the female gametophyte in the ovules of seed plants. Unlike most plant cells in which growth occurs by modification of the existing wall and the insertion of new material throughout its surface, pollen tubes extend strictly at their apex, undergoing a specialized type of growth called tip growth. Pollen tubes are thought to derive from the haustoria by which the primitive microgametophytes fed on the host sporophyte. They may grow extremely rapidly, with rates up to 1 cm/h like in lily, and their growth is often oscillatory (de Graaf et al. 2001; Holdaway-Clarke and Hepler 2003). Most studies on pollen tubes are performed with in vitro cultures of bicellular pollen (e.g. lily and tobacco) that supposedly grow autonomously. But even in optimized media, pollen tubes growing in vitro never achieve the high speed and length of those growing *in planta* as they lack the biochemical, physiological and physical environment of the pistil.

Pollen tubes bear similarity to other tip-growing cells like root hairs and moss and fern protonemal cells, with which they share a general cytoplasmic organization. Perhaps one of the most striking features of these cells is their growth mode that depends on polarized exocytosis at the growing tip and incorporation of new wall material. Quantitative data in pollen tubes has revealed, however, that the quantity of membrane delivered by exocytosis is clearly in excess for the cell growth rate (Steer and Steer 1989), indicating that, coupled to secretion, an underlying recycling process must take place. With the significant advances in molecular analysis and fluorescent probes, the mechanisms that drive this vesicle trafficking start to be unveiled. Here we review recent data and technological advances in the study of endocytosis/exocytosis in pollen tubes and discuss how they help us to dissect old hypotheses and raise new questions.

2
Exocytosis and Membrane Retrieval in Pollen Tubes

Until live imaging of secretory vesicles (SVs) became possible, mapping and quantification of endocytosis/exocytosis relied on the analysis of static electron microscopic images. Two main approaches were used and, together, they established the core information we now discuss and test.

The first approach was that of Morré and van der Woude (1974), who calculated the number of SVs needed for growth from the vesicle volume and the corresponding increase in wall volume for *Lilium* pollen tubes. This approach assumes a high similarity in the SVs, similar densities of SV content and the secreted wall, and negligible effects of processing for electron microscopy. Their estimation suggested identical vesicle requirements for both wall and membrane production. This, however, can be true only if the vesicle diameter is very high in comparison with the wall diameter. SVs with diameters smaller than or near the wall thickness necessarily will carry much more membrane material than required for tip expansion. A similar approach was used for tobacco by Derksen et al. (1995) but using rapid freeze fixing and substitution, which minimizes ultrastructural artefacts. They showed that the SV content and the wall at the tip exhibited similar electron densities (Fig. 1), although behind the tip the primary wall became thinner and appeared to lose wall material. The SV diameter was clearly smaller than the width of the wall at the tip.

The other approach was that by Picton and Steer (1981). They calculated the production rate of SV from the increase in SV numbers at the Golgi stacks by inhibiting their export from the Golgi with cytochalasin D. This approach omits possible effects of processing for electron microscopy, as only SV numbers need to be counted and is limited by the time exposure to the drug (prolonged cytochalasin D treatment perturbs SV production). Another drawback is that secretion depends not only on production rates but also on transport and fusion rates with the complete exclusion of other secretion sites, i.e. for the formation of the secondary wall. The procedure was used for *Tradescantia* pollen tubes after different treatments and was confirmed by

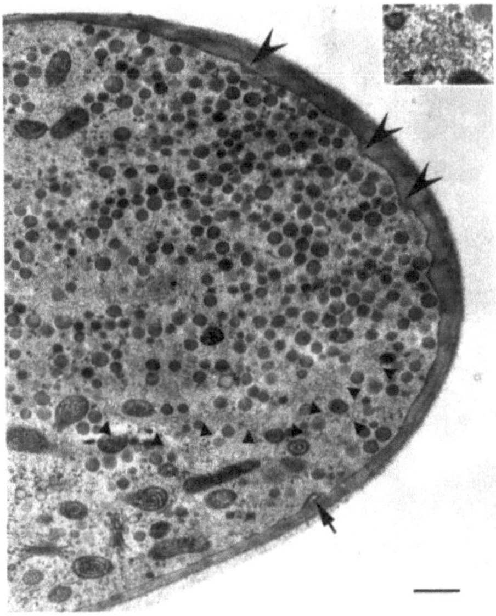

Fig. 1 Transmission electron microscopic image of a tobacco pollen tube grown in vitro. The image depicts a section closer to the apex. Many secretory vesicles (*SVs*) are present that show a similar electron density as the secreted wall. At the site of recently fused SVs, recognized from the slightly inside curved wall, small seemingly coated spots are visible (*split arrows*). On the membrane, vesicular structures are visible (*arrow*) that might represent remnants of in excess secreted membrane. In between and alongside the SVs in the tip trails of putative endoplasmic reticulum (*ER*) almost reach the site of exocytosis (*pointers*). *Inset* High magnification of part of the putative ER, the structures are not globular but are typical for strongly curved tubules. *Scale bar* 1 μm

estimating SV accumulations in the tip after fluoresceine isothiocyanate inhibition of SV fusion at the tip (Picton and Steer 1985). Despite all their pros and cons, the observations on *Tradescantia* and tobacco indicate a huge, up to 9 times, excess in membrane material delivered by the SVs at the tip.

Retrieval of membrane material secreted at the tip has been estimated from electron microscopic images as well. In frozen-fixed tobacco pollen tubes, Derksen et al. (1995) observed the presence of a collar of coated pits (CPs) immediately behind the apex. The coated vesicles (CVs) derived from the CP were too small, not more than 170 nm, to allow pinocytosis (Fig. 2a–c). Since these cells were grown in vitro (which excludes receptor-mediated CP formation), the data suggested that the CVs observed in the subapex were mainly involved in membrane retrieval. The calculations on possible endocytosis by a CP showed it to be nearly sufficient for retrieval of excess secreted membrane material. This estimation relied completely on an average 30-s CV maturation time; spatial and temporal differences in maturation time will considerably influence the real figure. However, the dependence of CP/CV formation on exocytosis is indicated by the almost absence of CPs

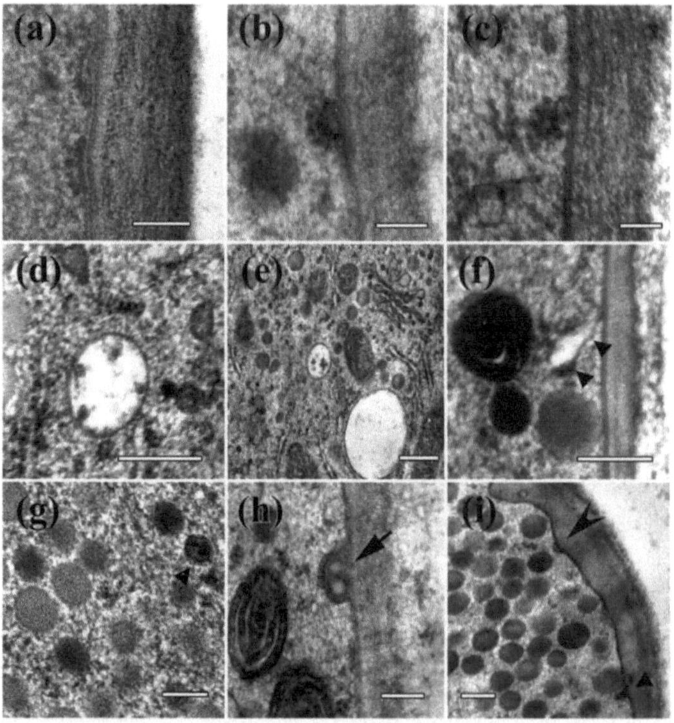

Fig. 2 Transmission electron microscopic images of possible endocytosis-related structures. **a, b** Typical clathrin-coated pits. **c** Typical coated vesicle. The clathrin decoration is shed after detachment from the membrane and becomes indistinguishable. **d, e** In the cytoplasm large multivesicular bodies (*MVBs*) (**d**) are found. Small MVBs occur also in the tip with the SVs (**e**). **f** Putative partially coated endosome structure at the plasma membrane near the tip. **g** Coated vesicles far from the wall in between the SVs. The coat is not typical for clathrin. **h** Vesicular structures on the membrane (*arrow*) also occur near the tip; part of the structure is clearly not coated with clathrin . **i** Clear finger-like invaginations occur regularly at the site of exocytosis only. The fast formation and the size of such structures make them difficult to observe and suggest they occur abundantly. *Scale bar* 100 nm

in brefeldin A treated pollen tubes (Rutten and Knuiman 1993). Markedly, species such as *Pinus sylvestris* (Derksen et al. 1999) and *Arabidopsis thaliana* (Derksen et al. 2002), with much slower growing pollen tubes, also show much lower CP densities and in caffeine-treated lily pollen tubes, CP areas occur at the dispersed sites of SV secretion (Lancelle et al. 1997).

Though the case for CP-/CV-mediated membrane retrieval is well argued for pollen tubes grown in vitro, the in vivo situation may be quite different. There, the growing rates generally are much higher, the wall may have different widths, SV production rates may be different and receptor-mediated CP/CV formation may occur as signals and cues for directional growth or incompatibility are present (Malhó et al. 2005). For example, the occurrence of numerous plasmatubules in *Nicotiana sylvestris* (Kandasamy et al. 1988), lily (Roy et al. 1997) and *Arabidopsis thaliana* (Lennon and Lord 2000) pollen tubes grown in vivo could reflect a possible alternative storage site for excessive membrane.

So far, the presence of alternative endocytic routes in pollen tubes has not been considered in electron microscopic studies (reliable probes were not available). However, target organelles for endocytosis, such as small vacuoles and small multivesicular bodies (MVBs) may occur (Figs. 1, 2c–e), although with a low abundance in the apex. In addition, larger MVBs and partially coated structures that resemble the endosomes described for root hairs are found in the cytoplasmic area only (Fig. 2f). Often, seemingly coated structures appear in the dome of the tip (Fig. 2g, h).

3 Mapping Endocytosis/Exocytosis in Living Pollen Tubes

Measurements of endocytosis/exocytosis in growing pollen tubes recently became possible with the introduction of FM dyes such as 1-43 [*N*-(3-triethylammoniumpropyl)-4-4-dibutylaminostyryl pyridinium dibromide] and 4-64 [*N*-(3-triethylammoniumpropyl)-4-6-4-diethylaminophenyl hexatrienyl pyridinium dibromide]. These probes are rendered fluorescent upon incorporation into the outer leaflet of the cell membrane and can be imaged with confocal and/or wide-field systems (Fig. 3; Šamaj, this volume). The changes in fluorescence recorded must then be critically interpreted because these probes do not directly report fusion/recycling events but simply membrane labelling. It is thus very important that analysis of FM dynam-

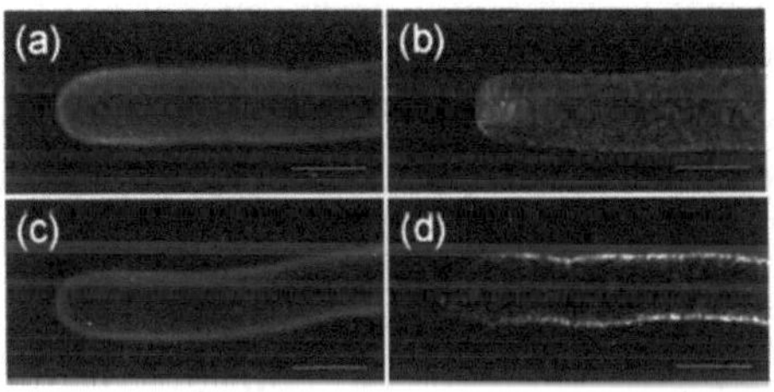

Fig. 3 Imaging of FM 1-43 (**a, b**) and 4-64 (**c, d**) in *Agapanthus umbellatus* pollen tubes using wide-field (**a, c**) or confocal (**b, d**) laser microscopy. *Scale bar* 10 μm. Given the high signal emitted by the plasma membrane, the use of wide-field systems (using a cool CCD camera) largely compromises the imaging of the FM signal in cytoplasmic regions. In contrast, thin sectioning using confocal microscopy allows the visualization of median cell sections without significant noise from the adjacent plasma membrane

ics is coupled to the analysis of spatial distribution, volumetric changes and rates of dye loss/bleaching. FM 1-43 and 4-64 have slightly different structures that could be the basis for the different observations recorded in pollen tubes. Using FM 1-43 and confocal microscopy, we observed that growing pollen tubes of *Agapanthus umbellatus* exhibit a fluorescence hotspot at the apex (Camacho and Malhó 2003). This hotspot has an inverted cone shape that confirms the labelling of the apical SVs by FM 1-43. When the tube reorients, the side of the dome to which the cell bent became more fluorescent, suggesting vesicle relocation and an asymmetric distribution of the fusion/recycling events. Furthermore, we found that if FM 1-43 is washed out from the growth medium, there is a progressive loss of internalized dye (Camacho and Malhó, unpublished data), possibly owing to the exocytic process (Richards et al. 2005). Similar observations have been made for synaptic vesicles (Zenisek et al. 2000), suggesting that FM 1-43 is a useful probe for the study of endocytosis and exocytosis dynamics. In *A. umbellatus* FM 4-64 was found to produce much stronger labelling in the plasma membrane and a very weak signal in the cytoplasm, making it more useful for studies of plasma membrane dynamics. In pollen tubes of *Nicotiana plumbaginifolia*, FM 4-64 showed a similar patter (Parton et al. 2003) but in *Lilium longiflorum*, the pattern for FM 4-64 fluorescence (Parton et al. 2001, 2003) was similar to that we described for FM 1-43, indicating that the usefulness and application of each probe must be evaluated for each biological model.

4
The Regulation of Endocytosis and Exocytosis by Ca^{2+} Ions

The cytoplasmic Ca^{2+} concentration ($[Ca^{2+}]_c$) is a key element in the regulation of pollen tube elongation and guidance. A tip-focused $[Ca^{2+}]_c$ gradient has been imaged with a high 1–3 μM Ca^{2+} concentration in the tip region and a low 0.2–0.3-μM Ca^{2+} concentration in the subapical and basal part of the tube (Malhó et al. 1995; Pierson et al. 1996; Messerli et al. 1997). This gradient is absent in nongrowing pollen tubes and its disruption leads to inhibition of tube growth. Both the intracellular $[Ca^{2+}]_c$ gradient and the extracellular Ca^{2+} influx suffer sinusoidal oscillations which are accompanied by oscillations in growth rate (Holdaway-Clarke and Hepler 2003) and there is strong evidence coupling Ca^{2+} with the control of cytoplasmic streaming. At basal levels (100–200 nM) streaming occurs normally; when the concentration is elevated to 1 μM or higher, streaming is rapidly but reversibly inhibited (Taylor and Hepler 1997). Pollen tubes show a marked inhibition of streaming when the intracellular $[Ca^{2+}]_c$ is elevated, and a recovery when the concentration returns to basal levels (Yokota et al. 1999). A high Ca^{2+} concentration is known to fragment F-actin (Yokota et al. 1998), which could explain why thick microfilaments are not visible in the tube apex. $[Ca^{2+}]_c$ at the tube apex is also sufficiently high to inhibit myosin motor activity and consequently streaming. Therefore, apical $[Ca^{2+}]_c$ may have an important role in the local regulation of vesicle docking and fusion with the plasma membrane (Battey et al. 1999). Indeed, the FM 1-43 data obtained in pollen tubes (Camacho and Malhó 2003) strongly resemble those obtained when cytosolic free ($[Ca^{2+}]_c$) was mapped during this process (Malhó and Trewavas 1996), indicating that $[Ca^{2+}]_c$ is a regulator of the coupling between growth and endo/exocytosis. Parton et al. (2001, 2003) also observed a correlation between growth rates and the dynamics of apical vesicle accumulation.

Modulation of intracellular GTP levels also resulted in a modification of apical secretion dynamics (Camacho and Malhó 2003) suggesting that cell growth is not strictly dependent of a Ca^{2+}-mediated stimulation of exocytosis. This is in agreement with the results of Roy et al. (1999) using the Yariv reagent who concluded that a Ca^{2+}-dependent exocytosis served mainly to secrete cell wall components. This is why rises in apical Ca^{2+} alone led to augmented secretion, reorientation of the growth axis but not to an increase in growth rates; this requires a strict coupling between exocytosis and endocytosis.

5 Molecular Regulators of Endocytosis and Exocytosis

Fusion and membrane recycling depends on Ca^{2+} but additionally on the presence of proteins such as annexins (Battey et al. 1999). Trotter et al. (1995) showed that relocation of annexins from the cytosol to the membranes started when $[Ca^{2+}]_c$ was elevated to 300 nM and at a concentration of 800 nM were no longer detected in the cytosol, reaching a maximum level in the membrane fraction. These $[Ca^{2+}]_c$ values are very close to those we measured in *A. umbellatus* pollen tubes (Malhó et al. 1995) for the body of the tube and tip region (180–220 and 700–1000 nM, respectively). Under such conditions annexins should be mainly associated with the plasma membrane in the apical region, and with the cytosol in the regions further back. Van der Hoeven et al. (1996) demonstrated the binding of an annexin-like peptide to a protein kinase C isoform and we obtained data showing a tip-high gradient of Ca^{2+}-dependent kinase activity in growing pollen tubes (Moutinho et al. 1998). This activity appears to be mainly associated with the plasma membrane of the apical region and cytosolic regions underneath. When pollen tubes reorient, the side towards which the cell bent showed a higher activity. These data suggest that an asymmetric activity of Ca^{2+}-dependent protein kinases within the growing dome could be involved in the regulation of secretory events. Ca^{2+} can achieve its role through cross talk with other molecules, namely calmodulin (CaM) and cyclic AMP (cAMP). Ca^{2+}–CaM is known to play a key role in the control of secretory activity in plant cells. Schuurink et al. (1996) and Rato et al. (2004) found that a decrease in CaM levels on one side of the apical dome results in a decrease of secretory activity and reorientation. Also, addition of W-12, a CaM antagonist, inhibited secretion, further suggesting a role for CaM in the control of endocytosis/exocytosis in pollen tubes. Diminishing cAMP levels mimicked this effect, while an increase of cAMP (which augments CaM activity) promoted secretion (Rato et al. 2004). In animal cells, cAMP has been reported to act on the secretory machinery, accelerating protein release in concert with a rise in $[Ca^{2+}]_c$ (Ammala et al. 1993).

GTP-binding proteins have also been claimed to regulate endocytosis/exocytosis (Li et al. 1999; Lin et al. 1996) and the relationships between exocytosis, endocytosis and the actin cytoskeleton in the growing pollen tube tip raise an interesting question to which no clear answer has been provided yet. The nonhydrolysable analogues of guanine nucleotide interfere with biological processes and it was found that GDPβS decreases growth rate, while GTPγS has the opposite effect (Ma et al. 1999; Camacho and Malhó 2003), thus confirming the importance of GTPases for tip growth. The release of GTPγS was also found to promote exocytosis, a process that seems to be more dependent on the active state of GTPases rather than on the cycling between the GTP/GDP bound state (Zheng and Yang 2000). Selective exposure to elevated Ca^{2+} concentrations alone is not sufficient to explain the selectivity of membrane retrieval (Smith et al. 2000) so it is likely that GTPases play an active role in endocytosis/exocytosis by coupling the actin cytoskeleton to the sequential steps underlying membrane trafficking at the site of exocytosis. Previous observations (Camacho and Malhó 2003) support a model in which Ca^{2+} and Rop GTPases act differentially, but in a concerted manner, in the sequential regulation of pollen tube secretion and membrane retrieval; Ca^{2+} would play a major role in the secretion of cell wall components, while Rop GTPases would be mainly involved in the fusion of docked vesicles and endocytosis.

In animal cells, phosphoinositides are central players in actin dynamics, vesicle trafficking and ion transport (Cremona et al. 1999; Stevenson et al. 2000); thus, they are excellent candidates for a key role in the regulation of vesicle secretion in tip growing cells (Malhó and Camacho 2004). Through phospholipase C (PLC), PIP_2 generates IP_3 and diacylglycerol (DAG), which can be converted to phosphatidic acid (PA) through DAG kinase (Munnik 2001). PIP_2 is also known to govern PLD activity, leading to elevated PA formation (Powner et al. 2002). Multiple PLD genes have been identified in plants and the corresponding proteins seem to be regulated by Ca^{2+} and G-proteins (Munnik 2001). PIP_2 has been shown to act in a common pathway to that of Rac GTPases (Kost et al. 1999). Recently, we have shown that caged-probe photorelease of minute levels of PIP_2 (Monteiro et al. 2005a) caused an increase in apical $[Ca^{2+}]_c$ and in FM 1-43 fluorescence, suggesting an increase in membrane recycling.

In pollen tubes, an IP_3-induced Ca^{2+} release seems to be required for transduction of signals from the apex to subapical regions of the cell (Malhó 1998) and it was found that, in contrast to PIP_2, IP_3 release induced a decrease in apical FM1-43 fluorescence (Monteiro et al. 2005a). This was accompanied by a significant increase in subapical $[Ca^{2+}]_c$ that led to disruption of the tip-focused gradient and growth perturbation. The different effects induced by photorelease of PIP_2 and IP_3 cannot be explained solely by a conversion of PIP_2 to IP_3. Clearly, the action of the two phosphoinositides hits different targets. One of the targets was suggested to be PA, since this phospholipid is a final product of PIP_2 hydrolysis via PLC or PLD. PA promotes membrane curvature and formation of SVs and has a crucial role in the structural integrity of the Golgi (Sweeny et al. 2002). In addition to a role in membrane curvature, PA has been reported to play a role in membrane transport (Kooijman et al. 2003). Supporting these observations, we found that antagonists of PA accumulation (e.g. 1-butanol) and inhibitors of PLC and PLD reversibly halted pollen tube polarity, caused dissipation of the $[Ca^{2+}]_c$ gradient and inhibited apical plasma membrane recycling (Monteiro et al. 2005a). Interestingly, reduction of PA levels was accompanied by an extensive formation of thick but nondirectional actin microfilaments, suggesting that PA participates in the correct anchoring and positioning of the actin cytoskeleton (Monteiro et al. 2005a). These multiple but coordinated effects of PA might thus explain the different responses induced by photolysis of caged PIP_2 or IP_3. An increase in PIP_2 is likely to result in its hydrolysis, leading to increased levels of IP_3 and PA. The effects of PA on membrane curvature are enhanced by Ca^{2+} (Kooijman et al. 2003); thus, simultaneously to the $[Ca^{2+}]_c$ rise, an increase in membrane turnover and actin-dependent trafficking might occur. On the other hand, increasing IP_3 levels lead to a rise in $[Ca^{2+}]_c$ and increased exocytosis but not equivalent the increases from membrane recycling. These three signalling molecules (PIP_2, IP_3 and PA) thus seem to have a concerted action modulating endocytosis and exocytosis.

6 Membrane Trafficking and the Actin Cytoskeleton

The actin cytoskeleton, along with the secretory apparatus and endocytic pathways (Šamaj et al., this volume), is a structural component in which most pollen tube signalling pathways seem to converge. The crucial role of the microfilaments in pollen germination and pollen tube growth has been reported in several studies (reviewed by Taylor and Hepler 1997). In the shank of the pollen tube, thick bundles of microfilaments are axially or sometimes helically oriented throughout the cortical and central cytoplasm of the tube cell. Towards the apex, these bundles of microfilaments branch into thinner, short-lived bundles and, in the subapical region, a dense network of fine, labile actin filaments is formed (Kost et al. 1998; Vidali et al. 2001). Dynamic short actin bundles, correlating with the oscillatory growth rate, were found to transiently penetrate into the extreme tip of tobacco pollen tubes (Fu et al. 2001), a process which was suggested to be due to the tip-focused Ca^{2+} gradient.

The typical F-actin array in the subapical region could act as a structural (compensating the weakness in the expanding apical cell wall) and molecular buffer to retain organelles in this area and to deliver the vesicles onwards to the apical dome. In addition, it may also play a role in the positioning of the endocytic machinery to a precise microdomain of the pollen tube plasma membrane and to define the tip region by restricting the exit of Ca^{2+} channels, SVs and other components necessary for regulated exocytosis (Staiger 2000). Such reorganizations may then promote topological changes in the transport of ions and metabolites across the plasma membrane within those regions (Derksen et al. 1995a). Thus, it is likely that in pollen tubes specific arrangements of the cortical F-actin may be induced in response to stimuli from extracellular cues. Several genes encoding small actin-bundling proteins (ABPs) have already been characterized from pollen (reviewed by Staiger 2000 and Malhó et al. 2005) and the proteins they code for are potent sensors of the pollen tube ionic and signalling environment (e.g. cofilin, profilin, actin depolymerizing factor). More recently, Cheung and Wu (2004) reported the presence of formins, which are actin-nucleating proteins that stimulate actin cable formation. A pollen villin was also isolated from *L. longiflorum* pollen tubes as a 135-kDa ABP (Yokota and Shimmen 1999) that colocalizes to microfilament bundles (Vidali et al. 1999). Ca^{2+} and CaM together inhibit the actin-bundling activity of this pollen villin (Yokota et al. 2000), adding villin to the group of Ca^{2+}-regulated ABPs. The pollen villin also includes a phosphoinositide-binding domain (Vidali et al. 1999). Profilin forms complexes with ADP-actin and promotes its phosphorylation to ATP-actin. These complexes are dissociated by phosphatidyl 4,5-bisphosphate (PIP_2) (Drøbak et al. 1994) so changes in profilin distribution and/or activity could change the physical state of the cytoskeletal network and thus modulate secretion.

7
Conclusions and Future Prospects

It was argued that a rapid endocytosis mechanism might occur in the apex of rapidly growing pollen tubes (Monteiro et al. 2005a, b). The rapid endocytosis mechanism is a Ca^{2+}-dependent process coupled to exocytosis that requires GTP hydrolysis and dynamin but not clathrin (Artalejo et al. 1995). The importance of the dynamin-like protein ADL1C for plasma membrane maintenance during pollen maturation has been demonstrated (Kang et al. 2003) and, as previously discussed, clathrin CVs are not abundant in the extreme apex of freeze-fixed rapidly growing pollen tubes (Lancelle and Hepler 1992; Derksen et al. 1995). Rapid endocytosis is characterized by the formation of a small and short-lived pore, which limits the size of particles that can be released/incorporated. Indeed, we found that without artificial permeabilization, growing pollen tubes of *A. umbellatus* were unable to incorporate dextrans larger than 4 kDa. A rapid endocytosis mechanism is also compatible with fast delivery of wall material and membrane recycling and fits the observations using contrast-enhanced video microscopy where full endocytic or exocytic events do not seem to occur.

This mechanism could operate in parallel to the CP/CV system where endocytosis and exocytosis are uncoupled. The uncoupling would be favoured when cell growth is arrested or it is slowed down (apical $[Ca^{2+}]_c$ is lower). In such a system, the formation of a full endocytic vesicle allows the incorporation of larger molecules and protoplasts obtained from pollen tubes (where $[Ca^{2+}]_c$ is low) indeed showed capacity to incorporate molecules up to 20 kDa. In barley protoplasts, Homann and Tester (1997) have also reported the existence of two exocytotic modes, a Ca^{2+}-dependent mode and a GTP-binding-dependent mode. The observations previously discussed from electron microscopic images of chemically fixed pollen tubes also allow us to speculate further on the existence of the two endocytic modes. With such a preservation method, growth is slowed down before coming to a complete halt and images of the cell apex reveal a larger number of CP/CV structures. The uncoupled mechanism could operate in subapical regions at $[Ca^{2+}]_c$ resting levels and at the apex in situations where the cell needs to interpret extracellular cues and reorient the growth axis. These invariably involve reduction of growth rates, localized changes in Ca^{2+}, redirectioning of the actin cytoskeleton and target sites for fusion. A putative rapid endocytosis mechanism would be confined to the extreme apex operating during the stages of fast-directed growth. To fully test this hypothesis and identify the structures involved, their target sites (organelles) and their molecular components is one of the challenges for future research in the study of plant endocytosis. Proteins involved in vesicle traffic events such as soluble *N*-ethylmaleimide-sensitive factor attachment protein receptors (SNAREs), soluble *N*-ethylmaleimide-sensitive factor attachment proteins (SNAPs), syntaxins and vesicle-associated membrane protein like proteins have been identified in the *Arabidopsis* genome, revealing that plant cells are equipped with molecular machinery equivalent to that of animals.

In addition to the regular delivery and recycling of membrane and wall material, this continuous process of endocytosis/exocytosis is likely to have another essential function: the interaction with the alien, sporophytic environment. To achieve fertilization, the pollen tube must continuously perceive extracellular signals and incorporate additional energy sources. Therefore, it is also foreseen that future work will require the use of in vivo systems in order to fully understand the secretory process.

Acknowledgements The work was supported by a UE TMR grant (TIPNET) and Fundação Ciência e Tecnologia, Lisboa, Portugal (grants nos. BCI/37555/2001; BCI/44148/2002; FEDER).

References

Ammala C, Ashcroft FM, Rorsman P (1993) Calcium-independent potentiation of insulin release by cyclic AMP in single beta-cells. Nature 363:356–358

Artalejo CR, Henley JR, McNiven MA, Palfrey HC (1995) Rapid endocytosis coupled to exocytosis in adrenal chromaffin cells involves Ca^{2+}, GTP and dynamin but not clathrin. Proc Natl Acad Sci USA 92:8328–8332

Battey NH, James NC, Greenland AJ, Brownlee C (1999) Exocytosis and endocytosis. Plant Cell 11:643–659

Camacho L, Malhó R (2003) Endo-exocytosis in the pollen tube apex is diferentially regulated by Ca^{2+} and GTPases. J Exp Bot 54:83–92

Cheung AY, Wu H-M (2004) Overexpression of an *Arabidopsis* formin stimulates supernumerary actin cable formation from pollen tube cell membrane. Plant Cell 16:257–269

Cremona O, Di Paolo G, Wenk MR, Lüthi A, Kim WT, Takei K, Daniell L, Nemoto Y, Shears SB, Flavell RA, McCormick DA, De Camilli P (1999) Essential role of phophoinositol metabolism in synaptic vesicle recycling. Cell 99:179–188

De Graaf BHJ, Derksen JWM, Mariani C (2001) Pollen and pistil in the progamic phase. Sex Plant Reprod 14:41–55

Derksen J, Knuiman B, Hoedemaekers K, Guyon A, Bonhomme S, Pierson ES (2002) Growth and cellular organization of *Arabidopsis* pollen tubes *in vitro*. Sex Plant Reprod 15:133–139

Derksen J, Rutten T, Lichtscheidl IK, de Win AHN, Pierson ES, Rongen G (1995) Quantitative analysis of the distribution of organelles in tobacco pollen tubes: implications for exocytosis and endocytosis. Protoplasma 188:267–276

Derksen J, Rutten T, van Amstel T, de Win A, Doris F, Steer M (1995a) Regulation of pollen tube growth. Acta Bot Neerl 44:93–119

Derksen J, van Wezel R, Knuiman B, Ylstra B, van Tunen AJ (1999) Pollen tubes of flavonol-deficient Petunia show striking alterations in wall structure leading to tube disruption. Planta 207:575–581

Drøbak BK, Watkins PAC, Valenta R, Dove SK, Lloyd CW, Staiger CJ (1994) Inhibition of plant plasma membrane phosphoinositide phospholipase C by the actin-binding protein, profilin. Plant J 6:389–400

Franklin-Tong VE, Drøbak BK, Allan AC, Watkins PAC, Trewavas AJ (1996) Growth of pollen tubes of *Papaver rhoeas* is regulated by a slow moving calcium wave propagated by inositol triphosphate. Plant Cell 8:1305–1321

Fu Y, Wu G, Yang Z (2001) Rop GTPase-dependent dynamics of tip-localized F-actin controls tip growth in pollen tubes. J Cell Biol 152:1019–1032

Holdaway-Clarke TL, Hepler PK (2003) Control of pollen tube growth: role of ion gradients and fluxes. New Phytol 159:539–563

Homann U, Tester M (1997) Ca^{2+}-independent and Ca^{2+}/GTP-binding protein-controlled exocytosis in a plant cell. Proc Natl Acad Sci USA 94:6565–6570

Kandasamy MK, Kappler R, Kristen U (1988) Plasmatubules in the pollen tubes of *Nicotiana sylvestris*. Planta 173:35–41

Kang B-H, Rancour DM, Bednarek SY (2003) The dynamin-like protein ADL1C is essential for plasma membrane maintenance during pollen maturation. Plant J 35:1–15

Kooijman EE, Chupin V, de Kruijff B, Burger KNJ (2003) Modulation of membrane curvature by phosphatidic acid and lysophosphatidic acid. Traffic 4:162–174

Kost B, Lemichez E, Spielhofer P, Hong Y, Tolias K, Carpenter C, Chua M-H (1999) Rac homologues and compartmentalized phosphatidilinositol 4,4-biphosphate act in a common pathway to regulate polar pollen tube growth. J Cell Biol 145:317–330

Lancelle SA, Hepler PK (1992) Ultrastructure of freeze-substituted pollen tubes of *Lilium longiflorum*. Protoplasma (1992) 167:215–230

Lancelle SA, Cresti M, Hepler PK (1997) Growth inhibition and recovery in freeze-substituted *Lilium longiflorum* pollen tubes: structural effects of caffeine. Protoplasma 196:21–33

Lennon KA, Lord EM (2000) The *in vivo* pollen tube cell of *Arabidopsis thaliana I*. Tube cell cytoplasm and wall. Protoplasma 214:45–56

Li H, Lin Y, Heath RM, Zhu MX, Yang Z (1999) Control of pollen tube tip growth by a Rop GTPase-dependent pathway that leads to tip-localized calcium influx. Plant Cell 11:1731–1742

Lin Y, Wang Y, Zhu J-K, Yang Z (1996) Localization of a Rho GTPase implies a role in tip growth and movement of the generative cell in pollen tubes. Plant Cell 8:293–303

Ma L, Xu X, Cui S, Sun D (1999) The presence of a heterotrimeric G protein and Its role in signal transduction of extracellular calmodulin in pollen germination and tube growth. Plant Cell 11:1351–1364

Malhó R (1998) The role of inositol(1,4,5)triphosphate in pollen tube growth and orientation. Sex Plant Reprod 11:231–235

Malhó R, Trewavas AJ (1996) Localized apical increases of cytosolic free calcium control pollen tube orientation. Plant Cell 8:1935–1949

Malhó R, Camacho L (2004) Signalling the cytoskeleton in pollen tube germination and growth. In: Hussey PJ (ed) The plant cytoskeleton in cell differentiation and development. Annual plant review series. Blackwell London, UK, pp 240–264

Malhó R, Read ND, Trewavas AJ, Pais MS (1995) Calcium channel activity during pollen tube growth and reorientation. Plant Cell 7:1173–1184

Malhó R, Camacho L, Moutinho A (2000) Signalling pathways in pollen tube growth and reorientation. Ann Bot 85 (Suppl A):59–68

Malhó R, Liu Q, Monteiro D, Rato C, Camacho L, Dinis A (2005) Signalling pathways in pollen germination and tube growth. Protoplasma (in press)

Messerli M, Robinson KR (1997) Tip localized Ca^{2+} pulses are coincident with peak pulsatile growth rates in pollen tubes of *Lilium longiflorum*. J Cell Sci 110:1269–1278

Monteiro D, Castanho Coelho P, Rodrigues C, Camacho L, Quader H, Malhó R (2005a) Modulation of endocytosis in pollen tube growth by phosphoinositides and phospholipids. Protoplasma (in press)

Monteiro D, Liu Q, Lisboa S, Scherer GEF, Quader H, Malhó R (2005b) Phosphoinositides and phosphatidic acid regulate pollen tube growth and reorientation through modulation of $[Ca^{2+}]_c$ and membrane secretion. J Exp Bot 56:1665–1674

Morré DJ, van der Woude WJ (1974) Origin and growth of cell surface components. In: Hay ED, King TJ, Papaconstantinou J (eds) Macromolecules regulating growth and development. Academic, New York, pp 81–111

Moutinho A, Trewavas AJ, Malhó R (1998) Relocation of a Ca^{2+}-dependent protein kinase activity during pollen tube reorientation. Plant Cell 10:1499–1510

Munnik T (2001) Phosphatidic acid: an emerging plant lipid second messenger. Trends Plant Sci 6:227–233

O'Luanaigh N, Pardo R, Fensome A, Allen-Baume V, Jones D, Holt MR, Cockcroft S (2002) Continual production of phosphatidic acid by phospholipase D is essential for antigen-stimulated membrane ruffling in cultured mast cells. Mol Biol Cell 13:3730–3746

Parton RM, Fischer-Parton S, Watahiki MK, Trewavas AJ (2001) Dynamics of the apical vesicle accumulation and the rate of growth are related in individual pollen tubes. J Cell Sci 114:2685–2695

Parton RM, Fischer-Parton S, Trewavas AJ, Watahiki MK (2003) Pollen tubes exhibit regular periodic membrane trafficking events in the absence of apical extension. J Cell Sci 116:2707–2719

Picton JM, Steer MW (1981) Determination of secretory vesicle production rates by dictyossomes in pollen tubes of *Tradescantia* using Cytochalasin D. J Cell Sci 49:261–272

Picton JM, Steer MW (1985) The effects of ruthenium red, lanthanum, fluorescein isothiocyanate and trifluoperazine on vesicle transport, vesicle fusion and tip extension in PT. Planta 163:20–26

Pierson ES, Miller DD, Callaham DA, van Aken J, Hackett G, Hepler PK (1996) Tip-localized calcium entry fluctuates during pollen tube growth. Dev Biol 174:160–173

Powner DJ, Wakelam MJO (2002) The regulation of phospholipase D by inositol phospholipids and small GTPases. FEBS Lett 531:62–64

Rato C, Monteiro D, Hepler PK, Malhó R (2004) Calmodulin activity and cAMP signalling modulate growth and apical secretion in pollen tubes. Plant J 38:887–897

Richards DA, Bai J, Chapman ER (2005) Two modes of exocytosis at hippocampal synapses revealed by rate of FM1-43 efflux from individual vesicles. J Cell Biol 168:929–939

Roy S, Eckard KJ, Lancelle S, Hepler PK, Lord EM (1997) High-pressure freezing improves the ultrastructural preservation of *in vivo*-grown lily pollen tubes. Protoplasma 200:87–98

Roy SJ, Holdaway-Clarke TL, Hackett GR, Kunkel JG, Lord EM, Hepler PK (1999) Uncoupling secretion and tip growth in lily pollen tubes: evidence for the role of calcium in exocytosis. Plant J 19:379–386

Rutten TLM, Knuiman B (1993) Brefeldin A effects on tobacco pollen tubes. Eur J Cell Biol 61:247–255

Šamaj J, Baluška F, Voigt B, Volkmann D, Menzel D (2005) Endocytosis and acto-myosin cytoskeleton (in this volume). Springer, Berlin Heidelberg New York

Schuurink RC, Chan PV, Jones RL (1996) Modulation of calmodulin mRNA and protein levels in barley aleurone. Plant Physiol 111:371–380

Smith RM, Baibakov B, Ikebuchi Y, White BH, Lambert NA, Kaczmarek LK, Vogel SS (2000) Exocytotic insertion of calcium channels constrains compensatory endocytosis to sites of exocytosis. J Cell Biol 148:755–767

Staiger CJ (2000) Signalling to the actin cytoskeleton in plants. Annu Rev Plant Physiol Plant Mol Biol 51:257–288

Steer MW, Steer JL (1989) Pollen tube tip growth. New Phytol 111:323–358

Stevenson JM, Perera IY, Heilmann I, Persson S, Boss WF (2000) Inositol signalling and plant growth. Trends Plant Sci 5:252–258

Sweeney DA, Siddhanta A, Shields D (2002) Fragmentation and re-assembly of the Golgi apparatus *in vitro*—a requirement for phosphatidic acid and phosphatidylinositol 4,5-bisphosphate synthesis. J Biol Chem 77:3030–3039

Taylor LP, Hepler PK (1997) Pollen germination and tube growth. Annu Rev Plant Physiol Plant Mol Biol 48:461–491

Trotter PJ, Orchard MA, Walker JH (1995) Ca^{2+} concentration during binding determines the manner in which annexin V binds to membranes. Biochem J 308:591–598

Van der Hoeven PCJ, Siderius M, Korthout HAAJ, Drabkin AV, De Boer AH (1996) A calcium and free fatty acid-modulated protein kinase as putative effector of the fusicoccin 14-3-3 receptor. Plant Physiol 111:857–865

Vidali L, McKenna ST, Hepler PK (2001) Actin polymerization is essential for pollen tube growth. Mol Biol Cell 12:2534–2545

Vidali L, Yokota E, Cheung AY, Shimmen T, Hepler PK (1999) The 135 kDa actin-bundling protein from *Lilium longiflorum* pollen is the plant homologue of villin. Protoplasma 209:283–291

Yokota E, Muto S, Shimmen T (1999) Inhibitory regulation of higher-plant myosin by Ca^{2+} ions. Plant Physiol 119:231–239

Yokota E, Muto S, Shimmen T (2000) Calcium-calmodulin suppresses the filamentous actin-binding activity of a 135-kilodalton actin-bundling protein isolated from lily pollen tubes. Plant Physiol 123:645–654

Yokota E, Shimmen T (1999) The 135-kDa actin-bundling protein from lily pollen tubes arranges F-actin into bundles with uniform polarity. Planta 209:264–266

Yokota E, Takahara K, Shimmen T (1998) Actin-bundling protein isolated from pollen tubes of lily: biochemical and immunocytochemical characterization. Plant Physiol 116:1421–1429

Zenisek D, Steyer JA, Almers W (2000) Transport, capture and exocytosis of single synaptic vesicles at active zones. Nature 406:849–854

Zheng Z-L, Yang Z (2000) The Rop GTPase switch turns on polar growth in pollen. Trends Plant Sci 5:298–303

Plant Cell Monogr (1)
J. Šamaj · F. Baluška · D. Menzel: Plant Endocytosis
DOI 10.1007/7089_018/Published online: 27 September 2005

Tip Growth and Endocytosis in Fungi

Jürgen Wendland (✉) · Andrea Walther

Junior Research Group: Growth Control of Fungal Pathogens,
Leibniz Institute for Natural Product Research and Infection Biology,
Hans Knöll Institute, Department of Microbiology, Friedrich Schiller University,
Hans-Knöll-Str. 2, 07745 Jena, Germany
juergen.wendland@uni-jena.de

Abstract Recent advances in molecular cell biology have provided new insights into different cellular processes that all turn out to contribute to polarized cell growth in a variety of model systems used to analyse growth, differentiation and development. Polarized cell growth, although a general feature of the living cell, can be found in a pronounced fashion during pollen tube outgrowth and root hair development in plants, during neurite outgrowth, and during filamentous hyphal growth. Filamentous fungi represent excellent model systems to analyse polarized cell growth owing to their genetic tractability and the ease of generating and keeping mutant strains. Contributing to this is the fact that already a number of fungal genomes have been sequenced, which allows the rapid analysis and comparison of gene function. This has led to the finding that polarized cell growth can be influenced by perturbations in different cellular pathways. Control of polarity establishment and the maintenance of polarized cell growth are exerted by a number of conserved GTP-binding proteins of the Ras/Rho subfamily and their specific regulators that organize the actin cytoskeleton. Hyphal tip growth requires coordination of vesicle transport using actin and microtubule cytoskeletons. Recent evidence has shown that hyphal growth not only depends on polarized secretion but also requires endocytosis, suggesting that the recycling of the membrane and sorting of vesicles is required for fast elongation of hyphal tubes. Key players on the molecular level that direct tip growth and endocytosis in the fungal hyphae based on differential regulation of the actin cytoskeleton are discussed.

1 Introduction

Endocytosis describes the general process of plasma membrane invagination and the generation of intracellular vesicles or endosomes. This process can be used for a variety of tasks such as the engulfment of extracellular particles or molecules and solutes known as phagocytosis and pinocytosis or the internalization of the plasma membrane with membrane-associated proteins (Engqvist-Goldstein and Drubin, 2003). Once internalized these compounds and membranes need to be delivered to specific cellular compartments, such as vacuoles for protein degradation, storage vesicles or recycling endosomes for reinsertion into the plasma membrane. Delivery needs to be a highly ordered process and regulation of vesicle flow is complicated by the secretory vesicles derived from the trans-Golgi network. *Neurospora crassa*, a fast-growing filamentous ascomycete, shows a tip extension rate of more than 30 μm/min sustained by more than 30 000 vesicles/min fused with the hyphal tip (Collinge and Trinci, 1974). Vesicle delivery, therefore, is a highly dynamic process. This vesicle flow requires intracellular tracks for targeted delivery composed of either the actin cytoskeleton or microtubules. Regulation of the assembly and organization of the actin cytoskeleton may provide crucial signals for the targeted transport of vesicles. Studies in other cell systems with highly polarized growth potential have shown that both exocytosis and endocytosis are required for cell expansion (reviewed by Mellman, 1996).

Even recently the mere existence of endocytosis in filamentous fungal systems was questioned (Cole et al., 1998; Torralba and Heath, 2002). Evidence, particularly based on the exploration of fungal genome sequences as well as on the analysis of mutant strains and time lapse microscopy monitoring the uptake of the membrane-selective lipophilic dye FM4-64, has shown that endocytosis occurs in filamentous fungi and is important for polarized growth (Yamashita and May, 1998; Fisher-Parton et al., 2000; Atkinson et al., 2002; Read and Kalkman, 2003; Walther and Wendland, 2004a, b).

This chapter summarizes genomic, genetic and microscopic evidence for endocytosis in filamentous fungi and merges these data from filamentous fungi and yeasts with our current knowledge on the regulation of cell polarity establishment and the maintenance of polarized hyphal growth via the actin cytoskeleton.

2
Endocytosis in Fungi

Endocytosis has been established as an essential process in a variety of eukaryotic cell systems and fulfils functions such as membrane retrieval and degradation, and internalization of membrane receptors and other signalling molecules (Ayscough, 2004). Endocytosis can generally either be clathrin-/lipid-raft-dependent or clathrin-/lipid-raft-independent. In mammalian systems and in the yeast *Saccharomyces cerevisiae* a large number of genes have been identified encoding proteins that are involved in these processes. The presence of homologues of these genes in filamentous fungi may therefore be considered bona fide evidence for the functional conservation of this process. A variety of fungal genome sequences are available and can be searched for homologues using, for example, the Fungal Blast tool (http://seq.yeastgenome.org/cgi-bin/nph-blast-fungal.pl). By searching the genome sequences of the filamentous ascomycetes *Ashbya gossypii* and *N. crassa* homologues of proteins involved in clathrin-mediated endocytosis could be found (Table 1). In yeasts, actin patch components and the Arp2/Arp3 complex are involved in endocytosis and most of the yeast proteins have corresponding homologues in filamentous fungi (Table 1). Some of these proteins, Abp1, Pan1 and Las17/Bee1 (encoding the yeast homologue of the human Wiskott–Aldrich syndrome protein, WASP) are activators of the Arp2/Arp3 complex (Winter et al., 1999; Duncan et al., 2001; Goode et al., 2001; Kaksonen et al., 2003). This results in the formation of branched actin

Table 1 *Saccharomyces cerevisiae* proteins involved in endocytosis and their homologs in filamentous ascomycetes

S. cerevisiae	Homology	*Ashbya gossypii*	*Neurospora crassa*
Clathrin			
Chc1	Clathrin heavy chain	AER359W	NCU02510.1
Clc1	Clathrin light chain	AGR309C	NCU04115.1
Apl1	β-Adaptin	ADR342C	NCU05232.1
Apl2	AP2-adaptor complex	AFR200W	NCU09721.1
Apl3		ADL302W	NCU03440.1
Apl4		ADR064C	NCU04120.1
Apm1	AP1-adaptor complex	ADL017C	NCU09688.1
Apm2		ABR047W	None
Apm3		AGL061W	NCU03998.1
Apm4		ADR315W	NCU09673.1
Aps1		AFR124W	None
Aps2		AFR370C	NCU07989.1
Aps3		AAL143W	NCU09461.1
Actin patch			
Dnm1	Dynamin	AAL174C	NCU09808.1
End3	Eps15 homology	AER416C	NCU06347.1
Ent1	Epsin	ACL157C	None
Ent2	Epsin	ACL157C	NCU04783.1
Ent3	Epsin	AER155C	NCU00725.1
Ent4	Epsin	ACL061C	None
Inp51	Synaptojanin	ADL002C	None
Inp52	Synaptojanin	AFL228W	NCU03792.1
Inp53	Synaptojanin	AFL228W	NCU03298.1
Las17/Bee1	WASP	AGR285W	NCU07438.1
Myo3	Type I myosin	AEL306C	NCU02111.1
Myo5	Type I myosin	AEL306C	NCU02111.1

Protein names in *S. cerevisiae* are according to the standard SGD designation. Nomenclature of the *A. gossypii* genes is derived from AGD (http://agd.unibas.ch/) and the annotation of *N. crassa* genes is based on the *N. crassa* genome database (http://www.broad.mit.edu/annotation/fungi/
neurospora/). Alias names for a yeast gene are indicated. *AP* adaptor protein, *WASP* Wiskott–Aldrich syndrome protein, *WIP* WASP-interacting protein, *CAP* adenylyl cyclase-associated protein.

Table 1 (continued)

S. cerevisiae	Homology	*Ashbya gossypii*	*Neurospora crassa*
Pan1	Eps15 homology	ADR018C	NCU06171.1
Rvs161/End6	Amphiphysin	AER193W	NCU01069.1
Rvs167	Amphiphysin	AFR140C	NCU04637.1
Sac6	Fimbrin	AGR069C	NCU03992.1
Sla2/End4	Talin-related	None	NCU01956.1
Srv2	CAP	AFR061W	NCU08008.1
Vrp1/End5	WIP	ABR038C	NCU03089.1
YAP1801	AP180	AEL209W	NCU02586.1
YAP1802	AP180	AEL209W	NCU02586.1
Arp2/3 complex			
Arp2		ADR316W	NCU07171.1
Arp3		AFR419C	NCU01756.1
Arc15		ADL061W	NCU03438.1
Arc18		AFR584C	NCU09572.1
Arc19		ACL092C	NCU01918.1
Arc35		AGL221W	NCU03050.1
Arc40		AFL022W	NCU02781.1

Protein names in *S. cerevisiae* are according to the standard SGD designation. Nomenclature of the *A. gossypii* genes is derived from AGD (http://agd.unibas.ch/) and the annotation of *N. crassa* genes is based on the *N. crassa* genome database (http://www.broad.mit.edu/annotation/fungi/
neurospora/). Alias names for a yeast gene are indicated. *AP* adaptor protein, *WASP* Wiskott–Aldrich syndrome protein, *WIP* WASP-interacting protein, *CAP* adenylyl cyclase-associated protein.

filaments (Madania et al., 1999; Engqvist-Goldstein and Drubin, 2003). The functional analyses of these homologues in filamentous fungi will, therefore, be of importance to understand endocytosis in these organisms and allow the comparison with yeasts.

Visualization of endocytosis can be achieved via in vivo time lapse fluorescence microscopy using the lipophilic dyes FM4-64 and Lucifer Yellow—the latter of which cannot cross the plasma membrane and thus serves as a marker for so-called fluid-phase endocytosis (Riezman, 1985; Vida and Emr, 1995; Šamaj, this volume; Baluška et al., this volume). Uptake of these dyes into the cells was reported for *Pisolithus tinctorius* and, for example, in *N. crassa* may be hampered by low permeability of the cell wall (Cole et al., 1998; Torralba and Heath, 2002). In consequence, failure in detecting the uptake of FM4-64 in a fungal species may not indicate the absence of endocytosis in these fungi and cell-wall-less protoplasts might be capable of internalizing these probes.

Endocytosis has been monitored in several fungal species, including *Magnaporthe grisea*, *Aspergillus nidulans* and *A. gossypii* (Yamashita and May,

1998; Atkinson et al., 2002; Walther and Wendland, 2004b). In vivo fluorescence time lapse microscopy in filamentous fungi is an important tool to analyse the uptake and transport processes within the hyphae and also includes imaging of vesicle fusion events (Fig. 1) and cytoplasmic streaming.

Endocytosis has been analysed on the molecular level in *A. nidulans* and *A. gossypii* (Yamashita and May, 1998; Walther and Wendland, 2004b). In *A. nidulans myoA* encodes the sole type I myosin. A *myoA* mutant was constructed in which a specific phosphorylation site (the "TEDS"-rule site) was changed into S371A or S371E. The phosphorylation mimic mutant (S371E) showed constitutive endocytosis on the basis of FM4-64 uptake and accumulated intracellular membranes. Moreover, this process was F-actin dependent since cells treated with the F-actin destabilizing drug cytochalasin D were

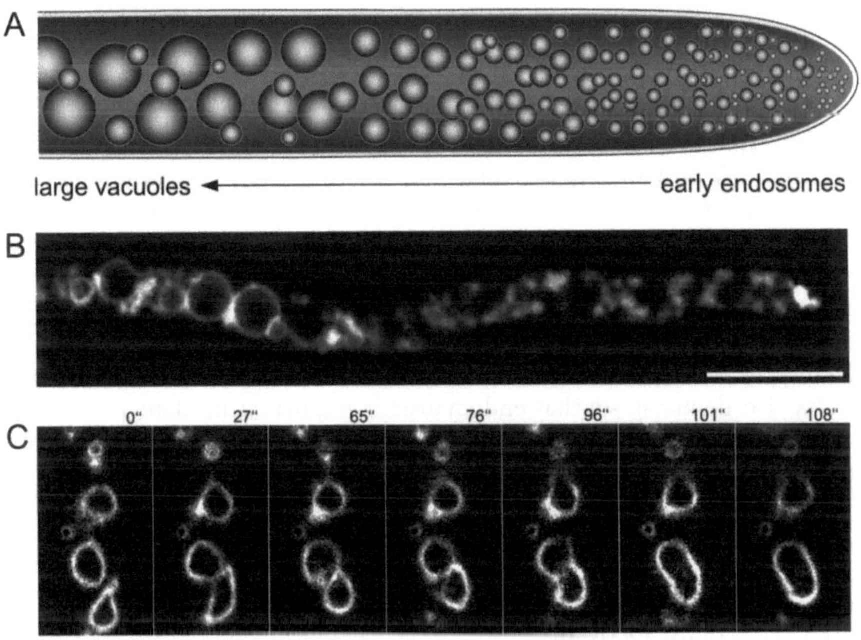

Fig. 1 Endocytosis and vacuole fusion in *Ashbya gossypii*. **a** The distribution of endocytic vesicles at the hyphal apex and larger vacuoles in subapical regions. **b** An *A. gossypii* hypha that was stained with the lipophilic dye FM4-64 (obtained from Molecular Probes and used at a concentration of 2 μM) for 30 min prior to microscopy. **c** In vivo time lapse imaging of vacuolar fusion in *A. gossypii*. A hyphal segment is shown in which one vacuole approaches another (at time point *0"*). The fusion of these two vacuoles is completed less than 108 s later (last time point), indicating that it is a very rapid event. The images were recorded with an Axioplan II imaging microscope (Zeiss, Jena) equipped with a MicroMax camera (Princeton Instruments/Roper Scientific, Trenton, NJ, USA) and processed using Metamorph software (Universal Imaging Corporation, Downingtown, PA, USA)

blocked in internalization (Yamashita and May, 1998). The actin cytoskeleton is of central importance for establishing and maintaining cell polarity as well as for the process of endocytosis. Particularly cortical actin patches are required for endocytosis in *S. cerevisiae* (Pruyne and Bretscher, 2000). In the dimorphic human fungal pathogen *Candida albicans* similar TEDS-rule mutants in the type I myosin (*CaMYO5*) were generated. Here S366D mutants allowed polarized hyphal growth even in the absence of polarized localization of cortical actin patches, whereas deletion of *CaMYO5* generated strains that were unable to form hyphae, indicating that type I myosin (and therefore endocytosis) is important for polarized morphogenesis (Oberholzer et al., 2002). In *S. cerevisiae* myosin I was found to play a role in endocytosis presumably at the step of fission of endocytic vesicles from the plasma membrane (Jonsdottir and Li, 2004).

Deletion of the WASP homologue *WAL1* in *A. gossypii* revealed several defects: (1) a severe reduction in the polarized growth rate; (2) uptake of FM4-64 was strongly reduced in *wal1*hyphae; (3) movement of large vacuoles was almost abolished. Interestingly, hyphal tips of *wal1* hyphae did not accumulate cortical actin patches and also did not contain endosomal vesicles, demonstrating a link between cortical actin patch positioning and endocytosis with polarized cell growth (Walther and Wendland, 2004b). The deletion of the *C. albicans* WASP, *WAL1*, showed very similar phenotypes as the *Camyo5* mutant in that *Cawal1* strains were defective in hyphal development (Walther and Wendland, 2004a). Similar analyses with the *A. gossypii cla4* mutant showed that in this strain movement of larger vacuoles was drastically reduced similarly as in the *wal1* mutant. This links the vacuolar distribution phenotypes with the cell polarity establishment machinery (Fig. 2). These results also demonstrate that endocytosis is important in filamentous fungi and that mutational analysis revealed a connection between polarized hyphal growth and endocytosis via Cdc42.

3
Cell Polarity Establishment and the Maintenance of Polarized Hyphal Growth

Polarized hyphal growth has largely been viewed as dealing with the problem of targeted delivery of secretory vesicles from the trans-Golgi network into the hyphal tip (Fig. 2). Designation of the hyphal tip as the endpoint of vesicle delivery requires a process known as polarity establishment. Maintenance of the axis of polarity then results in continuous polarized secretion, growth and tube elongation. It has been established in *S. cerevisiae* that the actin cytoskeleton plays an essential role in polarity establishment, whereas the microtubule cytoskeleton is not involved in this initial step (Wedlich-

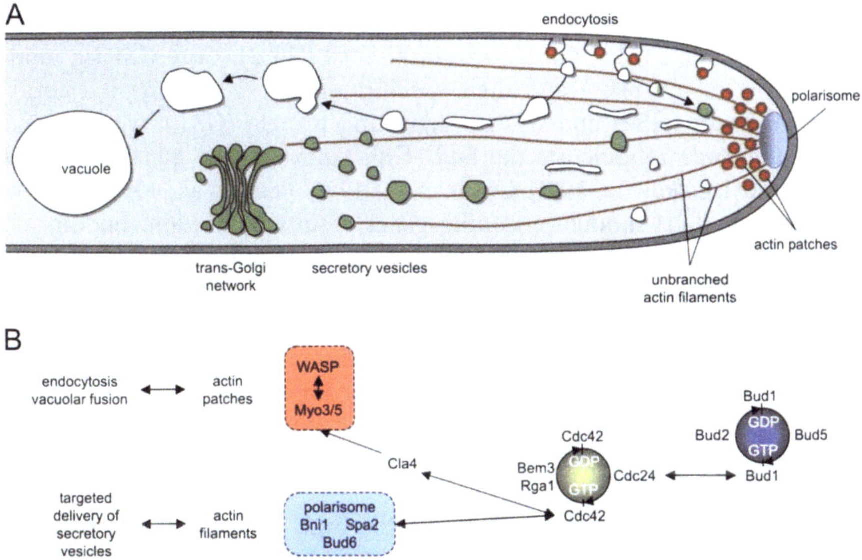

Fig. 2 Potential role of Cdc42 in coupling endocytosis and polarized growth in *A. gossypii*. **a** Schematic overview of vesicle trafficking in the hyphal apex of filamentous fungi. At the hyphal tip the polarisome complex generates linear actin filaments. Secretion from the trans-Golgi network occurs via secretory vesicles that are transported along either actin filaments (as shown) or microtubules. Endocytosis generates early endocytic vesicles that may share one of two fates: recycling and transport as secretory vesicles along cytoskeleton components to the hyphal tip or fusion with late endosomes and transport and fusion with vacuoles. **b** Signalling pathways involved in endocytosis and exocytosis. At the hyphal tip several GTPase modules are involved in establishment of the axis of growth and cell polarity. Cdc42 activates the polarisome. Particularly the formin homologue Bni1 is required for actin cable assembly. Actin cables provide the tracks for the targeted delivery of secretory vesicles and endosomes. Cdc42 may also activate the p21-activated kinase homologue Cla4 which phosphorylates type I myosins (Myo3/Myo5 in *Saccharomyces cerevisiae*). Myo3 together with the Wiskott–Aldrich syndrome protein (*WASP*) homologue activate Arp2/Arp3-dependent actin assembly required for endocytosis and vacuolar fusion

Söldner et al., 2004). Similar results were reported for the initiation and tip growth of plant root hairs previously (Baluška et al., 2000, reviewed by Šamaj et al. 2004). Also, the key role of the Rho-type GTPase Cdc42 in the organization of the actin cytoskeleton and in polarity establishment has been demonstrated in *S. cerevisiae* (Johnson, 1999). In *A. gossypii* loss of *CDC42* still allows spores to germinate and form germ cells. This indicates that Cdc42 is not required for breaking the dormancy of resting spores. However, germ tube formation is abolished in *cdc42* germ cells and the actin cytoskeleton cannot be polarized to a potential site of germ tube emergence. On the other hand, the nuclear cycle still continues in *cdc42* germ cells as also occurs in yeasts (Wendland and Philippsen, 2001). In *S. cerevisiae* bud-site selection depends on spatial landmark proteins that signal to the Ras-like GTPase Bud1 (Casamayor and Snyder, 2002; Kozminski et al., 2003). Core GTPase modules consist of a GTPase and their specific regulatory guanine nucleotide exchange factor (GEF) and GTPase activating protein (GAP); which in the case of the Bud1 module are the Bud1-GEF Bud5 and the Bud1-GAP Bud2 (Chant and Herskowitz, 1991; Chant et al., 1991; Park et al., 1993). Loss of either of the Bud1-module encoding genes results in random budding in *S. cerevisiae* which was suggested to be initiated by stochastic activation of Cdc42 (Wedlich-Soldner et al., 2003). Interestingly, deletion of the *A. gossypii BUD1* homologue resulted in alternating phases of polarized growth and pausing of growth that were coincident on the molecular level with localization and delocalization of the polarisome component Spa2 at the hyphal tip (Bauer et al., 2004). Disappearance of Spa2 from the hyphal tip resulted in aberrant growth and the formation of random bulges along the hyphal tube.

In *S. cerevisiae*, a direct link between the Bud1 and Cdc42 modules was established via the Cdc42-GEF Cdc24. Cdc24 interacts with Bud1-GTP, while Bem1, a protein involved with bud emergence, interacts with Bud1-GDP and also with the Cdc24/Cdc42 complex (Park et al., 1997; Gulli et al., 2000). Other Rho-protein modules are also involved in actin cytoskeleton organization, for example, Rho1/Rho2 are required for the activation of the cell wall integrity pathway and Rho3/Rho4 may share a function in the activation of the formin homologues Bni1/Bnr1 which are required for actin filament assembly (Bickle et al., 1998; Dong et al., 2003; Evangelista et al., 2003). Coordinate regulation of Cdc42 and Rho1 during yeast budding may involve the *S. cerevisiae* paxillin homologue Pxl1 (Gao et al., 2004), whereas genetic and functional overlaps were detected between Cdc42 and Rho3 (Robinson et al., 1999; Adamo et al., 2001).

In *A. nidulans*, screens of a collection of temperature-sensitive mutant strains resulted in the identification of *swo*, *pod* and *hyp* mutants defective in either polarity establishment, for example, *swoC*, *swoD* and *podB*, or polarity maintenance, for example, *swoA* (Harris et al., 1999; Momany et al., 1999). SwoC is homologous to the ribosomal RNA pseudouridine synthase Cbf5p of *S. cerevisiae* (Lin and Momany, 2003). SwoA encodes the protein *O*-mannosyltransferase pmtA (Shaw and Momany, 2002). Interestingly, the functional analysis of PMT genes in the dimorphic pathogen *C. albicans* revealed specific roles for growth, morphogenesis and antifungal resistance (Prill et al., 2005). A large-scale screen for conditional morphology mutants in *N. crassa* identified over 900 mutants with a diverse range of defects. Cloning of several genes also revealed mutants, for example, in the organization of the actin and microtubule cytoskeleton and in the secretory pathway. Some of the genes identified share overlapping functions with genes required for pollen tube growth (Seiler and Plamann, 2003).

4
The Spitzenkörper and its Role in Tip Growth

The Spitzenkörper was described as an apical body rich in cytoplasmic vesicles (Girbardt, 1957; Grove and Bracker, 1970). Polarized hyphal growth was suggested to arise from a gradient of wall-building vesicles generated by a vesicle supply centre (VSC) or Spitzenkörper (Gierz and Bartnicki-Garcia, 2001). Some evidence that the Spitzenkörper determines the shape of the fungal hyphae came from microscopic analyses of temporally displaced Spitzenkörper and computer simulations (Bartnicki-Garcia et al., 1995; Riquelme et al., 1998). Using FM4-64 to stain the Spitzenkörper allowed the localization of the VSC during germination of the rust fungi *Uromyces vignae*and *Puccinia graminis* f. sp. *tritici* grown on glass surfaces. These analyses demonstrated that a Spitzenkörper can be found during germ tube formation. The Spitzenkörper disappears during appressorium formation, which requires cell swelling (isotropic growth), but reforms upon penetration peg formation, which requires polarized growth again (Dijksterhuis, 2003).

While microtubules are not strictly required for the establishment of cell polarity, for example, in the filamentous ascomycete *A. nidulans* they nevertheless provide the tracks used for vesicle delivery and, thus, disruption of the microtubule integrity using benomyl results in a drastic decrease in the growth rate (Horio and Oakley, 2005). Similarly, disruption of the gene encoding the minus end directed microtubule motor protein dynein,*DYN1*, in *C. albicans* leads to the inability to promote hyphal growth upon induction owing to a failure of correct nuclear distribution into the hyphal tip (Martin et al., 2004). The general theme, therefore, is that polarity establishment is a function of Rho proteins that are master regulators of the actin cytoskeleton. The function of these Rho proteins is indispensable for tip extension. The role of the microtubule cytoskeleton in polarized growth is dependent on the biological system but is most important for nuclear positioning and vesicle delivery. In growing neurites, microtubules are essential (Dent and Gertler, 2003). In *Arabidopsis thaliana* root hair growth directionality was found to depend on the microtubule cytoskeleton (Bibikova et al., 1999; Bao et al., 2001). Loss of growth directionality was also observed in dynein/dynactin mutants of *N. crassa*. These mutant strains display a "ropy" (curly) morphology but were still able to grow in a polarized manner (Riquelme et al., 2000). In conifer pollen tubes of *Picea abies* transport of vesicles is dependent on the microtubule cytoskeleton and disassembly of microtubules results in a cessation of growth (Justus et al., 2004). Tip-localized cell growth in root hairs and pollen tubes is governed by the actin cytoskeleton (Miller et al., 1999; Baluška et al., 2000; Vidali and Hepler, 2001). Interestingly, the subcellular localization of the mitogen-activated protein kinase SIMK was shown to be dependent on a polymerized actin cytoskeleton at the growing tip of root hairs of *Medicago sativa* (alfalfa) (Šamaj et al., 2002). As with fungal systems, plant tip growth was shown to rely on targeted delivery of secretory vesicles and endocytosis (Voigt et al., 2005; Homann, this volume). Morphological defects in root hair mutants were shown to be caused by mutations of genes encoding Arp2/Arp3 complex components or proteins regulating actin filament turnover (Dong et al., 2001; Mathur et al., 2003).

5 Ca^{2+} Gradient and its Role in Tip Growth

Polarized hyphal growth bears another similarity with polarized growth in plant cells: the presence of a tip-high Ca^{2+} gradient (Torralba and Heath, 2001). In *N. crassa* the source of the cytosolic Ca^{2+} gradient appears to be the intracellular stores in organelles as, in contrast to plant systems, an appreciable tip-localized Ca^{2+} influx could not be observed (Felle and Hepler, 1997; Pierson et al., 1996; Silverman-Gavrila and Lew, 2001). In *N. crassa* a low conductance IP_3-activated Ca^{2+} channel localized to the endoplasmic reticulum and the plasma membrane regulates hyphal tip growth (Silverman-Gavrila and Lew, 2002). A similar mechanism may exist in plant cells (Holdaway-Clarke and Hepler, 2003). Also, IP_3 is involved in release of Ca^{2+} from vacuoles in *N. crassa*, *S. cerevisiae* and *C. albicans* (Cornelius et al., 1989; Belde et al., 1993; Calvert and Sanders, 1995). Recently, a calcium channel (Mid1) has been identified in *S. cerevisiae* that allows uptake of Ca^{2+} ions from the medium stimulated by the mating pheromone and in response to Ca^{2+} depletion from the endoplasmic reticulum (Tokes-Fuzesi et al., 2002; Yoshimura et al., 2004).

Why is such a calcium gradient so important for polarized growth in these systems? It has been shown in a plant system that elevated Ca^{2+} concentrations can stimulate the rate of exocytosis (Zorec and Tester, 1992). Another report using an animal system showed that the disassembly of actin filaments at the apical membrane also stimulated exocytosis (Muallem et al., 1995). A link between both processes may be direct as it was shown in pollen tubes that Ca^{2+} can induce the disassembly of actin filaments (Kohno and Shimmen, 1987). Ca^{2+}-calmodulin-sensitive actin binding proteins have been identified as villins and profilins. The monomeric actin binding protein profilin stimulates actin filament assembly, while the actin depolymerizing factor/cofilin controls filament turnover (Lappalainen and Drubin, 1997; Sagot et al., 2002a). The role of other ions as well as that of pH gradients for polarized hyphal growth need further attention in the future.

6
Model of Balanced Endocytosis and Exocytosis Integrated by Rho-Protein Signalling to Promote Polar Growth

Polarized hyphal growth is a highly dynamic process and it is obvious that vesicle sorting and delivery need to be tightly regulated. Several processes run simultaneously, for example, the targeted delivery of proteins and their plasma membrane insertion and tipward directed secretion during polarized hyphal growth as well as membrane turnover. In filamentous fungi, in contrast to yeast-like fungi, tip-directed secretion as well as secretion directed to septal sites during cytokinesis may occur at the same time in a hyphal tip cell. Furthermore, integrity and fluidity of the cell membrane need to be maintained as well as the spatial distribution of membrane proteins, for example, in lipid rafts. Outward-bound traffic meets inward-bound traffic of endocytic vesicles that also have different fates and may either be delivered to the vacuole or be reinserted into the membrane and may thus join the pool of secretory vesicles.

The actin cytoskeleton has a key role in endocytosis and exocytosis in filamentous fungi aided by the delivery of vesicles via the microtubule cytoskeleton. The role of WASP (resulting in Arp2/Arp3-dependent actin assembly of branched filaments) was shown to play a major role in endocytosis (Naqvi et al., 1998; Chang et al., 2003; Merrifield et al., 2004; Walther and Wendland, 2004a, b). The formation of branched actin filaments is contrasted by the assembly of linear actin filaments by a highly conserved group of eukaryotic proteins called formins which do not require Arp2/Arp3 complex activity (Evangelista et al., 2002, 2003; Sagot et al., 2002a, b). Formins are part of a protein complex termed the polarisome (Bidlingmaier and Snyder, 2004). This complex is tip-localized and contains the Spa2 and Bud6 proteins (Sharpless and Harris, 2002; Bauer et al., 2004). Deletion of *BUD1* in *A. gossypii* resulted in aberrant hyphal growth in which phases of growth and halts of growth alternated. During growth pauses the Spa2p-GFP signal at the hyphal tip was lost, indicating that the polarisome became unstable and delocalized from the tip. This demonstrates that the polarisome has a key role in maintaining polarized secretion at the hyphal tip and that activation of the polarisome requires G protein signalling.

WASP and Arp2/Arp3-mediated filaments are generated at sites of endocytosis as part of actin patches. Actin patches, therefore, mark sites of active endocytosis. In *S. cerevisiae* it was shown that cortical actin patches form at endocytic sites to foster the internalization process (Kaksonen et al., 2003; Huckaba et al., 2004). Our observations in *A. gossypii* revealed that in an *Ashbya* WASP deletion mutant actin patches are not accumulated in the hyphal tip but assemble in subapical regions. Concomitantly, endocytosis did not occur in the hyphal tip but was also shifted backwards (Walther and Wendland, 2004b).

Since myosin I and WASP mutants in *C. albicans* and *A. gossypii* showed decreased polarized growth rates, a molecular connection between endocytosis and exocytosis can be established (Oberholzer and Whiteway, 2002; Walther and Wendland, 2004a, b). Several arguments can be brought forward to explain why defects in endocytosis could slow down growth:

1. Compensatory endocytosis of the plasma membrane may be required to fuel hyphal tip growth. This is based on the assumption that many more vesicles are needed to deliver cell wall material than the membrane.
2. Endocytosis of the plasma membrane in subapical regions close to the hyphal tip may be required to reinforce the gradient of tip-polarized proteins involved in the maintenance of polarized hyphal growth. This could help to prevent diffusion of such factors that could otherwise lead to isotropic growth or swelling of the hyphal tip. Such a role may be attributed to proteins involved in bud emergence, for example, Bem1 and Bem2 in *S. cerevisiae*. In yeasts Bem1 was suggested to be part of a positive feedback loop that takes part in polarity establishment (Irazoqui et al., 2003). Swollen mutants have also been analysed in *A. gossypii* and *A. nidulans*. In the *A. gossypii bem2* mutant polarized growth is drastically slowed down and germ cells and hyphal tips are swollen enormously. *A. nidulans swo* mutants show several defects both in the establishment and in the maintenance of polarity (Momany et al., 1999; Wendland and Philippsen, 2000).
3. Balanced endocytosis and exocytosis may be required for maximal polar growth rates and a block in vesicle flow owing to defects in endocytosis perhaps causing a backlog in the secretory pathway.
4. Endocytic vesicles need to be sorted, for example, either for targeted degradation in the vacuole or for recycling to the plasma membrane. Such a sorting mechanism of vesicle flow may be directly connected and include the secretory machinery.

Thus, if endocytosis contributes substantially to this secretory flow, defects in endocytosis will automatically slow down the polar growth rate.

Importantly, both endocytosis and polarized cell growth are governed by actin cytoskeleton dynamics. Since Rho-protein modules are key regulators of the actin cytoskeleton, their interplay and cycling can generate the plasticity needed to respond to various extracellular and intracellular signals that may influence cell polarity establishment and polarized cell growth. In fact, both processes may be activated simultaneously by the highly conserved Rho-GTPase Cdc42. In *S. cerevisiae* Cdc42-GTP can trigger several feedback loops that can reinforce sites of polarized growth (1) by activating downstream effector proteins such as formins which generate unbranched actin filaments to steer targeted delivery of secretory vesicles, and (2) by activating p21-activated kinases (PAKs) that activate type I myosins (Myo3 and Myo5 in *S. cerevisiae*) which could recruit WASP and result in the assembly of branched actin filaments required for endocytosis (Lechler et al., 2000; Evangelista et al., 2000).

7
Conclusions and Future Prospects

Endocytosis is an essential eukaryotic process that is also found in filamentous fungi. Evidence for the presence of endocytic pathways stems from the presence of conserved proteins required for this process, from studying the uptake of endocytic markers in vivo, for example, using fluorescence time lapse microscopy methods, and by generating mutants that are defective in the endocytic pathway. It has become evident that endocytosis contributes substantially to polarized tip growth in filamentous fungi. Therefore, endocytosis could also be used as a target process for antifungal therapy. Endocytosis requires the temporal and spatial organization of the actin cytoskeleton. Key players involved in polarity establishment and the maintenance of polarized growth may also play a central role in regulating endocytosis. The elucidation of the interplay of endocytosis and secretion in filamentous fungi will be a challenging task in the future. This will help to explain the flow of vesicles and membranes that are bound to a diverse set of intracellular destinations.

Acknowledgements This work was supported by the Deutsche Forschungsgemeinschaft (We2634/2-1).

References

Adamo JE, Moskow JJ, Gladfelter AS, Viterbo D, Lew DJ, Brennwald PJ (2001) Yeast Cdc42 functions at a late step in exocytosis, specifically during polarized growth of the emerging bud. J Cell Biol 155:581–592

Atkinson HA, Daniels A, Read ND (2002) Live-cell imaging of endocytosis during conidial germination in the rice blast fungus, *Magnaporthe grisea*. Fungal Genet Biol 37:233–244

Ayscough KR (2004) Endocytosis: actin in the driving seat. Curr Biol 14:R124–R126

Bao Y, Kost B, Chua NH (2001) Reduced expression of alpha-tubulin genes in *Arabidopsis thaliana* specifically affects root growth and morphology, root hair development and root gravitropism. Plant J 28:145–157

Baluška F, Salaj J, Mathur J, Braun M, Jasper F, Šamaj J, Chua N-H, Barlow PW, Volkmann D (2000) Root hair formation: F-actin-dependent tip growth is initiated by local assembly of profilin-supported F-actin meshworks accumulated within expansin-enriched bulges. Dev Biol 227:618–632

Bartnicki-Garcia S, Bartnicki DD, Gierz G, Lopez-Franco R, Bracker CE (1995) Evidence that Spitzenkörper behavior determines the shape of a fungal hypha: a test of the hyphoid model. Exp Mycol 19:153–159

Bauer Y, Knechtle P, Wendland J, Helfer H, Philippsen P (2004) A Ras-like GTPase is involved in hyphal growth guidance in the filamentous fungus *Ashbya gossypii*. Mol Biol Cell 15:4622–4632
Belde PJ, Vossen JH, Borst-Pauwels GW, Theuvenet AP (1993) Inositol 1,4,5-trisphosphate releases Ca2+ from vacuolar membrane vesicles of *Saccharomyces cerevisiae*. FEBS Lett 323:113–118
Bibikova TN, Blancaflor EB, Gilroy S (1999) Microtubules regulate tip growth and orientation in root hairs of *Arabidopsis thaliana*. Plant J 17:657–665
Bickle M, Delley PA, Schmidt A, Hall MN (1998) Cell wall integrity modulates RHO1 activity via the exchange factor ROM2. EMBO J 17:2235–2245
Bidlingmaier S, Snyder M (2004) Regulation of polarized growth initiation and termination cycles by the polarisome and Cdc42 regulators. J Cell Biol 164:207–218
Calvert CM, Sanders D (1995) Inositol trisphosphate-dependent and -independent Ca2+ mobilization pathways at the vacuolar membrane of *Candida albicans*. J Biol Chem 270:7272–7280
Casamayor A, Snyder M (2002) Bud-site selection and cell polarity in budding yeast. Curr Opin Microbiol 5:179–186
Chang FS, Stefan CJ, Blumer KJ (2003) A WASp homolog powers actin polymerization-dependent motility of endosomes in vivo. Curr Biol 13:455–463
Chant J, Herskowitz I (1991) Genetic control of bud site selection in yeast by a set of gene products that constitute a morphogenetic pathway. Cell 65:1203–1212
Chant J, Corrado K, Pringle JR, Herskowitz I (1991) Yeast BUD5, encoding a putative GDP-GTP exchange factor, is necessary for bud site selection and interacts with bud formation gene BEM1. Cell 65:1213–1224
Cole L, Orlovich DA, Ashford AE (1998) Structure, function, and motility of vacuoles in filamentous fungi. Fungal Genet Biol 24:86–100
Collinge AJ, Trinci AP (1974) Hyphal tips of wild-type and spreading colonial mutants of *Neurospora crassa*. Arch Microbiol 99:353–368
Cornelius G, Gebauer G, Techel D (1989) Inositol trisphosphate induces calcium release from *Neurospora crassa* vacuoles. Biochem Biophys Res Commun 162:852–856
Dent EW, Gertler FB (2003) Cytoskeletal dynamics and transport in growth cone motility and axon guidance. Neuron 40:209–227
Dijksterhuis J (2003) Confocal microscopy of Spitzenkörper dynamics during growth and differentiation of rust fungi. Protoplasma 222:53–59
Dong CH, Kost B, Xia G, Chua NH (2001) Molecular identification and characterization of the *Arabidopsis AtADF1, AtADFS* and *AtADF6* genes. Plant Mol Biol 45:517–527
Dong Y, Pruyne D, Bretscher A (2003) Formin-dependent actin assembly is regulated by distinct modes of Rho signaling in yeast. J Cell Biol 161:1081–1092
Duncan MC, Cope MJ, Goode BL, Wendland B, Drubin DG (2001) Yeast Eps15-like endocytic protein, Pan1p, activates the Arp2/3 complex. Nat Cell Biol 3:687–690
Engqvist-Goldstein AE, Drubin DG (2003) Actin assembly and endocytosis: from yeast to mammals. Annu Rev Cell Dev Biol 19:287–332
Evangelista M, Klebl BM, Tong AH, Webb BA, Leeuw T, Leberer E, Whiteway M, Thomas DY, Boone C (2000) A role for myosin-I in actin assembly through interactions with Vrp1p, Bee1p, and the Arp2/3 complex. J Cell Biol 148:353–362
Evangelista M, Pruyne D, Amberg DC, Boone C, Bretscher A (2002) Formins direct Arp2/3-independent actin filament assembly to polarize cell growth in yeast. Nat Cell Biol 4:260–269
Evangelista M, Zigmond S, Boone C (2003) Formins: signaling effectors for assembly and polarization of actin filaments. J Cell Sci 116:2603–2611

Felle HH, Hepler PK (1997) The cytosolic Ca^{2+} concentration gradient of Sinapis alba root hairs as revealed by Ca^{2+}-selective microelectrode tests and fura-dextran ratio imaging. Plant Physiol 114:39–45
Fischer-Parton S, Parton RM, Hickey PC, Dijksterhuis J, Atkinson HA, Read ND (2000) Confocal microscopy of FM4-64 as a tool for analysing endocytosis and vesicle trafficking in living fungal hyphae. J Microsc 198:246–259
Girbardt M (1957) Der Spitzenkörper von Polystictus versicolor. Planta 50:47–59
Gao XD, Caviston JP, Tcheperegine SE, Bi E (2004) Pxl1p, a paxillin-like protein in *Saccharomyces cerevisiae*, may coordinate Cdc42p and Rho1p functions during polarized growth. Mol Biol Cell 15:3977–3985
Gierz G, Bartnicki-Garcia S (2001) A three-dimensional model of fungal morphogenesis based on the vesicle supply center concept. J Theor Biol 208:151–164
Goode BL, Rodal AA, Barnes G, Drubin DG (2001) Activation of the Arp2/3 complex by the actin filament binding protein Abp1p. J Cell Biol 153:627–634
Grove SN, Bracker CE (1970) Protoplasmic organization of hyphal tips among fungi: vesicles and Spitzenkörper. J Bacteriol 104:989–1009
Gulli MP, Jaquenoud M, Shimada Y, Niederhauser G, Wiget P, Peter M (2000) Phosphorylation of the Cdc42 exchange factor Cdc24 by the PAK-like kinase Cla4 may regulate polarized growth in yeast. Mol Cell 6:1155–1167
Harris SD, Hofmann AF, Tedford HW, Lee MP (1999) Identification and characterization of genes required for hyphal morphogenesis in the filamentous fungus *Aspergillus nidulans*. Genetics 151:1015–1025
Holdeway-Clarke TL, Hepler PK (2003) Control of pollen tube growth: role of ion-gradients and fluxes. New Phytol 159:539–563
Horio T, Oakley BR (2005) The role of microtubules in rapid hyphal tip growth of *Aspergillus nidulans*. Mol Biol Cell 16:918–926
Huckaba TM, Gay AC, Pantalena LF, Yang HC, Pon LA (2004) Live cell imaging of the assembly, disassembly, and actin cable-dependent movement of endosomes and actin patches in the budding yeast, *Saccharomyces cerevisiae*. J Cell Biol 167:519–530
Irazoqui JE, Gladfelter AS, Lew DJ (2003) Scaffold-mediated symmetry breaking by Cdc42p. Nat Cell Biol 5:1062–1070
Kohno T, Shimmen T (1987) Ca^{2+}-induced fragmentation of actin filaments in pollen tubes. Protoplasma 141:177–179
Johnson DI (1999) Cdc42: An essential Rho-type GTPase controlling eukaryotic cell polarity. Microbiol Mol Biol Rev 63:54–105
Jonsdottir GA, Li R (2004) Dynamics of yeast myosin I: evidence for a possible role in scission of endocytic vesicles. Curr Biol 14:1604–1609
Justus CD, Anderhag P, Goins JL, Lazzaro MD (2004) Microtubules and microfilaments coordinate to direct a fountain streaming pattern in elongating conifer pollen tube tips. Planta 219:103–109
Kaksonen M, Sun Y, Drubin DG (2003) A pathway for association of receptors, adaptors, and actin during endocytic internalization. Cell 115:475–487
Kozminski KG, Beven L, Angerman E, Tong AH, Boone C, Park HO (2003) Interaction between a Ras and a Rho GTPase couples selection of a growth site to the development of cell polarity in yeast. Mol Biol Cell 14:4958–4970
Lappalainen P, Drubin DG (1997) Cofilin promotes rapid actin filament turnover in vivo. Nature 388:78–82
Lechler T, Shevchenko A, Li R (2000) Direct involvement of yeast type I myosins in Cdc42-dependent actin polymerization. J Cell Biol 148:363–373

Lin X, Momany M (2003) The *Aspergillus nidulans* swoC1 mutant shows defects in growth and development. Genetics 165:543–554

Madania A et al. (1999) The *Saccharomyces cerevisiae* homologue of human Wiskott-Aldrich syndrome protein Las17p interacts with the Arp2/3 complex. Mol Biol Cell 10:3521–3538

Martin R, Walther A, Wendland J (2004) Deletion of the dynein heavy-chain gene DYN1 leads to aberrant nuclear positioning and defective hyphal development in *Candida albicans*. Eukaryot Cell 3:1574–1588

Mathur J, Mathur N, Kirik V, Kernebeck B, Srinivas BP, Hulskamp M (2003) *Arabidopsis CROOKED* encodes for the smallest subunit of the ARP2/3 complex and controls cell shape by region specific fine F-actin formation. Development 130:3137–3146

Mellman I (1996) Endocytosis and molecular sorting. Annu Rev Cell Dev Biol 12:575–625

Merrifield CJ, Qualmann B, Kessels MM, Almers W (2004) Neural Wiskott Aldrich syndrome protein (N-WASP) and the Arp2/3 complex are recruited to sites of clathrin-mediated endocytosis in cultured fibroblasts. Eur J Cell Biol 83:13–18

Miller DD, de Ruijter NCA, Bisseling T, Emons AMC (1999) The role of actin in root hair morphogenesis: Studies with lipochito-oligosaccharide as a growth stimulator and cytochalasin as an actin perturbing drug. Plant J 17:141–154

Momany M, Westfall PJ, Abramowsky G (1999) *Aspergillus nidulans swo* mutants show defects in polarity establishment, polarity maintenance and hyphal morphogenesis. Genetics 151:557–567

Muallem S, Kwiatkowska K, Xu X, Yin HL (1995) Actin filament disassembly is a sufficient final trigger for exocytosis in nonexcitable cells. J Cell Biol 128:589–598

Naqvi SN, Zahn R, Mitchell DA, Stevenson BJ, Munn AL (1998) The WASp homologue Las17p functions with the WIP homologue End5p/verprolin and is essential for endocytosis in yeast. Curr Biol 8:959–962

Oberholzer U, Marcil A, Leberer E, Thomas DY, Whiteway M (2002) Myosin I is required for hypha formation in *Candida albicans*. Eukaryot Cell 1:213–228

Park HO, Chant J, Herskowitz I (1993) BUD2 encodes a GTPase-activating protein for Bud1/Rsr1 necessary for proper bud-site selection in yeast. Nature 365:269–274

Park HO, Bi E, Pringle JR, Herskowitz I (1997) Two active states of the Ras-related Bud1/Rsr1 protein bind to different effectors to determine yeast cell polarity. Proc Natl Acad Sci USA 94:4463–4468

Pierson ES, Miller DD, Callaham DA, van Aken J, Hackett G, Hepler PK (1996) Tip-localized calcium entry fluctuates during pollen tube growth. Dev Biol 174:160–173

Prill SK, Klinkert B, Timpel C, Gale CA, Schroppel K, Ernst JF (2005) PMT family of *Candida albicans*: five protein mannosyltransferase isoforms affect growth, morphogenesis and antifungal resistance. Mol Microbiol 55:546–560

Pruyne D, Bretscher A (2000) Polarization of cell growth in yeast. J Cell Sci 113:571–585

Read ND, Kalkman ER (2003) Does endocytosis occur in fungal hyphae? Fungal Genet Biol 39:199–203

Riezman H (1985) Endocytosis in yeast: several of the yeast secretory mutants are defective in endocytosis. Cell 40:1001–1009

Riquelme M, Reynaga-Pena CG, Gierz G, Bartnicki-Garcia S (1998) What determines growth direction in fungal hyphae? Fungal Genet Biol 24:101–109

Riquelme M, Gierz G, Bartnicki-Garcia S (2000) Dynein and dynactin deficiencies affect the formation and function of the Spitzenkörper and distort hyphal morphogenesis of *Neurospora crassa*. Microbiology 146:1743–1752

Robinson NG, Guo L, Imai J, Toh EA, Matsui Y, Tamanoi F (1999) Rho3 of *Saccharomyces cerevisiae*, which regulates the actin cytoskeleton and exocytosis, is a GTPase which interacts with Myo2 and Exo70. Mol Cell Biol 19:3580–3587

Sagot I, Klee SK, Pellman D (2002) Yeast formins regulate cell polarity by controlling the assembly of actin cables. Nat Cell Biol 4:42–50

Sagot I, Rodal AA, Moseley J, Goode BL, Pellman D (2002) An actin nucleation mechanism mediated by Bni1 and profilin. Nat Cell Biol 4:626–631

Šamaj J, Ovecka M, Hlavacka A, Lecourieux F, Meskiene I, Lichtscheidl I, Lenart P, Salaj J, Volkmann D, Bogre L, Baluška F, Hirt H (2002) Involvement of the mitogen-activated protein kinase SIMK in regulation of root hair tip growth. EMBO J 21:3296–3306

Šamaj J, Baluška F, Voigt B, Schlicht M, Volkmann D, Menzel D (2004) Endocytosis, actin cytoskeleton and signalling. Plant Physiol 135:1150–1161

Šamaj J (2005) Methods and molecular tools to study endocytosis in plants—an overview (in this volume). Springer, Berlin Heidelberg New York

Šamaj J, Baluška F, Voigt B, Volkmann D, Menzel D (2005) Endocytosis and acto-myosin cytoskeleton (in this volume). Springer, Berlin Heidelberg New York

Seiler S, Plamann M (2003) The genetic basis of cellular morphogenesis in the filamentous fungus *Neurospora crassa*. Mol Biol Cell 14:4352–4364

Sharpless KE, Harris SD (2002) Functional characterization and localization of the Aspergillus nidulans formin SEPA. Mol Biol Cell 13:469–479

Shaw BD, Momany M (2002) *Aspergillus nidulans* polarity mutant swoA is complemented by protein O-mannosyltransferase pmtA. Fungal Genet Biol 37:263–270

Silverman-Gavrila LB, Lew RR (2001) Regulation of the tip-high [Ca^{2+}] gradient in growing hyphae of the fungus *Neurospora crassa*. Eur J Cell Biol 80:379–390

Silverman-Gavrila LB, Lew RR (2002) An IP3-activated Ca^{2+} channel regulates fungal tip growth. J Cell Sci 115:5013–5025

Tokes-Fuzesi M, Bedwell DM, Repa I, Sipos K, Sumegi B, Rab A, Miseta A (2002) Hexose phosphorylation and the putative calcium channel component Mid1p are required for the hexose-induced transient elevation of cytosolic calcium response in Saccharomyces cerevisiae. Mol Microbiol 44:1299–1308

Torralba S, Heath IB (2001) Cytoskeletal and Ca^{2+} regulation of hyphal tip growth and initiation. Curr Top Dev Biol 51:135–187

Torralba S, Heath IB (2002) Analysis of three separate probes suggests the absence of endocytosis in *Neurospora crassa* hyphae. Fungal Genet Biol 37:221–232

Vida TA, Emr SD (1995) A new vital stain for visualizing vacuolar membrane dynamics and endocytosis in yeast. J Cell Biol 128:779–792

Vidali L, Hepler PK (2001) Actin and pollen tube growth. Protoplasma 215:64–76

Voigt B, Timmers A, Šamaj J, Hlavacka A, Ueda T, Preuss M, Nielsen E, Mathur J, Emans N, Stenmark H, Nakano A, Baluška F, Menzel D (2005) Actin-based motility of endosomes is linked to the polar tip growth of root hairs. Eur J Cell Biol 84:609–621

Walther A, Wendland J (2004) Apical localization of actin patches and vacuolar dynamics in *Ashbya gossypii* depend on the WASP homolog Wal1p. J Cell Sci 117:4947–4958

Walther A, Wendland J (2004) Polarized hyphal growth in *Candida albicans* requires the Wiskott-Aldrich syndrome protein homolog Wal1p. Eukaryot Cell 3:471–482

Wedlich-Soldner R, Altschuler S, Wu L, Li R (2003) Spontaneous cell polarization through actomyosin-based delivery of the Cdc42 GTPase. Science 299:1231–1235

Wedlich-Soldner R, Wai SC, Schmidt T, Li R (2004) Robust cell polarity is a dynamic state established by coupling transport and GTPase signaling. J Cell Biol 166:889–900

Wendland J, Philippsen P (2000) Determination of cell polarity in germinated spores and hyphal tips of the filamentous ascomycete *Ashbya gossypii* requires a rhoGAP homolog. J Cell Sci 113 (Pt 9):1611–1621

Wendland J, Philippsen P (2001) Cell polarity and hyphal morphogenesis are controlled by multiple rho-protein modules in the filamentous ascomycete *Ashbya gossypii*. Genetics 157:601–610

Winter D, Lechler T, Li R (1999) Activation of the yeast Arp2/3 complex by Bee1p, a WASP-family protein. Curr Biol 9:501–504

Yamashita RA, May GS (1998) Constitutive activation of endocytosis by mutation of myoA, the myosin I gene of *Aspergillus nidulans*. J Biol Chem 273:14644–14648

Yoshimura H, Tada T, Iida H (2004) Subcellular localization and oligomeric structure of the yeast putative stretch-activated Ca^{2+} channel component Mid1. Exp Cell Res 293:185–195

Zorec R, Tester M (1992) Cytoplasmic calcium stimulates exocytosis in a plant secretory cell. Biophys J 63:864–867

Subject Index

MIX
Papier aus verantwortungsvollen Quellen
Paper from responsible sources
FSC® C105338

If you have any concerns about our products,
you can contact us on
ProductSafety@springernature.com

In case Publisher is established outside the EU,
the EU authorized representative is:
Springer Nature Customer Service Center GmbH
Europaplatz 3, 69115 Heidelberg, Germany

Printed by Libri Plureos GmbH
in Hamburg, Germany